Privacy-Preserving Machine Learning

A use-case-driven approach to building and protecting
ML pipelines from privacy and security threats

Srinivasa Rao Aravilli

Privacy-Preserving Machine Learning

Group Product Manager: Niranjan Naikwadi
Publishing Product Manager: Sanjana Gupta
Senior Editor: Sushma Reddy
Book Project Managers: Farheen Fathima & Shambhavi Mishra
Technical Editor: Kavyashree K S
Copy Editor: Safis Editing
Language Support Editor: Safis Editing
Project Coordinators: Farheen Fathima & Shambhavi Mishra
Proofreader: Sushma Reddy
Indexer: Rekha Nair
Production Designer: Alishon Mendonca
Marketing Coordinator: Vinishka Kalra

First published: April 2024

Production reference: 2100524

Published by Packt Publishing Ltd.
Grosvenor House
11 St Paul's Square
Birmingham
B3 1RB, UK.

ISBN 978-1-80056-467-1

www.packtpub.com

To my mother, Jaya, and my father, Rama Sarma, for their sacrifices and for exemplifying the power of determination. To my wife, Uma Madhavi, for being my loving partner throughout our joint life journey, and my son, Atchuta Ram, and daughter, Akhila, for their love, support, and inspiration.

– Srinivasa Rao Aravilli

Foreword

In an era defined by an abundance of data and the transformative power of machine learning, the need to safeguard privacy and ensure the security of sensitive information has never been more critical.

In this groundbreaking book, *Privacy-Preserving Machine Learning*, Srinivasa Rao Aravilli presents a comprehensive exploration of the intersection between privacy and machine learning. Srinivasa, an esteemed expert in the field, offers a unique perspective on addressing the privacy challenges in machine learning and deep learning applications. The book takes us on a compelling journey into the world of privacy-preserving machine learning, providing a deep understanding of the underlying principles, methodologies, and cutting-edge techniques. Srinivasa elucidates the potential risks and vulnerabilities inherent in data pipelines, shedding light on the potential consequences of privacy breaches. Drawing on his extensive experience, he guides us through the intricate landscape of privacy preservation, offering invaluable insights and practical solutions.

He explores the latest techniques such as federated learning, federated learning benchmarks, homomorphic encryption, and differential privacy, providing a glimpse into the future of privacy-preserving machine learning on secure enclaves to protect data from insider threats. With the insights and methodologies presented in this book, readers will be empowered to strike the delicate balance between data utility and privacy preservation, fostering trust among individuals and organizations alike.

I commend Srinivasa for his meticulous research, practical insights using open source privacy-preserving ML and DL frameworks, and dedication to advancing the field of privacy-preserving machine learning.

This book serves as a valuable resource for data scientists, machine learning engineers, privacy professionals, and decision-makers seeking to unlock the transformative potential of machine learning while upholding the privacy rights of individuals.

Sam Hamilton

Head of Data and AI

Visa Inc.

Contributors

About the author

Srinivasa Rao Aravilli boasts 27 years of extensive experience in technology and leadership roles, spearheading innovation in various domains such as information retrieval, search, ML/AI, Generative AI, distributed computing, network analytics, privacy, and security. Currently serving as a senior director of machine learning engineering at Capital One, Bangalore, he has a proven track record of driving new products from conception to outstanding customer success. Prior to his tenure at Capital One, Srinivasa held prominent leadership positions at Visa, Cisco, and Hewlett Packard, where he led product groups focused on large-scale data, privacy, machine learning, and Generative AI. He holds a master's degree in computer applications from Andhra University, Visakhapatnam, India.

I would like to thank the people who have been close to me and supported me, especially my family, parents, and my colleagues and team members.

About the reviewers

Dr. Nitin Agrawal is currently a privacy engineer at Snap Inc. Previously, he served as an applied scientist at Amazon in Alexa Privacy Science. Holding a doctoral degree from the University of Oxford, his research focuses on privacy-preserving machine learning and designing solutions for industry compliance with international privacy regulations, optimizing business utility. His work includes designing secure data-driven systems and advancing privacy protections through privacy-enhancing primitives including secure multi-party computation and homomorphic encryption. Recently, Dr. Agrawal has been actively involved in the privacy auditing of machine learning models, with a specific emphasis on privacy-aware Generative AI.

Dr. Agrawal acknowledges the unwavering support of his family, mentor, and colleagues in the production of this book.

Nandita Rao Narla is the head of technical privacy and governance at DoorDash. Previously, she was a founding team member of a data profiling startup and held various leadership roles at EY, where she helped Fortune 500 companies build and mature privacy, cybersecurity, and data governance programs. She is a Senior Fellow at Future of Privacy Forum and serves on the advisory boards and technical standards committees for IAPP, Ethical Tech Project, X Reality Safety Initiative, Institute of Operational Privacy Design, ISACA, and NIST. Nandita holds a BTech in computer science from JNT University, an MS in Information Security from Carnegie Mellon University, and certifications including FIP, CIPP/US, CIPT, CIPM, CDPSE, CISM, CRISC, and CISA.

Akshat Gurnani is a qualified professional in the field of computer science and machine learning, with a Master's degree. His expertise covers a variety of machine learning techniques, including natural language processing, computer vision, and deep learning. Akshat's significant contributions to academia are evident through his prolific publications in top-tier journals and conferences. His dedication to continuous learning ensures he remains at the forefront of the latest technological developments, seeking to drive forward advancements in artificial intelligence.

Table of Contents

2

Machine Learning Phases and Privacy Threats/Attacks in Each Phase   27

Part 2: Use Cases of Privacy-Preserving Machine Learning and a Deep Dive into Differential Privacy

3

Overview of Privacy-Preserving Data Analysis and an Introduction to Differential Privacy   79

4

Overview of Differential Privacy Algorithms and Applications of Differential Privacy   119

5

Developing Applications with Differential Privacy Using Open Source Frameworks   147

Part 3: Hands-On Federated Learning

6

Federated Learning and Implementing FL Using Open Source Frameworks   211

7

Federated Learning Benchmarks, Start-Ups, and the Next Opportunity   261

Part 4: Homomorphic Encryption, SMC, Confidential Computing, and LLMs

8

Homomorphic Encryption and Secure Multiparty Computation   283

Preface

In today's interconnected world, the vast amounts of data generated by individuals and organizations have become a valuable resource for developing powerful machine learning models. These models have the potential to revolutionize industries, improve services, and unlock unprecedented insights. However, this tremendous opportunity comes with a significant challenge: preserving the privacy and security of sensitive data.

As data breaches and privacy concerns continue to make headlines, individuals and organizations are increasingly aware of the potential risks associated with sharing and analyzing their data. There is a growing demand for innovative solutions that can harness the power of machine learning while simultaneously protecting the privacy of individuals and safeguarding sensitive information.

This book, *Privacy-Preserving Machine Learning*, aims to address these pressing concerns and explore the latest techniques and methodologies designed to reconcile the power of machine learning with the imperative of data privacy. We delve into the intricate world of privacy-preserving techniques, algorithms, and frameworks that enable organizations to unlock the full potential of their data while adhering to stringent privacy regulations and ethical considerations.

Throughout the pages of this book, we provide a comprehensive overview of the field, covering both fundamental concepts and advanced techniques. We discuss various privacy threats and risks associated with machine learning, including membership inference attacks and model inversion attacks. Moreover, we explore the legal and ethical aspects of privacy in machine learning, shedding light on regulations such as the **General Data Protection Regulation (GDPR)** and the **California Consumer Privacy Act (CCPA)**.

One of the central themes of this book is the exploration of privacy-enhancing technologies that enable secure and private machine learning. We delve into differential privacy, homomorphic encryption, secure multiparty computation, and federated learning, among others. We examine their underlying principles, strengths, and limitations, providing you with the necessary tools to choose the most appropriate techniques for your specific privacy requirements.

As the fields of artificial intelligence and data science continue to advance, it is imperative to ensure that privacy remains at the forefront of innovation.

This book aims to serve as a valuable resource for researchers, practitioners, and policymakers interested in the intersection of privacy and machine learning. By understanding the challenges, solutions, and emerging trends in privacy-preserving machine learning, we can collectively shape a future where privacy and innovation coexist harmoniously. Together, let us embark on a journey through the world of privacy-preserving machine learning and unlock the transformative potential of AI while upholding the rights and privacy of individuals and organizations.

Who this book is for

This book is intended for a wide range of readers who are interested in the intersection of privacy and machine learning. The target audience includes the following:

- **Data scientists and machine learning practitioners**: Professionals who work with data and develop machine learning models will find this book invaluable. It provides insights into privacy-preserving techniques and frameworks that can be integrated into their existing workflows, enabling them to build secure and privacy-aware machine learning systems.

- **Researchers and academics**: Researchers and academics in the fields of computer science, data science, artificial intelligence, and privacy will benefit from the comprehensive coverage of privacy-preserving machine learning techniques. The book explores the latest advancements and challenges in the field, offering a solid foundation for further research and exploration.

- **Privacy professionals and data protection officers**: Privacy professionals responsible for ensuring compliance with privacy regulations and protecting sensitive data will find this book highly relevant. It covers legal and ethical aspects of privacy in machine learning, providing guidance on incorporating privacy-enhancing technologies into organizational practices.

- **Policymakers and government officials**: Policymakers and government officials who are involved in shaping privacy regulations and guidelines can gain valuable insights from this book. It explores the regulatory landscape and discusses the implications of privacy-preserving machine learning for policy development and implementation.

- **Industry leaders and decision-makers**: Executives, managers, and decision-makers in various industries will find this book beneficial in understanding the importance of privacy in machine learning. It offers practical examples and use cases that demonstrate the benefits of privacy-preserving techniques, enabling informed decision-making regarding data protection strategies.

- **Privacy advocates and activists**: Individuals and organizations advocating for privacy rights and data protection will find this book useful in understanding the technical aspects of privacy-preserving machine learning. It equips them with the knowledge to engage in informed discussions and contribute to the development of privacy-friendly practices and policies.

Regardless of your level of expertise in machine learning or privacy, this book provides a comprehensive introduction to the subject and gradually builds upon foundational concepts. It offers both theoretical insights and practical applications, making it accessible and valuable to a diverse audience seeking to navigate the challenges and opportunities presented by privacy-preserving machine learning.

What this book covers

Chapter 1, Introduction to Data Privacy, Privacy Breaches, and Threat Modeling, serves as an introduction to various aspects of data privacy. We begin by exploring the concept of data privacy and distinguishing between sensitive data and personal sensitive data. Additionally, we delve into the realm of data privacy regulations, highlighting their significance in safeguarding individuals' information. The chapter also introduces the concept of privacy by design, emphasizing its importance in ensuring privacy throughout the data life cycle. Furthermore, we examine the real-world implications of privacy breaches by discussing notable cases and the resulting fines imposed on major enterprise companies. These examples shed light on the consequences of failing to protect sensitive data adequately. The chapter then delves into privacy threat modeling using the LINDDUN framework. We explain the concepts of linkability and identifiability threats, providing illustrative examples to enhance understanding. By covering these topics comprehensively, this chapter sets the foundation for a deeper exploration of privacy-preserving techniques and methodologies discussed throughout the book. It equips you with the necessary knowledge to understand the importance of data privacy, the risks associated with privacy breaches, and the strategies employed to mitigate these risks while enabling data analysis and utilization.

Chapter 2, Machine Learning Phases and Privacy Threats/Attacks in Each Phase, provides an overview of various types of machine learning, including supervised, unsupervised, and reinforcement learning, along with an exploration of the machine learning phases and pipeline. It provides various formats to persist ML models and challenges in model persistence as well. Additionally, it highlights the importance of privacy considerations at each phase of the machine learning process. We delve into the privacy needs associated with different phases of machine learning, namely, training data privacy, input data privacy, model privacy, and inference/output data privacy. The chapter proceeds by examining privacy attacks specific to each phase. We focus on the threats posed to training data, model persistence, and inference processes. We delve into model inversion attacks, model inference attacks, and training data extraction attacks, providing detailed examples to illustrate how these attacks work using open source frameworks.

Chapter 3, Overview of Privacy-Preserving Data Analysis and Introduction to Differential Privacy, serves as an introduction to privacy-preserving data analysis, privacy-enhanced technologies, and the concept of differential privacy. These topics are explored to provide a foundation for understanding and implementing privacy-preserving measures in data analysis and machine learning. This chapter also covers reconstruction attacks on SQL, explores practical use cases, and discovers how to prevent such attacks using the Open Diffix framework, which provides robust privacy protection. Specifically, the concept of differential privacy is explored in detail including privacy loss, privacy budgets, differential privacy mechanisms, and local/global differential privacy.

Chapter 4, Differential Privacy Algorithms and Limitations of Differential Privacy, deep dives into various algorithms (Laplace, Gaussian, count, sum, mean, variance, standard deviation, and thresholding algorithms) used in differential privacy and the limitations of differential privacy.

Chapter 5, Developing Applications with Differential Privacy Using Open Source Frameworks, provides a deep dive into **differential privacy (DP)** using open source frameworks and the implementation of a fraud detection use case using ML and DL open source frameworks. This chapter also provides an overview of real-world applications that make use of DP.

Chapter 6, Need for Federated Learning and Implementing Federated Learning Using Open Source Frameworks, covers the importance of **federated learning (FL)** and addresses the privacy concerns associated with sending data to a central server for model training. It explores the concepts of **independent and identically distributed (IID)** and non-IID datasets, along with the different categories of non-IID data. Understanding these data characteristics is crucial for effectively implementing FL. Furthermore, it discusses FL techniques (FedAvg, FedYogi, FedSGD, etc.) and introduces available open source frameworks that support FL implementations. It also implements a use case in the financial domain with FL using the Flower open source framework.

Chapter 7, Federated Learning Benchmarks, Start-Ups, and Next Opportunities, focuses on FL datasets and benchmarks, and a comparison of both. It delves into the available FL benchmarks and provides insights into how they can be used for evaluating FL algorithms and techniques. Additionally, it discusses the process of selecting the most appropriate FL benchmarks for your specific project, considering factors such as data characteristics and evaluation criteria. It also explores the state-of-the-art research in FL, highlighting the latest advancements, methodologies, and challenges in this field, and sheds light on start-ups that are actively working on FL and their specific focus areas.

Chapter 8, Homomorphic Encryption and Secure Multiparty Computation, explores various privacy-enhancing techniques, including encryption, anonymization, and de-identification. It discusses the principles and limitations of these techniques, understanding their effectiveness in safeguarding sensitive data while preserving data utility. The concept of **homomorphic encryption (HE)** and its mathematical foundations are covered along with an exploration of how HE can be applied in machine learning scenarios, allowing computations to be performed directly on encrypted data without compromising privacy. Furthermore, we discuss **secure multiparty computation (SMC)** and its use cases and present a use case implementation using the **private set interaction (PSI)** SMC technique. At the end of the chapter, we provide a high-level overview of **zero-knowledge proof (ZKP)**, a cryptographic protocol that enables one party to prove knowledge of certain information without revealing the information itself.

Chapter 9, Confidential Computing – What, Why, and Current State, delves into privacy and security attacks that target data stored in memory. We discuss the vulnerabilities and potential risks associated with such attacks, highlighting the importance of protecting data throughout its life cycle. We introduce the concept of confidential computation, focusing on **trusted execution environments (TEEs)**, and explore the concept of attestation of source code and how it aids in mitigating insider threat attacks. By verifying the integrity and authenticity of source code, organizations can establish trust and ensure that malicious actors cannot compromise the security of their systems. Additionally, we compare the support for secure enclaves in major cloud service providers such as AWS, Azure, GCP, and Anjuna. We assess the capabilities, features, and security measures offered by these providers, enabling you to make informed decisions when choosing a platform for deploying applications that require secure enclaves.

Chapter 10, Privacy Preserving in Large Language Models, introduces generative AI and the fundamentals of LLMs, the privacy vulnerabilities associated with them, and the technologies and approaches to preserve privacy while using these models. This chapter covers developing LLM applications using open source LLMs and protecting them from privacy attacks (prompt injection attacks, membership inference attacks, etc.) and ends with state-of-the-art privacy research on LLMs.

To get the most out of this book

While knowledge of **artificial intelligence (AI)**, **machine learning (ML)**, and **generative AI (GenAI)** is not required, having some familiarity with either will help grasp some of the concepts covered in this book. Knowledge of Python will help to execute the example source code provided in each chapter.

Software/hardware covered in the book	Operating system requirements
Python 3.7 or higher	Windows, macOS, or Linux
Jupyter Notebook	

To help go into depth on some of the more intricate concepts of this book, we have created Jupyter Notebook files on GitHub (details follow).

If you are using the digital version of this book, we advise you to type the code yourself or access the code from the book's GitHub repository (a link is available in the next section). Doing so will help you avoid any potential errors related to the copying and pasting of code.

Download the hands-on labs and example code files

You can download the hands-on labs and example code files for this book from GitHub at `https://github.com/PacktPublishing/Privacy-Preserving-Machine-Learning`. If there are any updates to the hands-on labs or any updates to any code, this will be updated in the GitHub repository referenced previously.

We also have other code bundles from our rich catalog of books and videos available at `https://github.com/PacktPublishing/`. Check them out!

Conventions used

There are a number of text conventions used throughout this book.

`Code in text`: Indicates code words in text, database table names, folder names, filenames, file extensions, pathnames, dummy URLs, user input, and Twitter handles. Here is an example: "Install Jupyter Notebook (pip install notebook)."

A block of code is set as follows:

```
import numpy as np

from sklearn.linear_model import LinearRegression

# Prepare sample input data with two features (feature 1, feature 2

X = np.array([[10, 10], [10, 20], [20, 20], [20, 30]])

# Assume that target feature has some relationship with the input
features with the formula y = 3 * x_1 + 5 * x_2 + 50

y = np.dot(X, np.array([3, 5])) + 50
```

When we wish to draw your attention to a particular part of a code block, the relevant lines or items are set in bold:

```
SELECT DEPTNO, DEPTNAME, MGRNO, MGRNAME, MGREMAILID,LOCATION FROM
DEPARTMENT;
```

Any command-line input or output is written as follows:

```
!pip3 install tmlt-analytics

!pip3 install onnx --user
```

Bold: Indicates a new term, an important word, or words that you see onscreen. For instance, words in menus or dialog boxes appear in **bold**. Here is an example: "The attacker's code aims to execute a model extraction attack by training a new **LogisticRegression** model (**extracted_model**) on the same dataset"

> **Tips or important notes**
> Appear like this.

Get in touch

Feedback from our readers is always welcome.

General feedback: If you have questions about any aspect of this book, email us at customercare@ packtpub.com and mention the book title in the subject of your message.

Errata: Although we have taken every care to ensure the accuracy of our content, mistakes do happen. If you have found a mistake in this book, we would be grateful if you would report this to us. Please visit www.packtpub.com/support/errata and fill in the form.

Piracy: If you come across any illegal copies of our works in any form on the internet, we would be grateful if you would provide us with the location address or website name. Please contact us at copyright@packt.com with a link to the material.

If you are interested in becoming an author: If there is a topic that you have expertise in and you are interested in either writing or contributing to a book, please visit authors.packtpub.com.

Share Your Thoughts

Once you've read *Privacy-Preserving Machine Learning*, we'd love to hear your thoughts! Scan the QR code below to go straight to the Amazon review page for this book and share your feedback.

https://packt.link/r/1-800-56467-8

Your review is important to us and the tech community and will help us make sure we're delivering excellent quality content.

Download a free PDF copy of this book

Thanks for purchasing this book!

Do you like to read on the go but are unable to carry your print books everywhere?

Is your eBook purchase not compatible with the device of your choice?

Don't worry, now with every Packt book you get a DRM-free PDF version of that book at no cost.

Read anywhere, any place, on any device. Search, copy, and paste code from your favorite technical books directly into your application.

The perks don't stop there, you can get exclusive access to discounts, newsletters, and great free content in your inbox daily

Follow these simple steps to get the benefits:

1. Scan the QR code or visit the link below

https://packt.link/free-ebook/9781800564671

2. Submit your proof of purchase

3. That's it! We'll send your free PDF and other benefits to your email directly

Part 1: Introduction to Data Privacy and Machine Learning

This part provides an introduction to the fundamental concepts of data privacy and the distinction between sensitive data and personal sensitive data, along with the importance of data privacy regulations. The concept of privacy by design is discussed, emphasizing the proactive integration of privacy measures into systems and processes. Additionally, notable privacy breaches in major enterprise companies are examined, highlighting the potential consequences and risks associated with such incidents. This introduction sets the foundation for understanding the significance of data privacy and the need for robust privacy measures. This part also covers privacy threat modeling using the LINDDUN framework in detail.

The second chapter in this part focuses on the different phases of the machine learning pipeline and the privacy threats and attacks that can occur at each stage. We will explore the phases of data collection, data preprocessing, model training, and inference. Within each phase, specific privacy threats and attacks, such as model inversion attacks and training data extraction attacks, are discussed in detail, providing illustrative examples. The importance of protecting training data privacy, input data privacy, model privacy, and inference/output data privacy is emphasized. This part highlights the potential risks and challenges associated with privacy in machine learning, underlining the need for robust privacy preservation techniques throughout the entire process. Exploration of privacy threats and attacks in each phase of the machine learning pipeline sheds light on the challenges of preserving privacy in machine learning systems.

This part has the following chapters:

- *Chapter 1, Introduction to Data Privacy, Privacy Breaches, and Threat Modeling*
- *Chapter 2, Machine Learning Phases and Privacy Threats/Attacks in Each Phase*

1

Introduction to Data Privacy, Privacy Breaches, and Threat Modeling

Privacy-preserving **machine learning** (**ML**) is becoming increasingly important in today's digital age, where the use of personal data is ubiquitous in various industries, including healthcare, finance, and marketing. While ML can bring many benefits, such as improved accuracy and efficiency, it also raises significant concerns about privacy and security. Many individuals are increasingly concerned about the risks associated with the use of their personal data, including unauthorized access, misuse, and abuse. Furthermore, there are regulations such as the **General Data Protection Regulation** (**GDPR**) and the **California Consumer Privacy Act** (**CCPA**) that require organizations to comply with strict privacy guidelines while processing personal data.

This book provides a comprehensive understanding of the techniques and tools available to protect individuals' privacy while enabling effective ML. This book will help researchers, ML engineers, software engineers, and practitioners to understand the importance of privacy and how to incorporate it into their ML algorithms and data processing pipelines. This book bridges the gap between the theoretical foundations of privacy and the practical implementation of privacy-preserving ML techniques, enabling data-driven decision-making without compromising individuals' privacy.

In this introductory chapter, we are going to learn about privacy, including data privacy; sensitive data versus personal sensitive data; data privacy regulations; **Privacy by Design** (**PbD**) concepts; and why data privacy is important. Once we have discussed these concepts, we will cover privacy threat modeling using the LINDDUN framework in detail and explain linkability and identifiability threats with an example. This chapter will help you to better understand privacy and why it is important. We will discuss key privacy regulations, such as the GDPR and CPRA, at a high level, as well as privacy threat modeling. At the end of this chapter, we will cover the need for privacy-preserving ML and a use case.

We will cover the following main topics:

- What do privacy and data privacy mean?
- Privacy by Design and a case study
- Privacy breaches
- Privacy threat modeling
- The need for privacy-preserving ML

What do privacy and data privacy mean?

Alan Westin's theory describes privacy as the control over how information about a person is handled and communicated to others. Irwin Altman added that privacy includes limiting social interaction and included regulating personal space and territory.

Personal data includes any information that by itself or in conjunction with other elements can be used to identify an individual, such as their name, age, gender, personal identification number, race, religion, address, email address, biometric data, device IDs, medical data, and genetics data, based on the regulations defined in the country where the orginated from.

Privacy refers to an individual's ability to keep their information, whether it is personal or non-personal data, to themselves and share data based on their consent. Privacy helps individuals maintain autonomy over their personal lives.

Data privacy focuses on the use and governance of personal data and policies to ensure that data is collected, processed, shared, and used/inferred in an appropriate way.

Privacy regulations

As per the latest statistics of the **United Nations Conference on Trade and Development** (**UNCTAD**), 71% of countries have their own privacy laws, which shows the importance of privacy and data protection across the world.

Most privacy laws deal with the collection of sensitive personal data, data processing, sharing data with other parties, and data subject rights. 137 out of 194 countries in the world have legal legislation to protect data and individuals' data privacy.

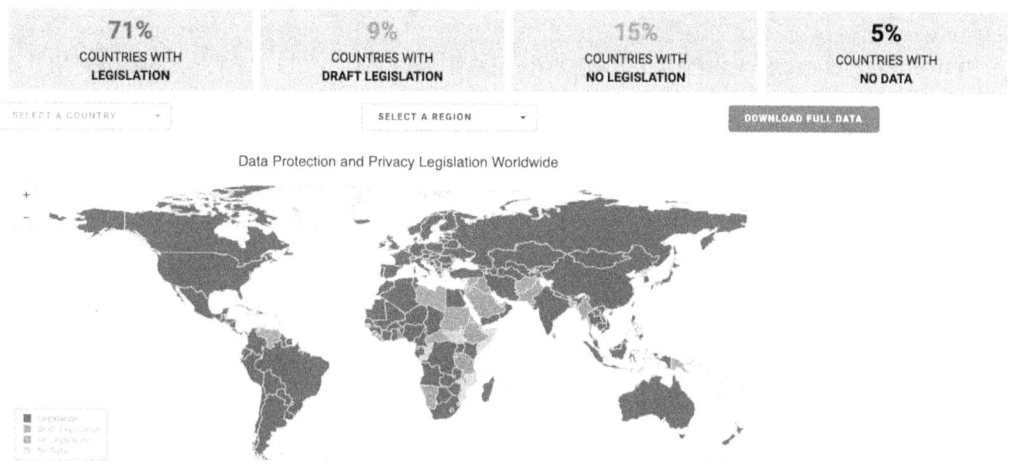

Figure 1.1 – Privacy legislation worldwide as of December 2021

Source: https://unctad.org/page/data-protection-and-privacy-legislation-worldwide

Out of these privacy regulations across the world, the most popular and widely implemented ones are the **GDPR** in Europe and the **CCPA** in the US.

As per the GDPR, personal data is defined as follows:

> *The data subjects are identifiable if they can be directly or indirectly identified, especially by reference to an identifier such as a name, an identification number, location data, an online identifier or one of several special characteristics, which expresses the physical, physiological, genetic, mental, commercial, cultural or social identity of these natural persons. In practice, these also include all data which are or can be assigned to a person in any kind of way. For example, the telephone, credit card or personnel number of a person, account data, number plate, appearance, customer number or address are all personal data.*

In this definition, the keywords are whether the person can be identified either *directly* or *indirectly* using an identifier mentioned as name. We will learn more about indirect identification, how individuals can be identified through indirect identifiers, and how their privacy is compromised in the *Privacy threat modeling* section.

The GDPR also defines sensitive personal data, which includes genetic, biometric, and health data, as well as personal data revealing racial and ethnic origin, political opinions, religious or ideological convictions, or trade union membership. Most regulations have articles/sections covering the following when working with personal data, non-personal data, and sensitive personal data:

- **Purpose, scope, and definition of personal data**: The purpose of the privacy law, specifying its scope and outlining the types of data and entities covered by the law. It clarifies the legal framework's intent and applicability and definitions of terms such as personal data and sensitive personal data.

- **Privacy enforcement authority**: Regulations define the role of the data protection authority or supervisory body responsible for overseeing compliance with the law, providing guidance, and handling complaints.

- **Fines and penalties**: Most laws have different fines/penalties based on the nature of privacy violations in that country. For example, the GDPR imposes a fine of 20 million euros, or up to 4% of the total global turnover of the preceding fiscal year of the company, whichever is higher for severe violations.

- **Rights**: Some countries define privacy as a fundamental right of people in that country. Each individual has rights to their data, that is, the right to know, remove, forget, and delete data.

The following table lists the data subject rights defined by popular privacy laws across the world:

Privacy Laws	Data Subject Rights
GDPR	Right to be informed
	Right to access
	Right to rectification
	Right to erasure/be forgotten
	Right to data portability
	Right to restrict data processing
	Right to withdraw consent
	Right to object processing
	Right to object automated decision-making

CCPA	Right to know
	Right to access
	Right to delete
	Right to opt-out
	Right to non-discrimination
	Right to correct
	Right to limit
LGPD (This is Brazil's General Personal Data Protection Law – Lei Geral de Proteção de Dados (LGPD))	Right to be informed
	Right to access
	Right to rectification
	Right to erasure
	Right to data portability
	Right to object processing
	Right to object automated decision-making

Table 1.1 – Data subject rights

We have gone through a high-level definition of privacy, privacy regulations in various countries, and data subject requests (rights of individuals).

Let's now learn more about PbD, what it is, and how it helps to protect data privacy.

Privacy by Design and a case study

The concept of PbD was created by Ann Cavoukian in the 1990s and presented in her 2009 presentation, *"Privacy by Design: The Definitive Workshop."* As Cavoukian states, the concept of PbD encompasses more than just technology.

PbD is a framework that promotes the integration of privacy and data protection principles into the design and development of systems, products, and services.

The PbD framework has seven foundational principles. The objective of these principles is to ensure that privacy is embedded in every stage of a system's development and that data subjects' privacy rights are protected:

- **Proactive not reactive measures**: PbD requires that privacy considerations be integrated into the design and development of a system from the outset, rather than being added as an afterthought.

- **Privacy as the default setting**: PbD requires that privacy settings be set to the highest level by default and that users must opt-in to more invasive settings.

- **End-to-end security**: PbD requires that privacy and security measures be integrated throughout the entire life cycle of a system, from design and development to deployment and decommissioning. I strongly suggest using a "begin with privacy" approach instead of shift-left privacy. In this way, privacy begins from the software requirements phase itself.

- **Full functionality**: PbD requires that privacy and data protection measures be integrated in a way that does not compromise the functionality of the system or product.

- **Visibility and transparency**: PbD requires that users are informed of the privacy risks associated with a system or product and that they have access to information about how their data is being collected, used, and shared.

- **Respect for user privacy**: PbD requires that users have control over their personal data and that their privacy preferences are respected.

- **Holistic approach**: PbD requires that privacy and data protection considerations are integrated into all aspects of a system or product, including its technical design, operational procedures, and business practices.

The PbD approach has become increasingly important in recent years, as privacy concerns have grown in response to the rapid expansion of data-driven technologies.

PbD is now widely recognized as a best practice for organizations that process personal data and is an important component of data protection regulations, such as the EU's GDPR.

Overall, PbD is a comprehensive framework that aims to ensure that privacy is an integral part of any system or product and that data protection is considered from the beginning of the development process, rather than as an afterthought.

Let's walk through an example to understand PbD in detail.

Example – Privacy by Design in a social media platform

PbD is a framework that advocates for embedding privacy considerations into the design and architecture of systems, products, and services from the very beginning. By incorporating privacy as a core component, organizations can proactively address privacy concerns and ensure the protection of user data. This example case study illustrates how a social media platform can implement PbD principles.

Case study description

Let's consider a hypothetical social media platform called "MyConnect," which aims to prioritize user privacy and data protection by implementing PbD principles throughout its development and operation. We will explore its principles one by one.

- **Minimized data collection**: MyConnect follows a privacy-focused approach by collecting only necessary user data. It only requests information that is directly relevant to providing the platform's core functionality. Unnecessary data points, such as excessive personal details or invasive tracking information, are deliberately avoided.

- **Privacy-oriented default settings**: MyConnect implements privacy-oriented settings to protect user privacy by default. For example, it sets user profiles to private, limiting the visibility of user information to only approved connections. Additionally, it enables opt-in consent for features such as location sharing, ensuring that users have to actively choose to share their location.

- **Granular privacy controls**: MyConnect offers granular privacy controls, empowering users to manage their privacy preferences. Users have control over who can view their posts, access their profile information, and send connection requests. The platform provides easy-to-use privacy settings that allow users to customize their privacy levels according to their preferences.

- **Secure data storage and encryption**: MyConnect prioritizes the security of user data by employing strong encryption mechanisms. User data, including personal information and communications, is stored securely and encrypted both at rest and during transmission. This ensures that even if a data breach occurs, the data remains unreadable and protected.

- **Regular security audits and updates**: MyConnect conducts regular security audits to identify potential vulnerabilities and address them promptly. It stays updated with the latest security measures and patches any identified security weaknesses to ensure the ongoing protection of user data.

- **Transparency and user education**: MyConnect maintains transparency with its users by providing clear and concise privacy policies and terms of service. It educates users about their rights, the data collected, and how it is used. The platform also offers user-friendly guides and resources to educate users about privacy best practices and how to protect their information. MyConnect also implements a new way of sharing the details of how it protects data privacy on its platform through "privacy data sheets."

The following mapping shows how the PbD principle is implemented at the social media platform company:

Privacy By Design – Principle	*MyConnect Implementation*
Proactive not reactive measures	Minimized data collection
Privacy as the default setting	Privacy-oriented default settings
Respect for user privacy	Granular privacy controls
End-to-end security	Secure data storage and encryption
Holistic approach	Regular security audits and updates
Visibility and transparency	Transparency and user education

Table 1.2 - MyConnect implementation

Benefits and outcomes

Implementing PbD principles in MyConnect yields several key benefits:

- **Enhanced user trust**: Users of MyConnect feel confident that their privacy is respected and their data is protected. The platform's commitment to privacy empowers users to engage and share content without undue concerns about their personal information being misused.

- **Compliance with privacy regulations**: By incorporating PbD principles, MyConnect ensures compliance with privacy regulations, such as the GDPR. This protects the platform from legal and reputational risks associated with privacy breaches.

- **Positive reputation and differentiation**: MyConnect gains a competitive advantage by promoting itself as a privacy-conscious social media platform. Its PbD approach can attract privacy-conscious users who prioritize the protection of their personal information.

- **Reduced privacy incidents and breaches**: PbD practices reduce the likelihood of privacy incidents and data breaches. By incorporating privacy considerations from the start of the project, MyConnect minimizes the potential vulnerabilities that could lead to unauthorized access or misuse of user data. We will go through privacy breaches in more detail in the next section.

MyConnect's implementation of PbD principles showcases the significance of considering privacy as a fundamental component in the design and operation of a social media platform by prioritizing minimized data collection, privacy-oriented defaults, granular privacy controls, secure data storage, regular audits, transparency, and user education.

Privacy breaches

What is a privacy breach?

A privacy breach, also known as a data breach, refers to an incident where unauthorized individuals or entities gain access to confidential or sensitive information without proper authorization. This breach of privacy can occur in various forms, such as hacking, theft, accidental exposure, or improper handling of data. It typically involves the unauthorized access, acquisition, disclosure, or use of personal information, which may include personally identifiable information (PII) such as names, addresses, social security numbers, financial details, and login credentials.

Privacy breaches can have serious consequences for individuals, organizations, and even society. They can lead to identity theft, financial fraud, reputational damage, loss of trust, legal implications, and emotional distress for those affected. Protecting personal data and maintaining privacy is essential to ensure the security and well-being of individuals and maintain trust in digital systems and services.

The following are some examples of privacy breaches: one involves a company utilizing web technologies in their product, while the other concerns a company incorporating **artificial intelligence (AI)** and ML into their products and services.

Equifax privacy breach

The Equifax privacy breach refers to a massive data breach that occurred in 2017, in which the personal information of approximately 147 million people was compromised. Equifax is one of the largest consumer credit reporting agencies in the United States, and the breach was one of the most significant data breaches in history.

The breach occurred when hackers exploited a vulnerability in Equifax's website software, allowing them to gain access to sensitive information such as names, social security numbers, birth dates, addresses, and, in some cases, driver's license numbers and credit card information.

The breach went undetected for several months, during which time the hackers were able to access and steal the information. The Equifax breach was a significant event, and it highlighted the importance of cybersecurity and the need for companies to take proactive measures to protect their customers' data.

The breach also resulted in numerous investigations, lawsuits, and settlements against Equifax, with the company ultimately agreeing to pay over $700 million in damages and penalties. In addition to the financial impact, the breach had serious consequences for the affected individuals, who were at risk of identity theft and other fraudulent activities. The breach highlighted the need for individuals to be vigilant about monitoring their credit reports, protecting their personal information, and taking steps to protect themselves from identity theft.

The attackers used a combination of techniques, including SQL injection and **cross-site scripting (XSS)**, to gain access to sensitive data stored in Equifax's databases. SQL injection is a type of attack in which an attacker injects malicious code into a SQL statement, allowing them to execute unauthorized actions on a database. In this case, the attackers used SQL injection to bypass Equifax's security controls and gain access to the personal information of millions of individuals.

The attackers also used XSS attacks, which involve injecting malicious code into a website to steal sensitive data from users. In this case, the attackers were able to inject malicious code into Equifax's website. Once the attackers gained access to Equifax's systems, they were able to extract large amounts of data over a period of several months without being detected. The data stolen included names, social security numbers, birth dates, addresses, driver's license numbers, and credit card information.

The Equifax breach highlights the importance of cybersecurity and the need for companies to take proactive measures to protect their systems and data from attackers. It also underscores the importance of ongoing monitoring and detection to quickly identify and respond to potential security threats.

Source: `https://en.wikipedia.org/wiki/2017_Equifax_data_breach`

Clearview AI Privacy breach

Clearview AI is a technology company that developed a controversial facial recognition system. The company's software was designed to match images of individuals with publicly available photos, scraping data from various sources on the internet, including social media platforms. The system gained widespread attention due to concerns over privacy and ethical implications.

In early 2020, Clearview AI found itself at the center of a major privacy breach. It was revealed that the company had amassed a massive database of billions of facial images without the knowledge or consent of the individuals involved. These images were collected from various online platforms, including Facebook, Instagram, and X (formerly Twitter).

The breach was brought to light by investigative reports and researchers who discovered that Clearview AI's database was accessible to law enforcement agencies and other organizations. It raised significant concerns about the potential misuse of the technology, as it could be employed for mass surveillance, tracking individuals, or invading people's privacy without their knowledge.

One of the primary concerns regarding Clearview AI's practices was the lack of transparency and consent. Individuals whose photos were included in the database had not given their permission or even been aware that their images were being used in this manner. Clearview AI's scraping of publicly available data bypassed many social media platforms' terms of service, further exacerbating the privacy issues.

The breach prompted legal and ethical debates about the use of facial recognition technology and the need for stronger regulations. Critics argued that Clearview AI's practices were an invasion of privacy, as people's faces were being used as biometric identifiers without their consent.

Additionally, there were concerns about the potential for racial bias and discrimination in the system, as facial recognition algorithms have shown to be less accurate for certain demographics. Following the revelation of the privacy breach, Clearview AI faced significant backlash from privacy advocates, technology experts, and the public. Several lawsuits were filed against the company, accusing it of violating privacy laws and regulations. As a result, Clearview AI was subject to investigations by various regulatory authorities.

In response to the backlash, Clearview AI made efforts to improve its practices and address privacy concerns. The company claimed to have implemented stricter policies regarding data access and established a verification system for potential clients. However, skepticism remains regarding the efficacy of these measures and the overall ethics of the company's operations.

The Clearview AI privacy breach serves as a cautionary tale about the potential dangers of unchecked facial recognition technology and the importance of safeguarding personal privacy. It has fueled discussions surrounding privacy laws, regulation of emerging technologies, and the ethical implications of mass surveillance. As the debate continues, it remains crucial to strike a balance between technological advancement and protecting individuals' rights and privacy.

Privacy threat modeling

In an increasingly digital world, privacy has become a paramount concern for individuals, organizations, and societies at large. With the widespread collection and processing of personal data, it is essential to assess and mitigate privacy threats effectively.

Privacy threat modeling – definition

Privacy threat modeling is a proactive process that aims to identify and understand potential threats to privacy before they materialize. By examining the system's architecture, data flows, and interactions, privacy threat modeling allows for the identification of vulnerabilities and risks that may compromise individuals' privacy. It helps organizations anticipate and address privacy concerns during the design and development stages, ensuring privacy protections are integrated into the system from the outset.

The importance of privacy threat modeling

Privacy threat modeling offers several key benefits, including the following:

- **Risk identification**: By systematically assessing potential privacy threats, organizations can identify and understand the risks they face. This knowledge enables them to prioritize privacy controls and allocate resources effectively.

- **PbD approach**: Privacy threat modeling encourages a privacy-centric approach to system design and development. By integrating privacy considerations early on, organizations can save time, effort, and costs that may otherwise be required for retrofitting privacy safeguards.

- **Compliance and accountability**: Privacy regulations and standards, such as the GDPR, require organizations to implement privacy measures. Privacy threat modeling helps organizations fulfill these requirements by identifying and addressing potential compliance gaps.

- **Stakeholder trust**: Demonstrating a commitment to privacy protection enhances stakeholder trust. Privacy threat modeling provides a systematic way to showcase an organization's dedication to safeguarding individuals' privacy, leading to increased confidence among users (internal and external) and customers. It builds a culture of responsible development.

Continuous privacy threat modeling helps clarify the requirements for privacy and allows organizations to move toward building standard privacy features and patterns. Focusing on proactive issue identification and fixing helps companies build a privacy-forward culture.

Privacy threat modeling's alignment to Privacy by Design principles

Privacy threat modeling involves identifying potential privacy risks and vulnerabilities within a system. While it doesn't directly encompass all PbD principles, it is an essential step in implementing those principles effectively.

Here's how privacy threat modeling aligns with different PbD principles:

- **Data minimization**: Helps identify areas where data collection might be excessive or unnecessary, leading to potential privacy risks

- **Purpose specification**: Identifies scenarios where collected data might be used for unintended purposes, helping to ensure that data use is appropriately specified

- **Consent mechanisms**: Highlights instances where data might be collected or used without proper user consent, assisting in designing effective consent processes

- **Access controls**: Identifies potential unauthorized access points, guiding the implementation of access controls to prevent unauthorized data exposure

- **Data encryption**: Reveals vulnerabilities in data storage or transmission that could lead to data breaches, informing the need for encryption

- **User empowerment**: Helps identify areas where users might lack control over their data, prompting the implementation of tools for user data management

- **Security measures**: Identifies potential security weaknesses that could compromise user data, contributing to the implementation of robust security measures

- **Regular audits and assessments**: Supports ongoing assessments by identifying areas of potential vulnerability that require regular monitoring and evaluation

While privacy threat modeling is not a direct substitute for PbD principles, it plays a crucial role in shaping the design and development process by identifying potential risks and vulnerabilities. The insights gained from threat modeling enable organizations to effectively apply PbD principles to address those risks and enhance the overall privacy posture of their systems.

Steps in privacy threat modeling

Performing an effective privacy threat assessment involves the following steps:

1. **Define the system**: Clearly define the scope of the system or application under assessment. Identify its components, data flows, and interfaces with other systems.

2. **Identify data types**: Determine the types of personal data the system processes and stores. Categorize the data based on sensitivity and regulatory requirements.

3. **Identify threat sources**: Enumerate potential threat sources, both internal and external, that may attempt to compromise the privacy of the system's data or users.

4. **Analyze threat scenarios**: Develop realistic threat scenarios by combining threat sources and system components. Consider scenarios that may exploit vulnerabilities in data handling, storage, transmission, or user interactions.

5. **Assess impact and likelihood**: Evaluate the potential impact and likelihood of each threat scenario materializing. Consider the potential harm to individuals, regulatory penalties, reputation damage, and other relevant factors.

6. **Identify controls**: Identify and implement appropriate privacy controls and safeguards to mitigate identified threats. Consider technical, organizational, and procedural measures to address vulnerabilities and protect privacy.

7. **Document and communicate**: Document the privacy threat assessment process, including identified threats, mitigations, and residual risks. Communicate the findings to stakeholders, such as system designers, developers, privacy officers, and management, to ensure collective awareness and buy-in.

8. **Review and update**: Regularly review and update the privacy threat assessment as the system evolves or new threats emerge.

Privacy threat modeling is an iterative process that should be integrated into the organization's ongoing privacy management practices

Privacy threat modeling frameworks

There are several privacy threat modeling frameworks available that provide structured methodologies and guidelines to assess and mitigate privacy risks effectively. Let's explore some of the widely recognized privacy threat modeling frameworks:

- **STRIDE (Microsoft)**: The STRIDE framework, originally developed by Microsoft, focuses on identifying threats to the security and privacy of a system. It stands for the following:

  - **Spoofing identity**: Unauthorized actors masquerade as legitimate users

  - **Tampering with data**: Unauthorized modification or destruction of data

  - **Repudiation**: Denial of actions or transactions by malicious actors

  - **Information disclosure**: Unauthorized access to sensitive information

  - **Denial of service**: Disruption or degradation of system availability

  - **Elevation of privilege**: Unauthorized escalation of privileges

The STRIDE framework helps identify potential privacy threats by considering how each threat category could impact the privacy of the system's users and data.

- **LINDDUN:** This is a privacy threat modeling framework. It provides a comprehensive approach to identify and address privacy concerns. The components of the LINDDUN framework are as follows:

  - **Linkability**: Assessing the potential for linking various data points to identify individuals

  - **Identifiability**: Evaluating the extent to which individuals can be identified or re-identified from the data

  - **Non-repudiation**: Ensuring that actions and transactions cannot be denied

  - **Detectability**: Assessing the ability to detect privacy breaches or unauthorized access

  - **Data disclosure**: Excessively collecting, storing, processing, or sharing personal data

  - **Unawareness**: Evaluating the level of user awareness and control over data collection and usage

  - **Non-compliance**: Identifying risks of non-compliance with privacy regulations and standards

  LINDDUN provides a holistic view of privacy threats and helps organizations analyze the impact of these threats on individuals' privacy.

- **PLOT4AI:** The **Privacy Library of Threats for Artificial Intelligence** (**PLOT4AI**) is a comprehensive resource meticulously crafted to tackle the intricate web of privacy concerns entwined with AI technology. Presently, the library comprises an assemblage of 86 unique threats, meticulously categorized into eight distinct groupings:

  - **Techniques and processes**: This category covers the potential downsides stemming from processes or technical maneuvers capable of detrimentally affecting individuals

  - **Accessibility**: This aims to rectify the deficiency in the accessibility and user-friendliness of AI systems for a diverse range of individuals

  - **Identifiability and linkability**: This casts a spotlight on threats and linking individuals to specific attributes or other individuals, coupled with apprehensions surrounding identification

  - **Security**: This zooms in on the potential perils arising from inadequately fortified AI systems and procedures against security vulnerabilities

  - **Safety**: This concentrates efforts on recognizing perils and shielding individuals from plausible harm or jeopardy

  - **Unawareness**: This confronts the issue of neglecting to inform individuals and extends to them the chance to intervene

- **Ethics and human rights**: This illuminates conceivable adverse effects on individuals or the harm borne out of an absence of consideration for values and principles

- **Non-compliance**: This directs attention toward threats emerging from the failure to adhere to data protection laws and other pertinent regulations

This repository stands as a robust arsenal to empower the AI community and stakeholders in safeguarding the paramount importance of privacy as AI continues to unfold its potential.

The library introduces a simplified four-phase **development life cycle** (**DLC**) approach, aligning with various methodologies, such as SEMMA, CRISP-DM, ASUM-DM, TDSP, and MDM. This streamlined approach ensures accessibility for non-technical stakeholders while maintaining alignment with established methodologies.

Data flow diagrams (**DFDs**) are employed as visual representations of systems under analysis. PLOT4AI emphasizes the importance of thorough threat modeling and suggests using both basic and detailed DFDs for different categories of threats.

Threats are presented in the form of cards, similar to LINDDUN GO, categorized by colors and icons representing the threat category and DLC phase. Some threats might have multiple category icons or DLC icons, reflecting their diverse impacts.

To practically apply PLOT4AI, it can be used as a card game in both physical and digital formats. Sessions are timeboxed to maintain engagement and focus, involving diverse stakeholders and a facilitator. In the sessions, participants are guided through each threat card's question, discussion, and potential recommendations.

By utilizing PLOT4AI, organizations can enhance their privacy practices, mitigate risks, streamline processes, and foster collaboration among stakeholders. The library's output can also contribute to data privacy impact assessments and facilitate compliance efforts.

While still in development, PLOT4AI offers benefits such as improved processes, reduced rework, clearer purpose, and alignment among stakeholders. The resource provides valuable insights into humanizing AI through the lens of privacy threat modeling.

The company provides an online assessment tool that, through responding to its queries, enables individuals to discern the threat model pertinent to the ML systems or products they are tasked with enhancing.

Here is a link to the assessment tool: `https://plot4.ai/assessments/`.

Let's deep dive into one of the frameworks, LINDDUN, with a detailed example to understand more about privacy threat modeling.

The LINDDUN framework

LINDDUN is a privacy threat modeling methodology that supports analysts in systematically eliciting and mitigating privacy threats in software architectures. This framework was developed by privacy experts at KU Leuven.

The LINDDUN framework consists of three main steps:

1. Model the system.
2. Elicit threats.
3. Mitigate threats.

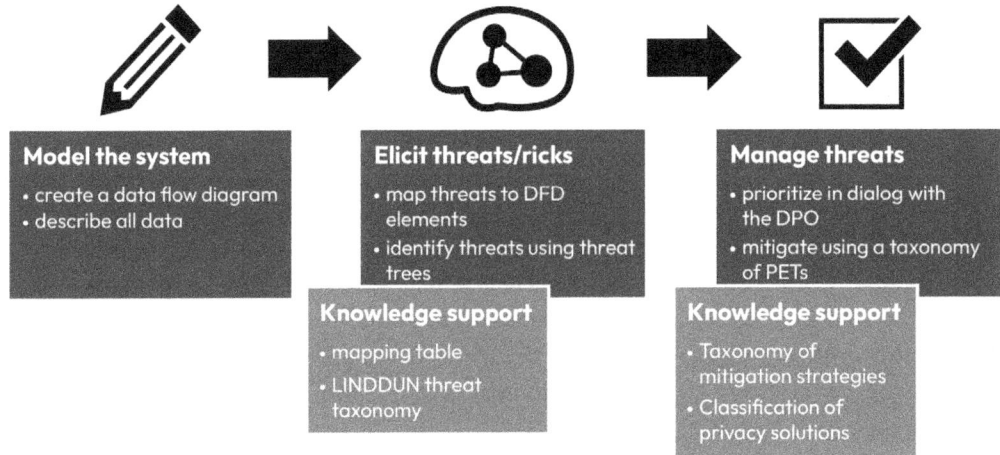

Figure 1.2 – LINDDUN framework steps

Step 1 – modeling the system

In this step, it is crucial to gain a comprehensive understanding of the system or product being developed, including detailed knowledge of data flows. This entails comprehending how data is collected, processed, utilized, retained, and shared, as well as identifying the system's users and their access methods. Additionally, it is essential to understand how the system interacts with other systems. To facilitate the analysis of privacy threats, LINDDUN employs DFDs as a means of capturing system or product knowledge, in the same way that STRIDE (Microsoft's security threat modeling method) is utilized for security threat modeling.

Step 2 – eliciting and documenting threats

LINDDUN encompasses seven distinct threat categories that enable the identification of threats through the utilization of threat trees. The framework provides the following high-level descriptions of these seven threat categories:

Threat Category	Privacy Threat Details
Linkability	This category involves linking two or more publicly available data sets to derive insights. Often, data subjects are unaware that their data can be linked with other datasets from the web, leading to the exposure of personal information, for instance, linking individuals with similar diseases or linking people who visit a specific mall.
Identifiability	Identifiability refers to the ability to identify specific personal information about an individual, such as their email, age, or gender, based on other available data through alternative means, for example, identifying a consumer based on their transaction data or identifying the readers of a particular website.
Non-repudiation	Non-repudiation ensures that a data subject cannot deny their involvement in a particular action. This can apply to instances such as social media comments or posts that can be attributed to a specific individual.
Detectability	Detectability involves the ability to determine whether a particular item of interest relating to a data subject exists. New personal information can often be derived based on the available data, for instance, by analyzing the type of social media posts to identify the author and extract further insights.
Disclosure of information	This category pertains to the ability to learn the content of a specific item of interest about a data subject. It involves accessing and obtaining personal information about an individual.
Unawareness	Unawareness refers to situations where data subjects are uninformed about the collection, processing, storage, or sharing of activities related to their personal data, including the corresponding purposes. Examples include collecting personal data without user consent, sharing it with third parties, or processing the data to generate insights, all without the knowledge of the data subjects.
Non-compliance	Non-compliance arises when systems collect and process personal data without adhering to privacy regulations. This can involve actions such as collecting data without consent, processing and sharing data without proper encryption, or retaining data for a longer period than necessary. These threat categories provide a framework for analyzing privacy threats within LINDDUN, enabling a comprehensive understanding of potential risks associated with personal data.

Table 1.3 – LINDDUN categories at a high level

Step 2a – map DFD elements to threat categories

Prepare a mapping table for each DFD item in the system and LINDDUN threats:

Data Flow Element	L	I	N	D	D	U	N
Entity	X						
Data store	X	X				X	X
Data flow			X				
Process	X	X		D			

Table 1.4 – Map between DFD and threat category

Put "X" marks where the DFD element has a potential privacy threat against the threat category.

Step 2b – elicit and document threats

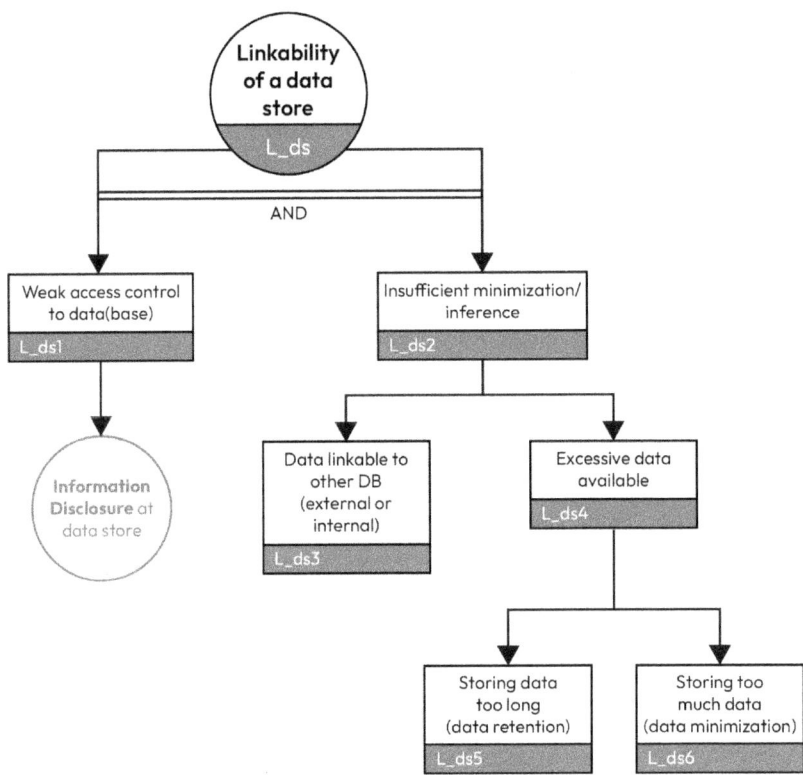

Figure 1.3 – Threat documentation steps

Use the LINDDUN trees to elicit and document the threats for each DFD category. In the preceding example, the **Linkability** category tree consists of the possible privacy tree model for data store linkability.

Step 3 – mitigating threats

In this step, identified privacy threats are handled appropriately.

Step 3a – prioritize the threats

Prioritize the threats to privacy as either low, medium, or high risk.

Step 3b – mitigation strategy

For each identified threat, plan for a mitigation strategy. The LINDDUN framework provides the following taxonomy of mitigation strategies for privacy threats.

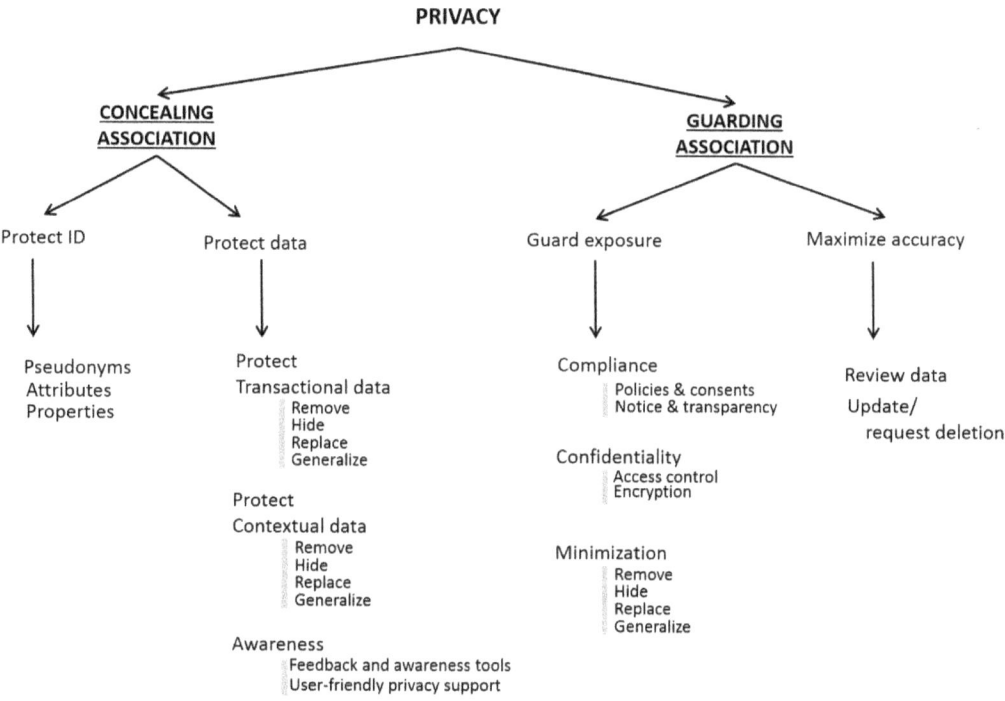

Figure 1.4 – Mitigation strategy for privacy threats

Source: LINDDUN mitigation strategies taxonomy (`https://www.linddun.org/linddun`)

Step 3c – select a privacy-enhancing solution

The final step of the framework involves planning a privacy-enhanced solution for each threat mitigation strategy.

Let's consider a practical example to explain the LINDDUN framework.

Example – social media platform (MyConnect) privacy threat modeling

In this example, we will use the LINDDUN framework to assess the privacy threats associated with the social media platform we looked at earlier in this chapter. The goal is to identify potential risks and vulnerabilities that may compromise user privacy:

- **Linkability**: Linkability refers to the ability to link different data points to identify individuals. In the case of a social media platform, linkability threats may include the following:

  - **User profile linking**: Assess the extent to which user profiles can be linked across different social media platforms or services, potentially revealing more information about individuals than intended

  - **Cross-device tracking**: Evaluate the risk of tracking users' activities across multiple devices to create a comprehensive profile and track their online behavior

- **Identifiability**: Identifiability focuses on the potential for individuals to be identified or re-identified from the data. Privacy threats related to identifiability may include the following:

  - **De-anonymization**: Assess the risk of adversaries being able to de-anonymize users' data by combining and analyzing various datasets

  - **Re-identification attacks**: Evaluate the likelihood of attackers being able to re-identify users by linking seemingly anonymous data with external datasets

- **Non-repudiation**: Non-repudiation ensures that actions and transactions cannot be denied. In the context of a social media platform, non-repudiation threats may include the following:

  - **User content alteration**: Assess the risk of unauthorized modifications or the tampering of user-generated content, potentially leading to false attributions or denial of user actions

  - **Account hijacking**: Evaluate the likelihood of malicious actors gaining unauthorized access to user accounts and performing actions on behalf of legitimate users, leading to the denial of user involvement

- **Detectability**: Detectability focuses on the ability to detect privacy breaches or unauthorized access. Potential threats in this category for a social media platform include the following:

  - **Unauthorized access monitoring**: Assess the platform's ability to detect and respond to unauthorized access attempts by malicious actors seeking to obtain user data

  - **Data leakage monitoring**: Evaluate the platform's capability to monitor and detect data leakage incidents, such as the unauthorized sharing of user data with third parties

- **Unawareness**: Unawareness refers to the level of user awareness and control over data collection and usage. Privacy threats related to unawareness may include the following:

  - **Data collection transparency**: Assess the platform's transparency in informing users about the types of data collected, the purposes, and the third parties with whom the data is shared

  - **Consent management**: Evaluate the effectiveness of the platform's consent mechanisms in obtaining explicit user consent for data processing activities

- **Non-compliance**: This involves identifying risks of non-compliance with privacy regulations and standards. For a social media platform, potential non-compliance threats may include the following:

  - **Inadequate privacy policies**: Assess the platform's privacy policies to ensure they comply with applicable privacy laws and provide clear information to users about data handling practices

  - **Data transfer across jurisdictions**: Evaluate the risks associated with data transfers to jurisdictions with differing privacy regulations and assess compliance with cross-border data transfer requirements

By applying the LINDDUN framework to the social media platform example, we can identify specific threats and vulnerabilities in each category. This allows the organization behind the platform to develop appropriate privacy controls and safeguards to mitigate the identified risks, enhancing user privacy protection and compliance with privacy regulations.

After exploring concepts such as data privacy, privacy threats, and threat modeling, it is time to delve into the central focus of this book: privacy-preserving ML. In this section, we will discover why it is necessary and explore the various technologies associated with privacy-preserving ML.

The need for privacy-preserving ML

Privacy-preserving ML has emerged as a response to the growing concerns about data privacy in AI and ML and the need to protect sensitive information while leveraging the power of ML algorithms. You will learn about ML and privacy-preserving ML techniques in the next chapter.

The following are some of the reasons why privacy-preserving ML is necessary:

- **Protection of sensitive data**: Privacy-preserving ML techniques enable organizations to utilize sensitive data without exposing it to potential breaches or unauthorized access. By implementing privacy safeguards, organizations can protect sensitive information such as PII, medical records, financial data, or proprietary business data. You will learn more in *Chapters 5 to 8* on how to protect sensitive data using different privacy-preserving technologies.

- **Compliance with privacy regulations**: Many jurisdictions have enacted strict privacy regulations and data protection laws, such as the European Union's **GDPR** and the **CCPA**. Privacy-preserving ML techniques help organizations comply with these regulations by ensuring that user data is processed in a privacy-conscious manner.

- **Preserving user trust**: Privacy breaches and data misuse incidents can severely damage user trust in organizations that handle personal data. By adopting privacy-preserving ML techniques, organizations demonstrate their commitment to protecting user privacy, thereby enhancing trust and fostering long-term relationships with customers or users.

- **Collaboration and data sharing**: Privacy-preserving ML techniques enable secure collaboration and data sharing among multiple organizations. These techniques allow organizations to combine their datasets without directly exposing sensitive information, facilitating joint research, and achieving collective insights while respecting data privacy.

- **Fairness and bias mitigation**: Privacy-preserving ML can also contribute to addressing issues related to fairness and bias in ML models. By applying privacy techniques, organizations can protect sensitive attributes and mitigate the risk of discrimination based on factors such as race, gender, or ethnicity, thus promoting fairness in ML applications.

- **Confidentiality in healthcare and research**: In domains such as healthcare and research, privacy-preserving ML techniques are crucial for maintaining the confidentiality of sensitive medical records and personal health information. These techniques allow healthcare providers and researchers to extract valuable insights from aggregated data while preserving patient privacy. You will learn more about aggregated datasets, privacy issues, and how to protect sensitivity and privacy of the individuals in *Chapters 3* and *4*.

- **Protection against insider threats**: Privacy-preserving ML techniques can help safeguard against insider threats by reducing the risk of unauthorized access to sensitive data by individuals within the organization. These techniques enable organizations to limit access to sensitive information and ensure that the privacy of users or customers is maintained, even within their own workforce. You will learn more about insider threats in *Chapter 9*.

Privacy-preserving ML techniques are essential for organizations to strike a balance between utilizing data for valuable insights and preserving the privacy of individuals.

By adopting these techniques, organizations can comply with privacy regulations, maintain user trust, foster collaboration, and promote fairness in ML applications while protecting sensitive information from unauthorized access or misuse.

Case study – privacy-preserving ML in financial institutions

Financial institutions handle large volumes of sensitive customer data, including personal and financial information. As ML techniques become more prevalent in the financial sector, the need to protect customer privacy while deriving insights from this data becomes critical. This case study examines a real-world scenario highlighting the necessity for privacy-preserving ML in financial institutions.

Case study description

ARDHA Bank (a fictional bank for illustration purposes), a leading global bank, aims to leverage ML algorithms to improve its fraud detection capabilities. The bank possesses a vast amount of transaction data, including credit card transactions, account activities, and customer profiles. However, ensuring the privacy and confidentiality of customer data is a top priority for ARDHA Bank. To address this concern, ARDHA Bank is exploring the adoption of the following privacy-preserving ML techniques:

- **Differential privacy**: ARDHA Bank adopts differential privacy techniques to protect customer privacy during the training of ML models. Differential privacy adds noise or perturbation to data to ensure that individual customer information remains obfuscated while still allowing accurate model training and analysis. You will learn more about differential privacy in *Chapters 3 to 5*.

- **Secure data aggregation**: To gather insights from customer data without exposing individual-level details, ARDHA Bank employs secure data aggregation techniques. This allows the bank to extract meaningful statistical information from the data while preserving customer privacy. Aggregated data can be used to train ML models without revealing sensitive information about specific customers. You will learn more about this in *Chapter 4*.

- **Federated learning**: ARDHA Bank implements federated learning, a privacy-preserving technique, to train ML models across multiple distributed data sources. Customer data remains decentralized and encrypted, allowing local models to be trained locally on each data source without sharing raw data. The trained models are then combined and aggregated to create a global model without exposing individual data. You will learn more about federated learning in *Chapters 6 and 7*.

- **Encrypted inference**: When performing fraud detection or other ML tasks in real time, ARDHA Bank employs encrypted inference techniques. By using homomorphic encryption or secure multi-party computation, the bank can execute ML models on encrypted customer data without decrypting it. This ensures that sensitive information remains protected during the prediction phase. You will learn more about this in *Chapter 8*.

Benefits and outcomes

ARDHA Bank has implemented privacy-preserving techniques to ensure the security of customer data. They use methods such as differential privacy, secure data aggregation, federated learning, and encrypted inference to protect individual customer information during the ML process. Despite this focus on privacy, the bank has developed effective fraud detection models by analyzing aggregated and anonymized data, enabling them to identify fraudulent patterns while respecting individual privacy. These efforts also ensure compliance with data protection regulations such as the GDPR and financial industry standards. By valuing customer privacy, ARDHA Bank builds trust, enhances its reputation, and strengthens customer relationships. The bank's approach showcases the significance of safeguarding customer data while making use of advanced analytics in the financial sector.

Privacy-preserving ML has equal significance in other domains as well, such as healthcare, education, and social networking.

Summary

To summarize, we have acquired a high-level understanding of privacy, data privacy, PbD concepts, and privacy threat modeling. Furthermore, we have explored the importance of privacy-preserved ML through a case study.

Moving forward, the next chapter will provide a brief overview of different types of ML (supervised and unsupervised) and the phases involved in ML (data extraction, data preparation, model development, model deployment, and inferencing). Additionally, we will examine the privacy threats and attacks associated with each phase.

Machine Learning Phases and Privacy Threats/Attacks in Each Phase

In this chapter, we will provide a quick refresher on the different types of **machine learning** (**ML**): supervised, unsupervised, and reinforcement learning. We will also review the essential phases or pipelines of ML. You may already be familiar with these; if not, this chapter will serve as a foundational introduction.

Subsequently, we will delve into the crucial topic of privacy preservation within each phase of the ML process. Specifically, we will explore the importance of maintaining privacy in training data, input data, model storage, and inference/output data. Additionally, we will examine various privacy attacks that can occur in each phase, such as training data extraction attacks, model inversion attacks, and model inference attacks. Through detailed examples, we will gain an understanding of how these attacks function and discuss strategies to safeguard against them.

We will cover the following main topics:

- ML types
- Overview of ML phases
- Privacy threats/attacks in the ML phases

ML types

Some of you may already be familiar with the different types of ML, namely supervised ML, unsupervised ML, and reinforcement learning. In the next sections, we will provide a quick refresher on these ML types, summarizing what you may have already learned.

Supervised ML

Supervised ML models involve the development of a mathematical model using a set of input data and corresponding actual output. The input data is known as the training data, while the output is referred to as the predicted output. These models employ mathematical functions to learn from the training data and aim to minimize the errors between the predicted output and the expected output using an optimal function. The training data, which consists of input examples, is typically represented in formats such as arrays, vectors, matrices, or tensors. This data is often referred to as feature data or feature vectors, where each attribute within the data is considered a feature.

Type	Example	Details
Scalar	1	A scalar is a single number.
Vector	[1 2 3 4]	A vector is an array of numbers or objects with different data types.
Matrix	$\begin{bmatrix} 1 & 2 & 3 \\ 4 & 5 & 6 \\ 7 & 8 & 9 \end{bmatrix}$	A matrix is an array of numbers arranged in rows and columns. In order to access the matrix, we need two indexes: column number and row number.
Tensor	[[1 2] [3 4] [5 6] [7 8] [9 0] [0 1]]	A tensor is an n-dimensional array with n>2.

Table 2.1 – Examples of scalar, vector, matrix, and tensor

In mathematical terms, supervised learning in ML can be represented by a model, with parameters, θ. This model acts as a mapping function between input x and output y, denoted as $y = f(x, \theta)$. In this context, x represents a vector of attributes or features with a dimensionality of n. The output or label y can vary in dimension depending on the specific learning task.

To train the model, a training set T, is utilized, which consists of data points in the form of T = {(x , y_i)}, where i ranges from 1 to n, representing the number of input–output pairs.

Supervised ML algorithms typically fall into two categories: regression and classification. These algorithms aim to learn patterns and relationships within the training data to make predictions or assign labels to new, unseen data.

Regression

A target variable in ML is the variable that we aim to predict or forecast. It is also often termed as the dependent variable. It is what the ML model is trained to predict using independent or feature variables. For example, in a house price prediction model, the house price would be the target variable.

Regression is a fundamental concept in ML that focuses on predicting continuous numerical values based on input variables. It is a supervised learning technique that involves analyzing the relationship between the input features and the target variable. The goal of regression is to build a mathematical model that can accurately estimate or approximate the value of the target variable when provided with new input data.

In regression, the target variable, also known as the dependent variable, is a continuous value. The input variables, also called independent variables or features, can be numerical or categorical. The regression model seeks to understand the relationship between these input variables and the target variable, enabling predictions to be made for unseen data.

The performance of a regression model is typically measured by evaluating the closeness of its predictions to the actual target values. Various regression algorithms exist, such as linear regression, polynomial regression, and more complex techniques such as support vector regression and random forest regression. These algorithms use mathematical optimization methods to fit a regression function that minimizes the difference between predicted values and the true values of the target variable.

Regression analysis finds applications in numerous fields, including finance, economics, healthcare, and social sciences, where it is used for tasks such as price prediction, demand forecasting, risk assessment, and trend analysis. By leveraging regression algorithms, valuable insights can be gained from data, allowing for informed decision-making and accurate predictions.

Regression model example

Here's a straightforward example of predicting a target variable using two input features. In this scenario, a model is trained on a set of historical data, typically representing the data from the last X days.

The purpose is to forecast or predict the target variable based on the provided input data. By analyzing patterns and relationships in the historical dataset, the trained model can make predictions about the target variable when presented with new input data. This process allows for forecasting future values or understanding potential outcomes based on the given inputs. The model's accuracy and performance are evaluated based on how well it can predict the target variable compared to the actual values. By leveraging historical data and utilizing ML algorithms, valuable insights can be gained, enabling accurate predictions and informed decision-making.

Feature 1 value	Feature 2 value	Target variable value
10	10	130
10	20	180
20	20	210
20	30	260
30	50	??

Table 2.2 – regression model example

This example was implemented using the scikit-learn library. I have used Python version 3.8 and scikit-learn version 1.2.1.

These are the steps followed in this example:

1. Install Jupyter Notebook (pip install notebook).
2. Open a Python Jupyter notebook (jupyter notebook).
3. Install sci-kit learn libraries (pip install -U scikit-learn).
4. Type the following code and run the model code.

You can directly execute the code using the Jupyter notebook provided in the GitHub location of this book—LinearRegression.ipynb—located at https://github.com/PacktPublishing/Privacy-Preserving-Machine-Learning/blob/main/Chapter%202/LinearRegression.ipynb:

```
import numpy as np
from sklearn.linear_model import LinearRegression

# Prepare sample input data with two features (feature 1, feature 2

X = np.array([[10, 10], [10, 20], [20, 20], [20, 30]])

# Assume that target feature has some relationship with the input
features with the formula y = 3 * x_1 + 5 * x_2 + 50

y = np.dot(X, np.array([3, 5])) + 50

# Run the model with the input data and output values
reg = LinearRegression().fit(X, y)
```

```
reg.score(X, y)
reg.coef_
reg.intercept_
reg.predict(np.array([[30, 50]]))
```

Output: array([390.])

In this example, target variable y is linearly dependent on input variables X_0 and X_1 with the linear equation $y= 3X_0+ 5X_1+50$ (50 is the intercept of the line).

The model uses the root mean square value as an optimal function and predicts the target variable. In this case, it predicts 100% accuracy because of the strong linear relationship between the features and the target variable.

Model persistence and retrieving the persisted model for inference

After developing and testing the model using training data and validating it with test data, the next step is to persist the model. This allows for easy sharing with other developers or engineers without revealing the training data and intricate model details. Additionally, if the model demonstrates sufficient accuracy during training, it can be deployed in production environments.

To persist the model, various formats are supported to store it in the disk or file system. The specific format utilized depends on the framework used to develop the ML or **deep learning** (**DL**) model. By employing these formats, the model can be stored and accessed efficiently, facilitating seamless integration into production systems or collaborations with other team members.

Persisting the model enables reproducibility and scalability, as it can be shared, reused, and deployed in different environments without the need to retrain it from scratch. It also helps protect proprietary information and intellectual property associated with the model, allowing organizations to safeguard their valuable research and development efforts.

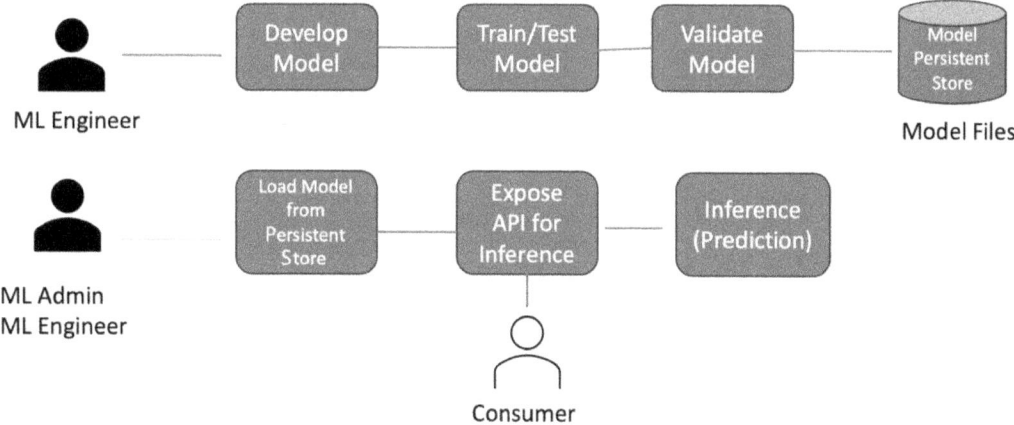

Figure 2.1 – Model persistence and retrieval

The following table shows some formats that are widely used and accepted in the community:

Framework	Model persistence format	Details
scikit-learn `https://scikit-learn.org/`	Joblib Pickle	The Joblib and pickle formats don't require any code changes. The pickle format has security issues, so most frameworks don't advise using the pickle format for model persistence because arbitrary code can be executed during unpickling.
TensorFlow/Keras `https://www.tensorflow.org/`	JSON YAML HDF5	This is model data stored in JSON format or YAML format, and these formats are text-based formats, so they are language-agnostic. Weights are saved in HDF5 format.
PyTorch `https://pytorch.org/`	state_dict Pickle	For neural network models to store weights and biases.
ONNX	Onnx	Models need to be converted to ONNX format and exported and loaded/executed using ONNX Runtime so that they can be run either on CPU- or GPU-based servers.

Table 2.3 – Examples of scalar, vector, matrix, and tensor

The following code shows how to store and retrieve a model in the Joblib format using Python:

> **What is Joblib?**
>
> Joblib (`https://joblib.readthedocs.io/en/latest/`) is a set of tools to provide lightweight pipelining in Python to persist (or serialize) the Python objects. Joblib version 1.2.0 is used in this code.

The Jupyter notebook for this example is LinearRegression_SaveModel.ipynb:

```python
# Persist the model in python Joblib file
# Retrieve the model weights and use it further prediction
import joblib
import numpy as np
from sklearn.linear_model import LinearRegression

X = np.array([[10, 10], [10, 20], [20, 20], [20, 30]])

# Assume that target feature has some relationship with the input
features with the formula y = 3 * x_1 + 5 * x2 + 50

y = np.dot(X, np.array([3, 5])) + 50# Run the model with the input
data and output values

reg = LinearRegression().fit(X, y)

# Persist the model in python Joblib file

filename = "sample_model.sav"

joblib.dump(reg, filename)

# regression model which is used earlier

# Share this file alone to others for deploying in to production and
for inferencing/predictions
```

Once the model is persisted in a file system or a file, then the file can be shared with other developers or engineers without sharing any of the training data or model details used in the code. Other developers/engineers can then load this file and use it for further predictions or deploy it in production for production usage. This is explained in *Figure 2.1*. This model is saved in the current directory and has the name sample_model.sav; you can make use of any extension as it doesn't matter which extension is used. The source code is in Linear_Regression_Load_Model.ipynb:

```python
import joblib

import numpy as np
```

```
filename = "sample_model.sav"

# Load the model from disk

loaded_model = joblib.load(filename)

result = loaded_model.predict(np.array([[30, 50]]))

print(result)
```

Output: [390.]

Classification

A classification model employs different algorithms to predict an output or dependent variable based on the relationship between the input variables. Classification algorithms are specifically designed to predict discrete values, such as spam/not spam, male/female, yes/no, and so on. Each of these predicted values is referred to as a label or class.

In binary classification scenarios, there are only two possible class labels, e.g., determining whether an email is spam or not spam. On the other hand, multi-label classification involves predicting multiple class labels simultaneously. An example could be classifying images into various categories such as cat, dog, and bird.

Classification models are trained using historical data that contains both the input variables and their corresponding class labels. The algorithms learn from this labeled data and establish patterns and relationships to make accurate predictions on new, unseen data. The performance of a classification model is evaluated based on metrics such as accuracy, precision, recall, and F1 score. These metrics assess how well the model can correctly assign the appropriate class labels to new instances based on their input features.

Classification models find extensive applications in various domains, including spam filtering, sentiment analysis, customer churn prediction, fraud detection, and medical diagnosis. By leveraging different classification algorithms, valuable insights can be gained from data, enabling informed decision-making and efficient problem-solving.

Classification type	Details	Examples	Algorithms
Binary	Predicts one of two classes based on the training data	Yes/no Spam/not spam Pass/fail Cancer/no cancer	Logistic regression K-nearest neighbors Decision trees Support vector machine Naive Bayes
Multi-class	Predicts one of more than two classes	Based on symptoms, e.g., cold, flu, or COVID-19	K-nearest neighbors Decision trees Naive Bayes Random forest Gradient boosting
Multi-label	Has two or more class labels	Prediction of the topic based on the content: finance, politics, science, language, or all of them	Multi-label decision trees Multi-label random forests Multi-label gradient boosting
Extreme	Classification task in which the number of candidate labels is huge	Amazon 3M dataset, where the number of labels is 2,812,281	DL algorithms More algorithms: `http://manikvarma.org/downloads/XC/XMLRepository.html`

Table 2.4 – Classification types and associated algorithms

Classification example

In this example, we will utilize the decision tree classification algorithm to determine the likelihood of a patient's survival based on two features: age and whether they have a pre-existing cancer condition.

The decision tree classification algorithm is a widely used technique in ML that constructs a tree-like model of decisions. It analyzes the provided data to create a structure that represents the decision-making process.

In our scenario, the age of the patient and their cancer status will be used as input features for classification. By examining a labeled dataset consisting of patient information, including age, cancer status, and survival outcome, the algorithm learns patterns and establishes decision rules.

Once the model is trained, it becomes capable of predicting the survival outcome for new patients who have not been previously encountered. By considering the age and cancer status of these patients, the model traverses the decision tree until reaching a leaf node that signifies the predicted outcome: whether the patient is expected to survive or not.

By employing the decision tree classification algorithm in this example, we aim to classify patients' survival probabilities based on their age and cancer status. This valuable insight can aid medical professionals in assessing patient prognosis and informing treatment decisions.

Age (years)	Has/had cancer (1 = yes, 0 = no)	Survived (1 = yes, 0 =no)
10	1	1
20	1	1
30	1	1
80	1	0
75	0	0
78	0	0
35	1	?? (predict)
78	1	?? (predict)

Table 2.5 – Toy dataset for classification example

In this toy dataset, the model needs to predict whether the last two patients survive or not (classification with two labels) based on the trained historical data of the model.

The source code is in `Classification_Example.ipynb`.

Scikit-learn provides various Python classes for classification algorithms. Since we have chosen the decision tree algorithm for this example, import the necessary classes and prepare the data in a format that the model accepts:

```
from sklearn import tree
X = [[10,1],[20,1],[30,1],[80,1],[75,0],[78,0]]
Y = [1,1,1,0,0,0]
clf = tree.DecisionTreeClassifier()
clf = clf.fit(X, Y)
clf.predict([[35,1]])
```

Output: array([1])

```
clf.predict([[78,1]])
```

Output: array([0])

In this case, the model predicted that the 35-year-old patient would survive but the 78-year-old patient would not survive based on the training data provided.

To understand more about decision trees and how the trees are split, let's look at the following line of code:

```
tree.plot_tree(clf)
```

This will plot the tree based on the input features and how the tree is split. This is useful when more features are in the training data and we need to know which feature has more importance:

```
[Text(167.4, 163.07999999999998, 'X[0] <= 52.5\ngini = 0.5\nsamples = 6\nvalue = [3, 3]'),
 Text(83.7, 54.360000000000014, 'gini = 0.0\nsamples = 3\nvalue = [0, 3]'),
 Text(251.10000000000002, 54.360000000000014, 'gini = 0.0\nsamples = 3\nvalue = [3, 0]')]
```

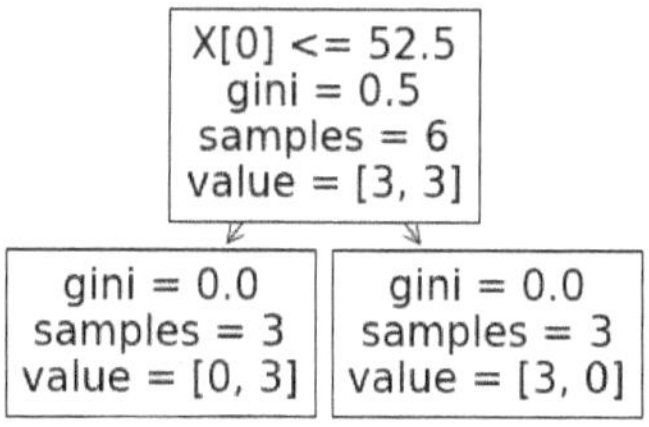

Figure 2.2 - Visualizing tree splitting

Once the model is trained and tested, it can be persisted in a file system or directly used for production.

In the last example, we persisted the model in the Joblib format.

Let's now try to persist the model with the ONNX format to learn more about it.

Model persistence using the ONNX format and executing the model

ONNX, short for **Open Neural Network Exchange**, is an open source format designed for ML and DL models. Its purpose is to facilitate the interoperability of models across different frameworks. It accomplishes this by providing an extensible computation graph model and defining a set of built-in operators and standard data types.

With ONNX, ML/DL models can be easily converted to the ONNX format, allowing for seamless deployment, export, loading, and execution using ONNX Runtime. ONNX Runtime is a powerful tool that enables high-performance execution of ML models on either CPU or GPU. Importantly, it does not rely on dependencies on the specific training framework used to develop the models. By leveraging ONNX and ONNX Runtime, developers can ensure that their models are portable across various frameworks and can be efficiently executed. More details about ONNX can be found at `https://github.com/onnx/onnx`.

Converting the sklearn sample model to the ONNX format

Converting the model to ONNX format requires the frameworks `onnx`, `onnxruntime`, and `skl2onnx` for scikit-learn. Install the frameworks in the following maner:

```
pip3 install onnx --user
pip3 install onnxruntime --user
pip3 install skl2onnx --user
```

Once the frameworks are installed, execute the following code to convert the model to ONNX format (the source code is in `Model_Persistence_Load_ONNX_Format.ipynb`):

```
import numpy as np

from sklearn.linear_model import LinearRegression

from skl2onnx import convert_sklearn

from skl2onnx.common.data_types import FloatTensorType

initial_type = [('float_input', FloatTensorType([None, 2]))]

onx = convert_sklearn(clf, initial_types=initial_type)

with open("survive.onnx", "wb") as f:
    f.write(onx.SerializeToString())
```

In this case, to convert the ML model that was developed using `sklearn` to ONNX format, first, the data types used in the training need to be provided:

```
initial_type = [('float_input', FloatTensorType([None, 2]))]
```

Later, use the methods provided to convert the model to ONNX format and specify the classifier that is used in `sklearn`. In our examples, we have used decision trees and named the model `clf`:

```
onx = convert_sklearn(clf, initial_types=initial_type)
```

Once the model is converted to ONNX format, store it in the disk and name the model file (in our example, `survive.onnx`):

```
with open("survive.onnx", "wb") as f:
    f.write(onx.SerializeToString())
```

Now the model is stored in ONNX format and it can be loaded and executed on any framework that supports ONNX Runtime.

Loading the ML model using ONNX format and executing the model

The following lines of code show how to load the model stored in ONNX format and how to use the model for inference. ONNX version 1.14.1 is used in this code (the source code is in `Model_Persistence_Load_ONNX_Format.ipynb`):

```
import onnxruntime as rt
import numpy as np

sess = rt.InferenceSession("survive.onnx")

input_name = sess.get_inputs()[0].name

# To test whether the patient will survive or not with 78 years age
and doesn't have prior cancer

X_test = np.array([[78.0, 0.0]])

pred_onx = sess.run(None, {input_name: X_test.astype(np.float32)})[0]

print(pred_onx)
```

Output: [0]

Unsupervised ML

In unsupervised ML, the model is trained using unlabeled training data, which means there are no target labels or classes provided. Instead, unsupervised ML models focus on understanding the inherent patterns and structures within the data. Unlike supervised learning, where the model learns from labeled examples, unsupervised machine learning models uncover hidden patterns and relationships within the data without any predefined class labels. This allows for the discovery of previously unknown insights and patterns that may not be readily apparent. By leveraging unsupervised ML techniques, analysts and data scientists can gain valuable insights from unlabeled data, uncover hidden structures, and make data-driven decisions based on the inherent patterns discovered in the dataset.

Clustering

Clustering is a primary technique used in unsupervised ML. It involves grouping similar data points together based on their intrinsic characteristics. By examining the data and identifying similarities, unsupervised models create clusters, which represent distinct groups or patterns within the dataset.

Clustering algorithms, such as k-means clustering, hierarchical clustering, or density-based clustering, are commonly employed in unsupervised ML to organize data into meaningful groups. These clusters can help in data exploration, anomaly detection, customer segmentation, and other data-driven tasks.

Clustering example

Let's consider a scenario where a company aims to offer transportation services to its employees and wants to cluster them based on their residential locations. To achieve this, the company can utilize a clustering model that takes the longitude and latitude coordinates of each employee's residence as input data. The ML model will cluster the employees based on the specified cluster size, which can be equal to the number of vehicles available for transportation. By analyzing the spatial data of employees' locations, the clustering model will group individuals who live in close proximity to one another. This grouping enables the company to efficiently allocate vehicles to each cluster.

Once the clustering model is trained and established, it can predict the appropriate cluster for new employees based on their residential coordinates. This allows the company to easily determine which cluster the new employee should join, facilitating seamless transportation arrangements.

By utilizing ML clustering techniques in this scenario, the company can effectively organize its transportation services and optimize resource allocation based on employees' residential locations.

Employee number	Latitude (o N)	Longitude (o E)
1	12.93	77.4472
2	12.32	77.4472
3	12.51	77.4472
4	12.62	77.4472
5	12.73	77.4472
6	12.84	76.4158
7	12.91	76.4158
8	12.41	76.4158
9	12.92	76.4158
10	12.55	76.4158

Table 2.6 – Toy dataset for clustering example

The `sklearn` framework supports various clustering algorithms, and we will use the K-means clustering algorithm in this example to cluster the given data.

The K-means algorithm is a centroid-based algorithm, where each cluster is associated with a centroid. The main aim of this algorithm is to minimize the sum of distances between the input data point and their corresponding cluster.

The source code is in `Clustering_Example.ipynb`.

Import the `sklearn` K-means clustering classes and prepare the training data as a numpy array format:

```
from sklearn.cluster import KMeans
import numpy as np
from sklearn.cluster import KMeans

import numpy as np

X = np.array([
            [12.93,77.4472],
            [12.32,77.4472],
            [12.51,77.4472],
            [12.62,77.4472],
```

```
            [12.73,77.4472],
            [12.84,76.4158],
            [12.91,76.4158],
            [12.41,76.4158],
            [12.92,76.4158],
            [12.55,76.4158],
          ])

kmeans = KMeans(n_clusters=2, random_state=0).fit(X)

kmeans.labels_

array([1, 1, 1, 1, 1, 0, 0, 0, 0, 0], dtype=int32)

      kmeans.cluster_centers_

array([[12.726 , 76.4158],
       [12.622 , 77.4472]])

kmeans.predict([[12.88, 76.88]])

      array([0], dtype=int32
```

In this case, the first five employees are assigned to cluster 1 and the remaining are assigned to cluster 0:

Employee number	Latitude (°N)	Longitude (°E)	Assigned cluster
1	12.93	77.4472	1
2	12.32	77.4472	1
3	12.51	77.4472	1
4	12.62	77.4472	1
5	12.73	77.4472	1
6	12.84	76.4158	0
7	12.91	76.4158	0
8	12.41	76.4158	0
9	12.92	76.4158	0
10	12.55	76.4158	0

Table 2.7 - Assigned cluster

The cluster model learned based on the input and formed two clusters. K-means is a clustering algorithm that finds the center of the cluster, divides the data into clusters, and predicts the new data based on the nearest cluster.

The following is the list of clustering algorithms supported by the `sklearn` framework:

- Affinity propagation
- Agglomerative clustering
- BIRCH
- DBSCAN
- K-means
- Mini-batch K-means
- Mean shift
- OPTICS
- Spectral clustering
- Mixture of Gaussians

Reinforced ML

Reinforcement learning (**RL**) is a type of ML technique that enables an agent to learn in an interactive environment by trial and error using feedback from its own actions and experiences. The agent learns a series of actions that lead to the final goal, maximizing its total rewards. RL differs from supervised learning in that the model learns from taking actions and observing the results, not from explicit teaching.

One classic use case of RL is in gaming, such as teaching a model to play and excel at chess. The model starts with no knowledge of the game but learns by making moves and seeing the outcome of the game it plays, with the aim of maximizing the reward (i.e., winning the game).

In the context of RL, exploration and exploitation are two strategies that an agent can use to navigate through the environment:

- **Exploration**: This is when the agent seeks to learn more about its environment. It means trying out different actions and gathering more information to learn about each possible action's reward. The agent aims to balance out the reward it gets from known information with the possibility of receiving an even higher reward from unknown areas. However, exploration might involve the risk of the agent making non-optimal choices.

- **Exploitation**: Here, the agent uses the information it has already learned to make the best action that will maximize its reward. It means using known information to maximize success instead of further exploring. The benefit of exploitation is that it allows for more assured, immediate rewards, but excessive exploitation can lead to suboptimal results as it may neglect even better options. The challenge lies in finding the right balance, as focusing too much on exploration might mean the agent will lose out on immediate rewards while focusing extensively on exploitation might prevent the agent from exploring options that could lead to larger rewards in the future. This trade-off is often referred to as the exploration–exploitation dilemma.

Example problem using RL—the multi-armed bandit problem

The multi-armed bandit problem is a classic problem in the field of RL that captures the fundamental trade-off between exploration and exploitation. The name is derived from a hypothetical experiment in which you face several slot machines (also known as "one-armed bandits") with different fixed payouts. Because of these differing payouts, your goal is to maximize your total payout over a certain number of attempts by figuring out which machines to play, how many times to play each machine, and in what sequence—hence, the "multi-armed bandit problem."

The primary challenge in the multi-armed bandit problem is balancing the immediate rewards from exploitative actions (playing the machine that you believe currently has the highest expected payout) with the possible benefits from exploration (trying out others that might have higher expected payouts but you're less certain about). This tension between exploration and exploitation is at the core of many reinforcement learning problems.

The following is example code for reinforcement learning—the source code is in `BandIt_RL_Example.ipynb`:

```
import numpy as np
# This example will use a 4-armed bandit.
# Initialization
num_trials = 1000
bandit_probabilities = [0.1, 0.15, 0.3, 0.35]  # The probability of
each bandit to give reward
reward_counts = np.zeros(4)
selected_bandit = 0
for i in range(num_trials):
    # Select a bandit
    selected_bandit = np.random.randint(0, 4)
    # Pull bandit's arm
    random_num = np.random.random()
```

```
    if random_num <= bandit_probabilities[selected_bandit]:
        reward = 1
    else:
        reward = 0
    reward_counts[selected_bandit] += reward
print("Most successful bandit:", np.argmax(reward_counts))
```

Most successful bandit: 3

In this example, the epsilon-greedy strategy is used, where epsilon is 1.

This is a very simplistic example and real-world RL problems require much more sophisticated algorithms (e.g., Q-learning, policy gradient, etc.) and are therefore implemented using specialized libraries.

In this section, we have covered the different types of ML and provided examples of how to save and load models for inference and prediction. Moving forward, the next section will delve into the various phases of ML, providing a detailed exploration.

Overview of ML phases

ML encompasses a variety of techniques and approaches, and it involves several distinct phases or stages in the process of developing and deploying ML models. These phases help guide engineers through the iterative and cyclical nature of ML projects, allowing them to build effective and accurate models.

The ML process typically consists of several key phases, each serving a specific purpose and contributing to the overall success of the project. These phases are not always strictly linear, and iterations may occur between them to refine and improve the models. The specific steps and terminology used may vary depending on the ML methodology employed, but the core phases remain consistent.

The ML phases provide a systematic framework for developing and deploying ML models, guiding practitioners through the complexities and challenges inherent in building effective solutions. By following these phases, practitioners can maximize their chances of success and create ML models that deliver valuable insights and predictions in a wide range of applications.

The main phases of ML

The following are the main phases of ML:

- **Data collection**: This phase involves gathering relevant data from various sources, such as databases, APIs, or manual collection. The data should be representative of the problem domain and cover a wide range of scenarios.

- **Data preparation**: In this phase, the collected data is preprocessed and transformed into a suitable format for analysis. This may include tasks such as cleaning the data, handling missing values, removing outliers, and normalizing or scaling the features.

- **Feature engineering**: Feature engineering involves selecting and creating relevant features from the available data that will enhance the model's predictive power. This phase requires domain knowledge and creativity to extract meaningful insights from the data.

- **Model development**: In this phase, a suitable ML algorithm or model is selected based on the problem at hand. The model is trained on the prepared data to learn patterns and relationships within the data.

- **Model evaluation**: The trained model is evaluated using appropriate evaluation metrics to assess its performance. This helps in understanding how well the model generalizes to unseen data and whether it meets the desired criteria for accuracy and reliability.

- **Model optimization**: If the model's performance is not satisfactory, this phase involves fine-tuning the model by adjusting hyperparameters or trying different algorithms to improve its performance. The optimization process aims to achieve the best possible results.

- **Model Deployment**: Once the model is trained and optimized, it is deployed in a production environment where it can make predictions on new, unseen data. This phase involves integrating the model into existing systems or creating an interface for users to interact with the model.

- **Model monitoring and maintenance**: After deployment, the model needs to be monitored to ensure it continues to perform well over time. Monitoring involves tracking performance metrics, identifying drift in data distribution, and updating the model if necessary. Regular maintenance is essential to keeping the model up to date and accurate.

These phases provide a systematic approach to building and deploying ML models, enabling organizations to leverage the power of data and make informed decisions.

Sub-phases in the ML process

The following diagram shows the phases of the ML process:

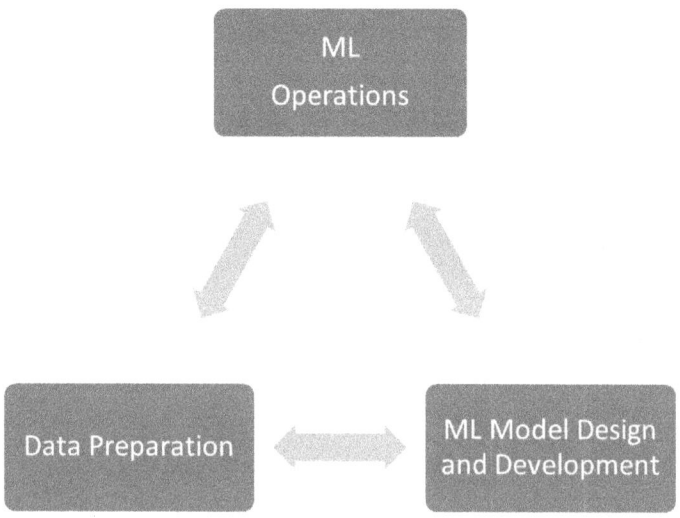

Figure 2.3 – ML phases

These are the main phases:

- Data preparation phase
- ML model phase (design and development)
- ML operations phase

Let's look at these in more detail.

Data preparation phase

The data preparation phase deals with data collection, extraction, and manipulation.

Phase	Sub-phases	Details
Data preparation	Data collection	Identify the data that needs to be analyzed
	Data extraction	Extract the data from the data source
	Data manipulation	Data transformation, missing data, duplicate data, noise, and data preprocessing
	Exploratory data analysis (EDA)	EDA and handling data

Table 2.8 - Data preparation phase

The following figure shows the data preparation sub-phases:

Data Preparation Phase

Figure 2.4 – ML data preparation sub-phases

ML model phase

This phase is subdivided into several phases to deal with feature engineering, actual model identification, the training and testing of models, and so on.

Phase	Sub-phases	Details
ML model	Model identification	This involves classification, clustering, re-enforcement, time series analysis, and so on.
	Feature engineering	In this phase, features are selected from the data.
	Input data preparation for the model	This involves data processed and data prepared in the format the model expects.
	Split the data (train, test, and validate)	Split the entire data into three parts—training data, test data, and validation data—to train, test, and validate the models.
	Train the model with the training dataset	In this phase, the model is trained with the training data.
	Test the model with the testing dataset	The ML model is tested with the test data to find out the accuracy of predictions.
	Version of data, model, model parameters, and results	Version control is applied to datasets used, as well as to the model and its parameters, along with the results of each experiment.
	Validate the dataset with the trained model	This is similar to test data but the samples are from the validation dataset.
	Predict results with new data (inference)	For inference, use the new data and find out the results.

Table 2.9 - ML model phase

The following figure shows the ML model sub-phases:

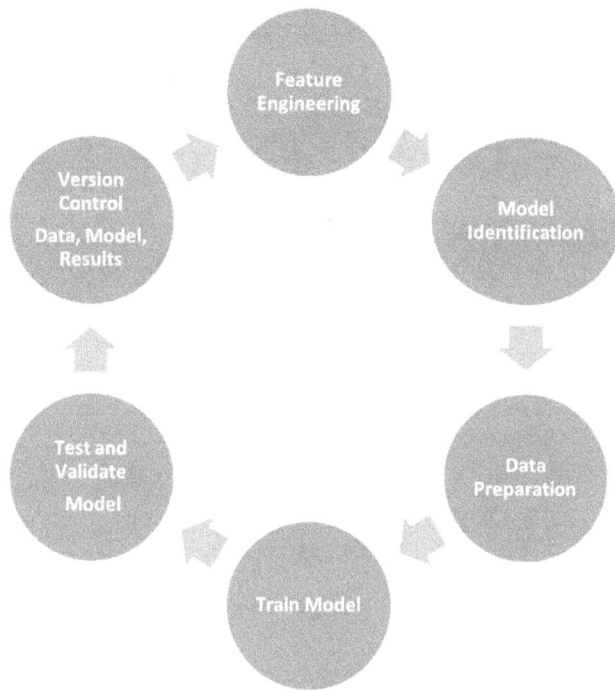

Figure 2.5 – ML model sub-phases

ML operations phase

This phase is mainly focused on the operations of the models in production.

Phase	Sub-phases	Details
ML operations (**MLOps**)	Package model artifacts	Persist the model (store the weights and biases) in ONNX format or other formats.
	Deploy model	This involves the production deployment of the model. (A/B testing, canary deployment, shadow models, etc.)
	Validate the inference results	
	Monitor model performance	Monitor the performance model, i.e, whether the accuracy stays constant or degrades over a period.
	Retrain the model and repeat the ML model life cycle	Retrain the model if the model performance degrades and handle model drift and data drift accordingly.

Table 2.10 - ML operations

The following figure shows the ML operations sub-phases:

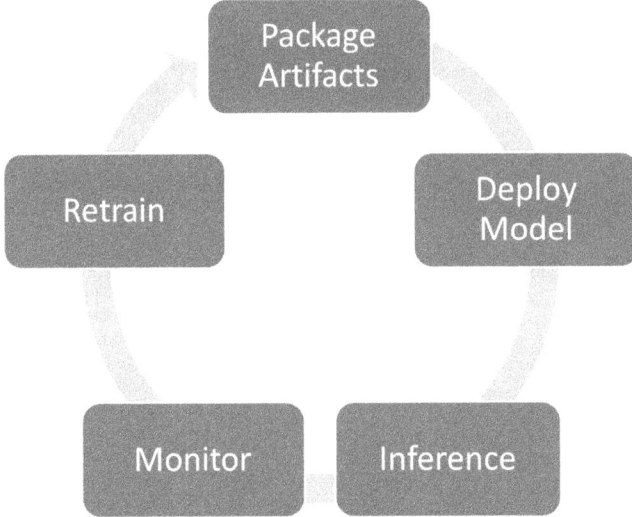

Figure 2.6 – ML operations sub-phases

We have thoroughly covered the development of ML models and the various phases involved in the process. In the upcoming section, our focus will shift to exploring the privacy threats and attacks that can occur in each phase of ML. We will delve into understanding these threats and discuss effective mitigation strategies to safeguard the privacy of the data and models involved. By addressing these privacy concerns at each stage, we can ensure the responsible and secure implementation of ML techniques.

Privacy threats/attacks in ML phases

ML projects are developed in collaboration with data engineers, ML engineers, and software engineers, and each one plays a different role in order to develop end-to-end systems to predict and provide insights.

Collaborative roles in ML projects

ML projects are collaborative efforts involving various roles such as data engineers, ML engineers, and software engineers. Each role contributes in different ways to develop end-to-end systems that can predict outcomes and provide valuable insights. Let's explore the roles and their responsibilities in the ML project life cycle:

- **Data engineer**: The data engineer primarily focuses on the data preparation phase. They are responsible for extracting data from one or multiple sources and ensuring its quality and suitability for the ML project. Data engineers work on tasks such as data cleaning, transformation, and feature selection to prepare the data for ML modeling.

- **ML engineer**: ML engineers play a crucial role in designing and developing ML models. They leverage the data provided by the data engineer to train and test the models. ML engineers are responsible for selecting appropriate algorithms or model architectures, tuning hyperparameters, and optimizing the models for accuracy and efficiency. They validate the model against validation test data and provide APIs for inference or export/deployment of the model into production.

- **Model consumer**: The model consumer can be an individual or another application that interacts with the ML model. They make API calls and provide input to the model for prediction or inference. Model consumers utilize the insights generated by the ML model to make informed decisions or take appropriate actions.

Privacy threats/attacks in ML

In the context of ML, an adversary refers to an entity or system that actively tries to undermine or exploit the machine learning model or system. The goal of an adversary is typically to manipulate the model's behavior, gain unauthorized access to sensitive information, or deceive the system by exploiting vulnerabilities.

Adversaries can take various forms and have different motives.

Here are a few examples:

- **Adversarial examples**: In this case, the adversary aims to create input samples (e.g., images or text) that are intentionally crafted to mislead or deceive the ML model. Adversarial examples are designed to exploit vulnerabilities in the model's decision-making process, leading to incorrect predictions or misclassifications.

- **Data poisoning**: An adversary may try to inject malicious or misleading data into the training dataset. By inserting carefully crafted samples, the adversary aims to manipulate the model's training process, leading to biased or compromised results. This can be particularly problematic in scenarios where the training data is collected from untrusted or unreliable sources.

- **Model inversion**: An adversary might attempt to extract sensitive information from a trained model. By providing specific input and observing the model's output, the adversary aims to infer confidential or private data that was used to train the model, such as **personally identifiable information (PII)** or proprietary knowledge.

- **Evasion attacks**: Adversaries can also launch evasion attacks, also known as adversarial attacks, during the deployment phase. In these attacks, the adversary tries to bypass or manipulate the model's defenses by carefully modifying input samples. For example, in the case of a spam email classifier, an adversary may add specific patterns or keywords to trick the model into classifying a malicious email as legitimate.

To mitigate the impact of adversaries, researchers and practitioners develop robust ML models and techniques, such as adversarial training, defensive distillation, and input sanitization. These approaches aim to enhance the resilience of ML systems against adversarial attacks and maintain their performance and reliability in the presence of potential threats.

Throughout the ML life cycle, privacy threats or attacks can occur, posing risks to the confidentiality of sensitive information. In the context of ML, adversaries attempt to gain unauthorized access to confidential data used in ML, the core ML model, or specific features of the data.

There are two primary types of attacks, white-box and black-box:

- **White-box attack**: A white-box attack assumes that the adversary has full knowledge and access to the ML model, including its architecture, input, output, and weights. The attacker exploits this information to extract confidential details.

- **Black-box attack**: In contrast, a black-box attack assumes that the attacker only has access to the input and output of the ML model. They have no knowledge of the underlying architecture or weights used in the ML/DL model. Despite the limited information, they aim to infer sensitive information from the model.

Privacy threats/attacks classification:

The following are the privacy attacks on the ML models for classification

- **Membership inference attack**: This attack aims to determine whether a particular data point was part of the training dataset used to train the ML model. The adversary tries to infer membership information by exploiting the model's responses.

- **Model extraction attack**: In this attack, the adversary attempts to extract the entire or partial ML model architecture, weights, or parameters. This attack allows the attacker to replicate the ML model for their own purposes, potentially leading to intellectual property theft.

- **Reconstruction attack**: This attack focuses on reconstructing sensitive information from the ML model's outputs. The attacker aims to infer private data or specific features that were used to generate the model's predictions. By understanding and addressing these privacy threats, ML practitioners can take appropriate measures to safeguard sensitive data and ensure the security of ML models throughout their life cycle.

We'll look at these in more detail in the following sections.

Membership inference attack

Assume a classification model is developed with certain input training data, X { x1, x2, x3, …. Xn} to predict certain labels, y, using a function, F.

A membership inference attack tries to determine whether an input sample, x, was used as part of the training dataset, X, or not. Basically, an adversary (attacker) needs to find out whether the data point at hand belongs to the original dataset that is used for training the ML model or not.

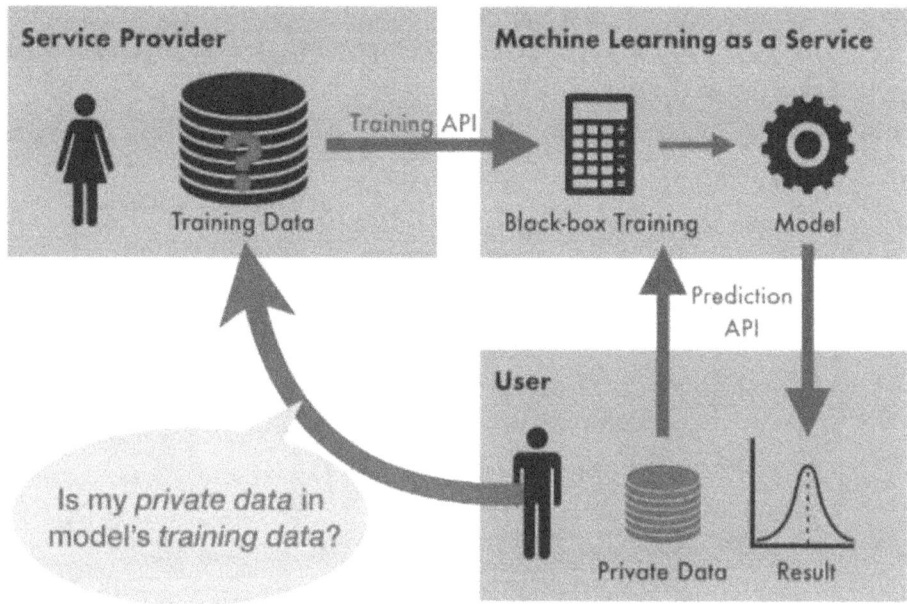

Figure 2.7 – Membership inference attack - Example

(Image source: Suha Hussain, PrivacyRaven, `https://blog.openmined.org/privacyraven-comprehensive-privacy-testing-for-deep-learning/`)

Let's look at an example. Suppose an adversary wants to determine whether a particular person's data was used in the training data or not without the knowledge of that person. Later, this data is used to derive insights on whether to approve that person's insurance policy or not.

This is the most popular category of attacks and was first introduced by Shokri et al.

Here is the reference to the full paper: *Reza Shokri, Marco Stronati, Congzheng Song, and Vitaly Shmatikov. 2017. Membership inference attacks against machine learning models. In 2017 IEEE Symposium on Security and Privacy (SP). IEEE, San Francisco, CA, USA, 3–18.*

This is a kind of black-box testing attack because the adversary doesn't have the details of the actual ML model; all they have is the set of input data and inference results from the model.

This is what the paper says about this approach:

> *"The attacker queries the target model with a data record and obtains the model's prediction on that record. The prediction is a vector of probabilities, one per class, that the record belongs to a certain class. This prediction vector, along with the label of the target record, is passed to the attack model, which determines whether the record was in or out of the target model's training dataset."*

The following figure also comes from the aforementioned paper (`https://arxiv.org/pdf/1610.05820.pdf`):

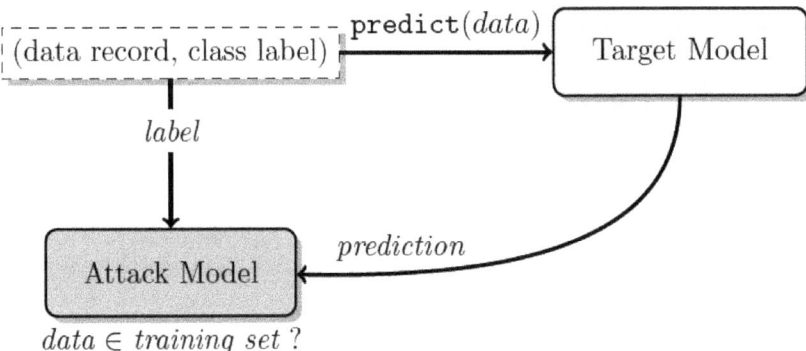

Figure 2.8 – Membership inference attack - Example

Membership inference attack—basic example

In a membership inference attack, an adversary attempts to determine whether a specific data point was used in the training set of an ML model, as stated earlier.

Here's an example using a simple decision tree classifier (the source code can be found in `Membership_Inference_basic_example.ipynb`):

```
from sklearn import datasets
from sklearn.model_selection import train_test_split
from sklearn.tree import DecisionTreeClassifier
# Load the dataset (e.g., Iris dataset)
iris = datasets.load_iris()
X = iris.data
y = iris.target
# Split the data into training and test sets
X_train, X_test, y_train, y_test = train_test_split(X, y, test_
size=0.5, random_state=42)

# Train a decision tree classifier on the training set
clf = DecisionTreeClassifier()
```

```
clf.fit(X_train, y_train)
for i in range (0,75):
    # Select a data point from the test set
    target_data = X_test[i]
    #Determine if the target data point was in the training set
    predicted_class = clf.predict([target_data])
    is_in_training_set = predicted_class == y_test[i]
    # Print the result
    if is_in_training_set:
        print(target_data, "Membership Inference Attack successful!
Target data point was in the training set.")
    else:
        print( target_data, "Membership Inference Attack unsuccessful.
Target data point was not in the training set.")
```

Output:
**[6.9 3.1 5.1 2.3] Membership Inference Attack successful! Target data
point was in the training set.**
**[6.2 2.2 4.5 1.5] Membership Inference Attack unsuccessful. Target
data point was not in the training set.**

In this example, we load the Iris dataset from scikit-learn and split it into a training set and a test set. We then train a decision tree classifier on the training set. Next, we select a data point (or set of points) from the test set and attempt to determine whether it was present in the training set by predicting its class. If the predicted class matches the actual class from the test set, we infer that the target data point was in the training set, indicating a successful attack. Remember, conducting membership inference attacks without proper authorization is unethical and often illegal.

Membership inference attack—advanced example

Let's consider a scenario where an adversary seeks to determine whether a specific individual's data is present in the training data without any prior knowledge of that person. This scenario involves discovering whether a person's name exists within a hospital's sensitive clinical data. The adversary intends to exploit this information to make decisions, such as granting or denying an insurance policy, based on the insights gained.

To illustrate this scenario, let's use a sample dataset (for illustrative purposes only) similar to the classification example we previously discussed. The dataset focuses on predicting whether a patient will live for the next 5 to 10 years or not based on factors such as age and existing diseases.

In this context, the adversary's goal is to identify whether the data related to a specific person, whose identity they are unaware of, is present in the training data. By discovering this information, the adversary can potentially manipulate decisions related to insurance policies based on the insights gained from the training data.

It is important to note that this example serves to highlight a potential privacy threat and does not aim to validate its accuracy or real-world applicability. The objective is to raise awareness about the importance of safeguarding sensitive data and implementing robust privacy measures to prevent unauthorized access and misuse.

Age (years)	Has/had cancer (1 = yes, 0 = no)	Survived (1 = yes, 0 = no)
10	1	1
20	1	1
30	1	1
80	1	0
75	0	0
78	0	0

Table 2.11 - Training data

The source code for this ML model can be found in `Membership_Inference_advanced_example.ipynb`:

```
from sklearn import tree
X = [[10,1],[20,1],[30,1],[80,1],[75,0],[78,0]]
Y = [1,1,1,0,0,0]
clf = tree.DecisionTreeClassifier()
clf = clf.fit(X, Y)
clf.predict([[35,1]])
```

Inference results with sample test data

The adversary creates synthetic test data with diverse inputs to evaluate the model's performance. Their objective is to determine whether the given patient data exists in the training dataset or not using the model's predictions:

```
testY=[
        [25,1],[25,0],[30,1],[30,0],[45,0],[45,1],
        [50,1],[50,0],[60,1],[60,0],[75,0],[75,1],
        [80,1],[80,0],[90,1],[90,0],[100,0],[100,1],
        [10,1],[20,1],[30,1],[78,0]
    ]
clf.predict(testY)


array([1, 1, 1, 1, 1, 1, 1, 1, 0, 0, 0, 0, 0, 0, 0, 0, 0, 0, 1, 1, 1,
0])
```

```
clf.predict_proba(testY)

array([[0., 1.],
       [0., 1.],
       [0., 1.],
       [0., 1.],
       [0., 1.],
       [0., 1.],
       [0., 1.],
       [0., 1.],
       [1., 0.],
       [1., 0.],
       [1., 0.],
       [1., 0.],
       [1., 0.],
       [1., 0.],
       [1., 0.],
       [1., 0.],
       [1., 0.],
       [1., 0.],
       [0., 1.],
       [0., 1.],
       [0., 1.],
       [1., 0.]])
```

The results are in the table format; 1 means the person has cancer and the predicted probability column shows the percentage of predicted probability:

Age	Cancer	Class predicted	Predicted probability
25	1	1	100
25	0	1	100
30	1	1	100
30	0	1	100
45	1	1	100
45	0	1	100
50	1	1	100
50	0	1	100
60	1	0	0
60	0	0	0
75	1	0	0

Age	Cancer	Class predicted	Predicted probability
75	0	0	0
80	1	0	0
80	0	0	0
90	1	0	0
90	0	0	0
100	1	0	0
100	0	0	0
10	1	0	100
20	1	0	100
30	1	0	100
78	0	1	0

Table 2.12 - Predicted probability

Next, the adversary proceeds to develop shadow models and a final attack model. These models are designed to predict whether a given data record was used in the training dataset. By utilizing each record and its predicted class, along with the corresponding predicted probabilities, the adversary infers whether the data record was part of the training set or not.

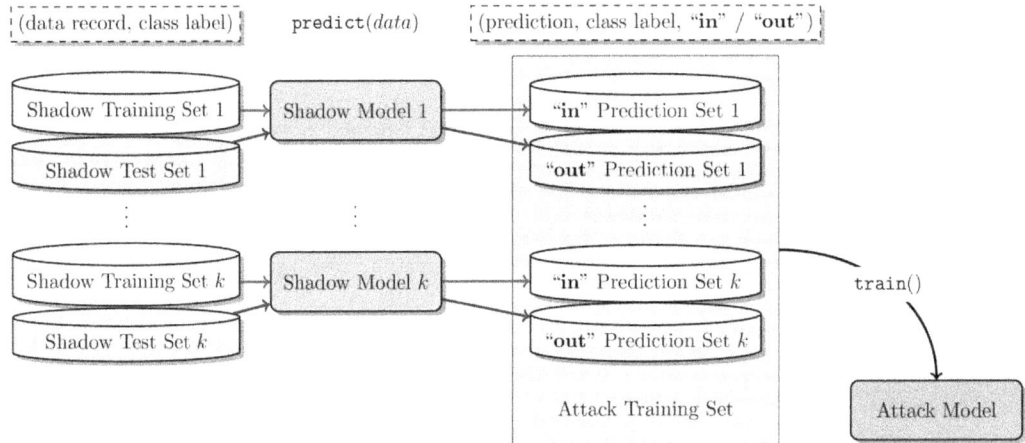

Figure 2.9 – Shadow models

The preceding figure was sourced from the paper at https://arxiv.org/pdf/1610.05820.pdf.

The attacker model makes use of the class label 'In' for the given input record used in the training, while 'out' means it is not used.

This is the training data for the final attack model:

Age	Cancer	Class predicted	Predicted probability	Record used in the training set (in = 1, out = 0)
25	1	1	100	1
25	0	1	100	1
30	1	1	100	1
30	0	1	100	1
45	1	1	100	1
45	0	1	100	1
50	1	1	100	1
50	0	1	100	1
10	1	0	100	1
20	1	0	100	1
30	1	0	100	1
60	1	0	0	0
60	0	0	0	0
75	1	0	0	0
75	0	0	0	0
80	1	0	0	0
80	0	0	0	0
90	1	0	0	0
90	0	0	0	0
100	1	0	0	0
100	0	0	0	0
78	0	1	0	0

Table 2.13 - Training data for final attack model

Membership inference attacks can be easily executed with remarkable accuracy in simple linear models, as shown in the preceding example. ML models hosted in the cloud are susceptible to such inference attacks, as researchers have successfully demonstrated membership attack models with accuracies exceeding 90%.

Techniques to mitigate membership inference attacks

Membership inference attacks can be a concern in ML models that involve sensitive data.

Some techniques to mitigate membership inference attacks are as follows:

- **Limit access to sensitive information**: One of the simplest ways to mitigate membership inference attacks is to limit access to sensitive information. By minimizing the amount of sensitive data that is exposed, you reduce the potential for attackers to perform membership inference.

- **Differential privacy**: Differential privacy is a technique that adds noise to the training data or the model's output, making it harder for an attacker to determine whether a specific record was part of the training set. Applying differential privacy mechanisms can help protect against membership inference attacks. We will learn about differential privacy in the next chapter.

- **Training set augmentation**: By augmenting the training set with additional synthetic or generated data, you can make it more difficult for attackers to distinguish between genuine training instances and potential members. Augmentation techniques such as data generation, perturbation, or adding noise can help to increase the privacy of the training set.

- **Regularization and dropout**: Applying regularization techniques such as L1 or L2 regularization and incorporating dropout layers in neural networks can improve model robustness and reduce overfitting. Regularization can help in reducing the memorization of training instances, making it harder for attackers to infer membership.

- **Model compression**: When sharing models or making predictions, consider using model compression techniques to reduce the amount of information leaked about the training data. Techniques such as quantization, pruning, or knowledge distillation can help reduce the model's sensitivity to the training set.

- **Ensemble methods**: Training an ensemble of multiple models with different architectures or using different algorithms can make it more difficult for attackers to perform accurate membership inference. Ensemble methods make it harder for an attacker to learn the specific patterns in the training data.

- **Secure aggregation**: If the model is trained using a distributed setting, secure aggregation protocols can be employed to ensure that individual contributions from different parties are protected and the membership information is not exposed.

- **Randomized response**: Randomized response techniques can be used to introduce noise into the model's outputs during inference, making it harder for an attacker to determine membership status. Randomized response mechanisms ensure plausible deniability for individual records.

- **Access control and authorization**: Implementing access control measures and strong authorization mechanisms can help restrict access to sensitive models and data, limiting the exposure to potential attackers.

- **Model monitoring**: Continuously monitoring the model's behavior for any unusual patterns or unexpected outputs can help detect potential membership inference attacks. Monitoring can involve techniques such as outlier detection, adversarial robustness checks, or statistical analysis of model outputs.

It's important to note that no single technique can provide complete protection against membership inference attacks. A combination of multiple techniques and a comprehensive approach to privacy and security is usually required to effectively mitigate these attacks.

Model extraction attack

A model extraction attack is a type of black-box attack in which the adversary aims to extract information, and possibly recreate a model, by creating a substitute model (denoted as f') that closely emulates the behavior of the original model being targeted (denoted as f).

Let's consider a scenario where we have developed an ML model specifically designed to predict whether a given post/tweet pertains to a disaster or not. We provide APIs to consumers, enabling them to access these prediction capabilities, and charge a fee for each API request made.

Figure 2.10 - Securing ML Model Integrity

The adversary takes advantage of the API provided and systematically submits thousands of input tweets in order to obtain their respective predictions. Subsequently, the adversary proceeds to construct a new ML model using these tweets, which were obtained by querying the API exposed by the original author. The predicted results obtained from the API serve as the class labels for this new model, indicating whether the tweets are classified as disaster-related or not.

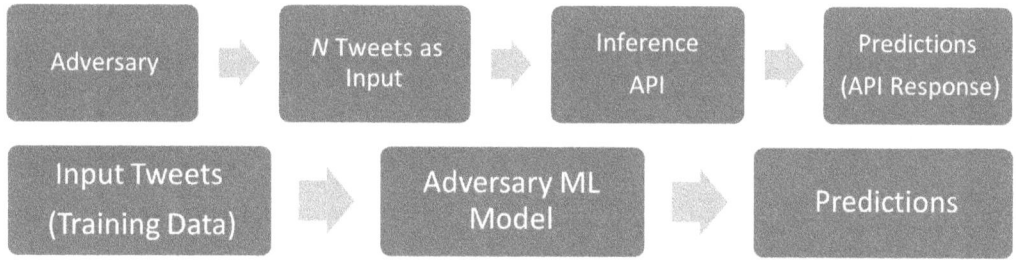

Figure 2.11 – Model extraction attack—adversary ML model

In certain instances, the ML model developed by the adversary may exhibit superior accuracy compared to the original author's ML model. As a consequence, this can significantly affect the revenue of the original author's company. The adversary may exploit this advantage by exposing similar inference APIs, charging substantially lower fees than the original author, and potentially engaging in the theft of intellectual property. Furthermore, the model extraction attack enables the adversary to gain access to private information associated with the ML model, further exacerbating the potential damages caused.

Example of a model extraction attack

The source code for this example can be found in `Model Extraction_Attack Example.ipynb`.

In this example, we start by creating a sample dataset that consists of two features (X) and their corresponding labels (y). Subsequently, we train a logistic regression model on this dataset using the `LogisticRegression` class from scikit-learn.

The attacker's code aims to execute a model extraction attack by training a new `LogisticRegression` model (`extracted_model`) on the same dataset. The attacker's objective is to replicate the original model's behavior without having direct access to its internal workings.

Once the extracted model is successfully generated, it can be utilized for unauthorized purposes, such as making predictions on new data (`new_data`) without requiring access to the original model. This unauthorized usage raises concerns regarding the security and integrity of the original model's functionality.

The following is the source code for the model extraction attack (`Model Extraction_Attack Example.ipynb`):

```
import numpy as np
from sklearn.linear_model import LogisticRegression

# Create a sample dataset for demonstration
X = np.array([[1, 1], [2, 2], [3, 3], [4, 4], [5, 5]])
y = np.array([0, 0, 0, 1, 1])

# Train a simple logistic regression model on the dataset
model = LogisticRegression()
model.fit(X, y)

# Attacker's code to perform model extraction attack
extracted_model = LogisticRegression()
extracted_model.fit(X, y)

# Extracted model can be further used for unauthorized purposes
# such as making predictions on new data without access to the
original model
new_data = np.array([[6, 6], [7, 7]])
predictions = extracted_model.predict(new_data)
print(predictions)
```

```
[1 1]
```

Techniques to mitigate model extraction attacks

Mitigating model extraction attacks, where an adversary tries to extract the underlying model's architecture, parameters, or functionality, is crucial for protecting intellectual property and maintaining the security of sensitive models.

Some of the techniques to mitigate model extraction attacks are as follows:

- **Model watermarking**: Embedding a unique watermark into the model's parameters or architecture can help identify the origin of the model and deter unauthorized extraction. Watermarking techniques can be designed to be resilient against removal attempts or modifications while remaining imperceptible to normal model operations.

- **Model obfuscation**: Applying obfuscation techniques to the model's code or architecture can make it harder for attackers to understand the internal workings of the model. Obfuscation can involve techniques such as code obfuscation, function renaming, control flow diversification, or encryption to protect the model's implementation details.

- **Secure model sharing**: When sharing models with authorized users or collaborators, it's important to employ secure sharing mechanisms. This can involve encryption during transit and at rest, strong access control measures, and secure authentication and authorization protocols to prevent unauthorized access to the model.

- **Model compression**: Using model compression techniques such as quantization, pruning, or knowledge distillation can make the model more compact and reduce the amount of information that can be extracted. Compressed models often have fewer parameters and structural details, making them more resistant to model extraction attacks.

- **Fine-grained access control**: Implementing fine-grained access control mechanisms can limit the exposure of sensitive models. This can involve providing access to only the necessary components or functionalities of the model based on user roles and permissions.

- **Secure execution environment**: Running the model in a secure execution environment can help protect against extraction attacks. Techniques such as secure enclaves (e.g., Intel SGX or AMD SEV), **trusted execution environments** (TEEs), and **secure multiparty computation** (MPC) can provide isolation and integrity guarantees for executing models, preventing unauthorized access to the model's internals. We will learn more about TEE in *Chapter 9*.

- **Model metadata protection**: Protecting the metadata associated with the model, such as the training data, hyperparameters, or training process details, can make it harder for attackers to extract meaningful information about the model. Techniques such as differential privacy or data perturbation can help preserve privacy in model metadata.

- **Monitoring for abnormal model usage**: Implementing model monitoring and anomaly detection mechanisms can help identify suspicious activities, such as repeated queries or excessive model interactions, which could indicate unauthorized extraction attempts. Monitoring can trigger alerts or initiate defensive actions when potential attacks are detected.

- **Legal and licensing measures**: Implementing legal protections, such as copyright, patent, or licensing agreements, can provide additional legal recourse and deter unauthorized model extraction and usage.

As we discussed with the membership inference attack, it's important to note that no single technique can provide complete protection against model extraction attacks; a combination of multiple techniques is usually required. The choice of mitigation techniques depends on the specific threat model, the sensitivity of the model, and the desired level of protection.

We have learned about membership inference attacks and model extraction attacks on ML models. Let's now explore the third type of privacy attack on ML models: the reconstruction attack.

Reconstruction attacks—model inversion attacks

Reconstruction attacks try to recreate one or more instances of training data and/or their respective class labels. The reconstruction may be partial or full, depending on the strength of the original model. A fully successful attack can generate more realistic training data and various samples to match exact class label predictions.

Model inversion or attribute inference are kinds of reconstruction attacks. They come under the black-box attack category because the attacker doesn't need to know the details of the model's structure or internal workings. They only need access to the model's output based on some input data. Using that, they can infer details about the data used to train the model.

A step-by-step example of creating a model inversion attack

In this example, we first create a simple dataset with two input features (X) and binary labels (y). We train a logistic regression model using this dataset. The model_inversion_attack function attempts to invert the model by finding an input that produces the desired output probability.

Please note that this is a basic example to illustrate the concept of model inversion attacks. In real-world scenarios, model inversion attacks can be more complex and require sophisticated techniques to handle larger and more complex models.

The full source code can be found in Model_Inversion_LR_Sample.ipynb:

```
from sklearn.linear_model import LogisticRegression
import numpy as np

# Create a sample dataset
X = np.array([[1, 2], [3, 4], [5, 6], [7, 8]])  # Input features
y = np.array([0, 0, 1, 1])  # Corresponding labels

# Train a logistic regression model on the dataset
model = LogisticRegression()
model.fit(X, y)
```

```
# Function to perform model inversion attack
def model_inversion_attack(model, output):
    # Generate a random input within a certain range
    input_range = np.arange(0, 10, 0.01)
    best_input = None
    best_loss = float('inf')

    # Find the input that minimizes the loss function
    for i in input_range:
        input_guess = np.array([[i, i]])
        predicted_output = model.predict_proba(input_guess)
        loss = abs(predicted_output[0][1] - output)

if loss < best_loss:
            best_input = input_guess
            best_loss = loss

    return best_input

# Perform model inversion attack on a specific output
target_output = 0.8
inverted_input = model_inversion_attack(model, target_output)
print("Inverted Input:", inverted_input)

target_output = 1
inverted_input = model_inversion_attack(model, target_output)
print("Inverted Input:", inverted_input)

target_output = 0.5
inverted_input = model_inversion_attack(model, target_output)
print("Inverted Input:", inverted_input)

target_output = 0
inverted_input = model_inversion_attack(model, target_output)
print("Inverted Input:", inverted_input)


Inverted Input: [[5.64 5.64]]
Inverted Input: [[9.99 9.99]]
Inverted Input: [[4.5 4.5]]
Inverted Input: [[0. 0.]]
```

Let's explore a more complex example to understand model inversion attacks.

Model inversion attacks in neural networks

Neural networks are a class of ML models inspired by the structure and functioning of the human brain. They are designed to recognize complex patterns and relationships in data. Neural networks consist of interconnected layers of artificial neurons, known as nodes or units, which collectively form a network.

Each neuron receives input signals, applies a mathematical operation to them, and produces an output signal. These signals are passed through the network, with weights assigned to the connections between neurons determining the strength of the signal. Neural networks are trained using a process called backpropagation, which adjusts the weights based on the errors between predicted and actual outputs.

The hidden layers of a neural network enable it to learn and represent intricate nonlinear relationships in the data, making it capable of solving highly complex tasks such as image and speech recognition, natural language processing, and even playing games. Popular neural network architectures include feedforward neural networks, **convolutional neural networks (CNNs)**, and **recurrent neural networks (RNNs)**.

Neural networks have achieved remarkable success in various fields, demonstrating state-of-the-art performance in many domains. They have become a fundamental tool in machine learning and continue to advance the boundaries of artificial intelligence by enabling sophisticated decision-making and pattern recognition capabilities.

We will not delve into the intricacies of neural networks, as this exceeds the scope of this book.

In this example, we will demonstrate how an adversary can generate input data using the output of a neural network model. The adversary's goal is to reconstruct the original input that led to a specific output prediction by leveraging the characteristics of the model's behavior. By reverse-engineering the relationship between the model's output and the corresponding input data, the adversary can gain insights into the original data points used for training the model.

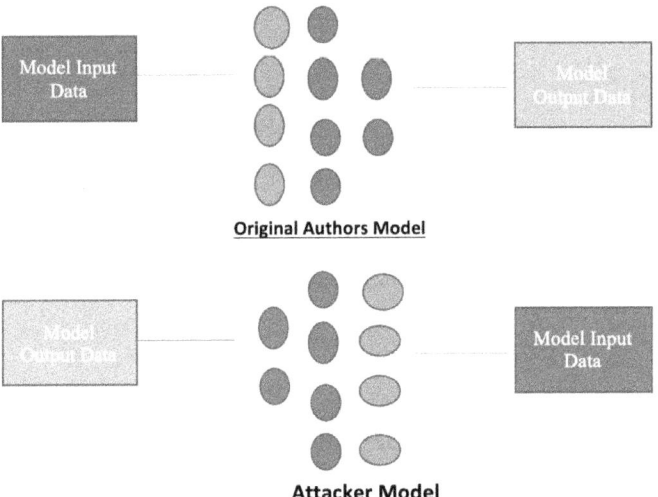

Figure 2.12 – Neural network models of the original author and the adversary

Input data

In this example, we utilize the **Modified National Institute of Standards and Technology (MNIST)** dataset to train a neural network model. The MNIST dataset comprises 60,000 grayscale images of handwritten single digits between 0 and 9. Each image is a small square with dimensions of 28 x 28 pixels.

Original authors model

The following code snippet demonstrates how to load the MNIST dataset from the PyTorch GitHub examples repository using the pickle format. It further splits the data into training and test datasets. In this example, the MNIST dataset is stored in a pickle file named `mnist_data.pkl`. We open the file in binary read mode and load the dataset using the `pickle.load()` function. The dataset is then split into the training and test sets, with the corresponding labels.

> **Note**
>
> Download the dataset from Github (`https://github.com/pytorch/tutorials/blob/main/_static/mnist.pkl.gz`) and keep it in the `data/mnist` directory.

The full source code can be found in `Model_Inversion_Attack_Example.ipynb`. We are using PyTorch version 1.13.1 here:

```
from pathlib import Path
import requests
import pickle
import gzip

DATA_PATH = Path("data")
PATH = DATA_PATH / "mnist"

PATH.mkdir(parents=True, exist_ok=True)

# 10.195.33.40 - Github
URL = "https://10.195.33.40/pytorch/tutorials/raw/main/_static/"
FILENAME = "mnist.pkl.gz"

with gzip.open((PATH / FILENAME).as_posix(), "rb") as f:
((x_train, y_train), (x_valid, y_valid), _) = pickle.load(f,
encoding="latin-1")
```

After loading the data, you can visualize one sample image using the Matplotlib library and obtain its shape. Here's the code snippet:

```
from matplotlib import pyplot
import numpy as np

pyplot.imshow(x_train[0].reshape((28, 28)), cmap="gray")
print(x_train.shape, y_train[0])
```

This results in the following output:

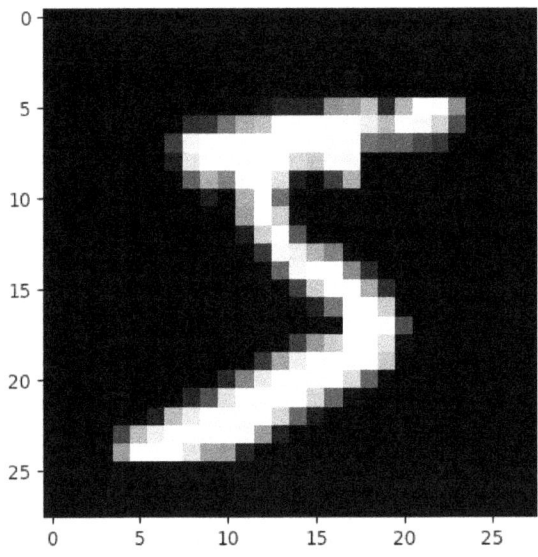

```
(50000, 784) 5
```

Now, convert the training samples into the tensor format in order to use the same in the neural network model as input:

```
import torch

x_train, y_train, x_valid, y_valid = map(
    torch.tensor, (x_train, y_train, x_valid, y_valid)
)
n, c = x_train.shape
print(x_train, y_train)
print(x_train.shape)
print(y_train.min(), y_train.max())
```

```
tensor([[0., 0., 0.,  ..., 0., 0., 0.],
        [0., 0., 0.,  ..., 0., 0., 0.],
        [0., 0., 0.,  ..., 0., 0., 0.],
        ...,
        [0., 0., 0.,  ..., 0., 0., 0.],
        [0., 0., 0.,  ..., 0., 0., 0.],
        [0., 0., 0.,  ..., 0., 0., 0.]])
tensor([5, 0, 4,  ..., 8, 4, 8])
```

```
print(x_train.shape)
```

```
torch.Size([50000, 784])
```

```
print(y_train.min(), y_train.max())
```

```
tensor(0) tensor(9)
```

Let's build a simple sequential neural network model with linear layers using **rectified linear unit (ReLU)** as an activation function:

```
from torch import nn, optim
class AuthorsNN(nn.Module):
  def __init__(self):
    super().__init__()
    self.first_sec = nn.Sequential(
                         nn.Linear(784, 450),
                         nn.ReLU(),
                     )
    self.second_sec = nn.Sequential(
                         nn.Linear(450, 450),
                         nn.ReLU(),
                         nn.Linear(450, 10),
                         nn.Softmax(dim=-1),
                     )

  def forward(self, x):
    return self.second_sec(self.first_sec(x))
```

AuthorsNN class extends the nn.Module class from PyTorch. This class represents a neural network model designed for a classification task. Here's a breakdown of the preceding code and its functionality:

- **The AuthorsNN class**: This class represents the neural network model designed by the author(s). It inherits from the nn.Module class, which is the base class for all neural network modules in PyTorch.

- **The __init__ method**: This method is the constructor of the `AuthorsNN` class and is called when an instance of the class is created. Inside this method, the architecture of the model is defined:

  - `self.first_sec` is a sequential module consisting of two layers:

    - `nn.Linear(784, 450)` represents a linear layer with 784 input features and 450 output features.

    - `nn.ReLU()` applies the ReLU activation function to introduce non-linearity.

  - `self.second_sec` is another sequential module consisting of three layers:

    - `nn.Linear(450, 450)` represents a linear layer with 450 input features and 450 output features.

    - `nn.ReLU()` applies the ReLU activation function.

    - `nn.Linear(450, 10)` represents a linear layer with 450 input features and 10 output features, corresponding to the number of classes.

    - `nn.Softmax(dim=-1)` applies the softmax activation function to convert the raw output scores into probabilities, ensuring they sum to 1 across classes.

- **The forward method**: This method defines the forward pass of the model, specifying how input data flows through the network. The input, x, is passed through `self.first_sec`, followed by `self.second_sec`, and the resulting output is returned.

The following code creates an instance of the `AuthorsNN` class named `auth_nn`. This instance represents the initialized neural network model:

```
auth_nn = AuthorsNN()
auth_nn
```

Printing `auth_nn` will display information about the model, such as its architecture and the number of trainable parameters:

```
AuthorsNN(
  (first_sec): Sequential(
    (0): Linear(in_features=784, out_features=450, bias=True)
    (1): ReLU()
  )
  (second_sec): Sequential(
    (0): Linear(in_features=450, out_features=450, bias=True)
    (1): ReLU()
    (2): Linear(in_features=450, out_features=10, bias=True)
    (3): Softmax(dim=-1)
  )
)
```

Next, we define a loss function in order to measure the error between the actual data versus the predicted data:

```
loss_func = nn.CrossEntropyLoss()
loss_func
```

CrossEntropyLoss()

To enhance the network, let's add the Adam optimizer function. This is an optimization algorithm that replaces **stochastic gradient descent (SGD)** for training DL models.

It combines the desirable aspects of the AdaGrad and RMSProp algorithms, making it suitable for handling sparse gradients in noisy problem scenarios:

```
from torch import optim
optimizer = optim.Adam(auth_nn.parameters(), lr = 0.01)
```

Here, we import the optim module from PyTorch. After instantiating the AuthorsNN class, we define the Adam optimizer using the optim.Adam() function. The optimizer is initialized with the model's parameters (auth_nn.parameters()), enabling it to optimize the model during training.

Next, print optimizer to provide details about the model's optimizer's configuration:

```
optimizer
```

This results in the following output:

```
Adam (
Parameter Group 0
    amsgrad: False
    betas: (0.9, 0.999)
    capturable: False
    differentiable: False
    eps: 1e-08
    foreach: None
    fused: False
    lr: 0.01
    maximize: False
    weight_decay: 0
)
```

Now, train the neural network model with the MNIST training dataset that we loaded earlier and wrap it in a Python function:

```
def train(num_epochs, ann):

    ann.train()

    for epoch in range(num_epochs):

            output = ann(x_train)

            loss = loss_func(output, y_train)

            # clear gradients for this training step
            optimizer.zero_grad()

            # backpropagation, compute gradients
            loss.backward()
            # apply gradients
            optimizer.step()

            print(epoch, loss.item())

    pass
```

Let's break this code down:

- **The train function**: This function trains the neural network model (ann) for a specified number of epochs (num_epochs). The function assumes the presence of input data (x_train) and corresponding target labels (y_train) used for training the model. Here, ann.train() is called to set the model in training mode, enabling functionalities such as dropout and batch normalization.

- **The training loop**: For each epoch in the range of num_epochs, the following steps are executed:

  - output = ann(x_train): Forward passes through the model, obtaining the output predictions.

  - loss = loss_func(output, y_train): Computes the loss between the predicted output and the ground truth labels.

  - optimizer.zero_grad(): Clears the gradients accumulated from the previous iteration.

  - loss.backward(): Performs backpropagation to compute the gradients of the model's parameters with respect to the loss.

- `optimizer.step()`: Updates the model's parameters by applying the computed gradients using the chosen optimizer.

- `print(epoch, loss.item())`: Prints the current epoch number and the `loss` value.

- The `pass` statement: A placeholder that does nothing in this context and can be removed if not needed.

Now, train the neural network model with the MNIST training dataset that we loaded earlier with the author's neural network model, which was built in the previous step with `100` epochs:

```
train(100,auth_nn)
```

```
0 1.6085695028305054
1 1.6047792434692383
2 1.59657621383667
....
100 1.4667187929153442
```

Once the model is trained, it can be used for further predictions. Now, we will construct the adversary attacker model with the objective of recreating the training data.

Adversary model to get the trained input data

Considering that the author's model has been trained on the MNIST dataset and we have access to the size `450` vector output from the model's first section (`first_sec`), we can utilize this information for our attack. Next, we will develop our adversary model. This model takes a size `450` vector as input, which corresponds to the output of the target's first section. The adversary model's objective is to generate a size `784` vector, matching the size of the original input data:

```
class Adversary(nn.Module):
  def __init__(self):
    super().__init__()
    self.layers= nn.Sequential(
                    nn.Linear(450, 800),
                    nn.ReLU(),
                    nn.Linear(800, 784),
                 )

  def forward(self, x):
    return self.layers(x)
```

Based on the information available, the authors' original model was trained on a dataset consisting of handwritten images. This knowledge provides us with an understanding of the model's training data source.

To train our adversary model, we can utilize the MNIST test data. Specifically, we will use the first 1,000 rows of the MNIST test data to train our adversary model. After training, we can evaluate the accuracy of the adversary model using the MNIST test data ranging from the 1,000th row to the 2,000th row.

Let's train the adversary model:

```
adversary = Adversary()
optimizer = optim.Adam(adversary.parameters(), lr=1e-4)

for i in range (0,1000):
    optimiser.zero_grad()
    #print(x_train[i])
    target_outputs   = auth_nn.first_sec(x_valid[i])
    adversary_outputs = adversary(target_outputs)
    #print(adversary_outputs)
    loss = ((x_valid[i] - adversary_outputs)**2).mean()
    #print(loss.item())
    loss.backward()
    optimiser.step()

Now, let's test the adversary model:
for i in range (1000,2000):
    target_outputs = auth_nn.first_sec(x_train[i])
    recreated_data = adversary(target_outputs)
    #print(recreated_data)
```

To assess the similarity between the recreated data and the original trained images from the training dataset, we can utilize the Matplotlib library to visualize the images. By plotting the recreated image, we can determine the level of resemblance it holds with the original trained images:

```
With torch.no_grad():
  pyplot.imshow(recreated_data.reshape((28, 28)), cmap="gray")
```

This results in the following output:

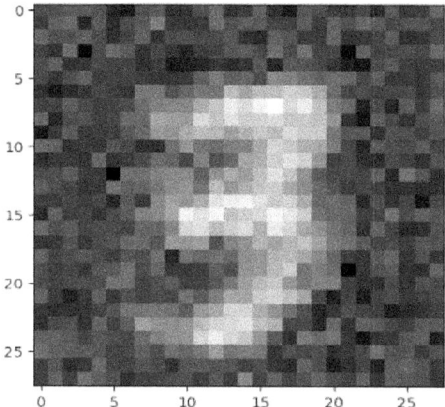

It is evident that the images generated using model inversion attacks closely resemble the training data. This example demonstrates the successful recreation of training data without requiring complete knowledge of the model details, thereby achieving a model inversion attack.

Techniques to mitigate model inversion attacks

Mitigating model inversion attacks, where an adversary tries to infer sensitive training data from a trained model's outputs, is crucial for preserving privacy and protecting sensitive information.

Some techniques to mitigate model inversion attacks include the following:

- **Differential privacy**: Applying differential privacy mechanisms during the training process can help protect against model inversion attacks. Differential privacy adds controlled noise to the training data or model's outputs, making it harder for an attacker to extract specific sensitive information from the model's predictions. We will learn more about differential privacy in the next two chapters.

- **Limit access to sensitive output**: Restricting access to sensitive model output or predictions can help mitigate model inversion attacks. By carefully controlling who has access to the output and under what circumstances, you can reduce the risk of an adversary inferring sensitive training data.

- **Preprocessing and postprocessing**: Applying preprocessing and postprocessing techniques to the data and model's output can help protect against model inversion attacks. For example, data anonymization, aggregation, or transformation techniques can be applied to remove or obfuscate sensitive information from the input or output. We will learn more about data anonymization and aggregation in the subsequent chapters.

- **Regularization**: Incorporating regularization techniques such as L1 or L2 regularization during the model training process can help improve privacy by reducing the model's reliance on specific sensitive features. Regularization can help prevent overfitting and limit the leakage of sensitive information through the model's predictions.

- **Generative adversarial networks** (**GANs**): Using generative models such as GANs can help protect against model inversion attacks. By generating synthetic data that preserves the statistical properties of the original data, GANs can provide alternative output for the attacker without revealing specific sensitive training instances.

- **Secure multi-party computation** (**MPC**): Leveraging secure MPC protocols can enable multiple parties to collaboratively train a model while keeping their individual training data private. Secure MPC ensures that no party can access the sensitive data of others, thereby mitigating model inversion attacks.

- **Secure aggregation**: In scenarios where models are trained using a distributed setting, secure aggregation protocols can be employed to prevent sensitive information leakage during the aggregation of model updates. This protects against model inversion attacks during the training process.

- **Access control and authorization**: Implementing access control measures and strong authorization mechanisms can help restrict access to sensitive model output, limiting the exposure to potential attackers. Only authorized entities should have access to sensitive predictions or output.

- **Synthetic data generation**: Instead of training models directly on sensitive data, using synthetic data generated from the original data can help mitigate model inversion attacks. Synthetic data retains the statistical characteristics of the original data but does not expose sensitive information.

- **Model monitoring**: Continuously monitoring the model's behavior for any unusual patterns or unexpected output can help detect potential model inversion attacks. Monitoring can involve techniques such as outlier detection, adversarial robustness checks, or statistical analysis of model predictions.

Like the previous two attacks, It's important to note that choosing mitigation techniques depends on the specific context, the sensitivity of the data, and the desired level of privacy protection. Multiple techniques can be combined to achieve stronger privacy guarantees against model inversion attacks.

Summary

To summarize, we have covered different types of ML (supervised and unsupervised) and explored how to save and execute models in various formats. Additionally, we delved into the different phases of ML (data extraction, data preparation, model development, model deployment, and inferencing) and discussed the privacy threats and attacks associated with each phase in detail.

In the next chapter, we will dive deeper into privacy-preserving data analysis and focus on understanding the concept of differential privacy. This will allow us to explore techniques and methodologies that ensure privacy while conducting data analysis tasks. By gaining a thorough understanding of differential privacy, we can better safeguard sensitive information and mitigate privacy risks in the context of ML.

Part 2: Use Cases of Privacy-Preserving Machine Learning and a Deep Dive into Differential Privacy

This part focuses on privacy-preserving data analysis and explores privacy-enhanced technologies, with a particular emphasis on differential privacy.

We introduce the concept of privacy-preserving data analysis and delve into techniques and methodologies that allow for the analysis of data while protecting individuals' privacy.

We highlight the risks associated with reconstruction attacks in SQL, where an adversary attempts to reconstruct sensitive information from seemingly anonymized data. We discuss various prevention methods and countermeasures that can be employed to mitigate such attacks and protect individuals' privacy.

This part also provides an overview of privacy-enhanced technologies, such as differential privacy, federated learning, secure multiparty computation (SMC), and homomorphic encryption.

We provide an introduction to differential privacy and machine learning using differential privacy and explore how differential privacy can be incorporated into machine learning algorithms to provide privacy guarantees for individuals whose data is used in the learning process.

We also provide a deep dive into various algorithms employed in differential privacy. These algorithms include Laplace, Gaussian, count, sum, mean, variance, standard deviation, and thresholding algorithms.

We look at developing differential privacy applications using the OpenMined PyDP framework. Here, we focus on the practical steps involved in developing differential privacy applications using the OpenMined PyDP framework.

We also cover deep learning with differential privacy and look at a fraud detection use case. This showcases how open source frameworks such as PyTorch and Opacus can be utilized to implement DP in deep learning models.

Finally, we provide an overview of real-world applications utilizing differential privacy. Here, we highlight the use cases where DP has been successfully employed to balance data analysis and privacy protection in fields such as healthcare, finance, and social sciences.

This part has the following chapters:

- *Chapter 3, Overview of Privacy-Preserving Data Analysis and Introduction to Differential Privacy*

- *Chapter 4, Differential Privacy Algorithms and Limitations of Differential Privacy*

- *Chapter 5, Developing Machine Learning Applications with Differential Privacy Using Open Source Frameworks*

3

Overview of Privacy-Preserving Data Analysis and an Introduction to Differential Privacy

In this chapter, we will explore the concept of privacy in the context of big data, along with the associated risks. We will delve into privacy in data analysis, focusing on the trade-off between privacy and utility. Furthermore, we will investigate various privacy-preserving techniques, such as anonymization, k-anonymity, t-closeness, and ℓ-diversity, while also discussing their limitations. Later on, we will introduce one of the key privacy-enhancing approaches, known as differential privacy. We will provide a high-level overview of differential privacy, covering essential concepts such as privacy loss, privacy budgets, and differential privacy mechanisms.

The main topics covered in this chapter include the following:

- Privacy in data analysis:

  - Privacy in data analysis, the need for privacy in data analysis, and the objectives of privacy in data analysis

- Privacy-preserving techniques:

  - Investigating various privacy-preserving techniques, including anonymization, k-anonymity, t-closeness, and ℓ-diversity

  - Data aggregation

  - Privacy attacks with data aggregation

  - Tools/framework to protect data privacy in SQL

- Privacy-enhancing technologies:

  - Introducing differential privacy as a privacy-enhancing technology and federated learning and homomorphic encryption

- Differential privacy:

  - Deep diving into differential privacy concepts, including privacy loss, privacy budgets, and differential privacy mechanisms

Privacy in data analysis

Privacy in data analysis is a crucial aspect that ensures sensitive information about individuals is not disclosed or misused. It involves implementing measures such as data anonymization and encryption to protect the identity and personal details of individuals while still allowing for meaningful data analysis.

The need for privacy in data analysis

Many enterprises, social networking companies, e-commerce platforms, networking companies, taxi/cab aggregators, food delivery services, and government organizations, among others, gather and process vast amounts of data – both personal and non-personal – to derive insights using machine learning and AI techniques. The data collected by these entities encompasses a wide range of information, such as browsing history, purchase records, social network interactions, health data, location data, content consumption patterns, device information, and more. It is important to note that this data often contains sensitive personal information. Sharing or retaining this data for longer durations than necessary (considering data retention laws based on data categories) can pose privacy risks and may lead to non-compliance with privacy and legal regulations. As we discussed in the first chapter, some privacy risks/breaches are related to the following:

- **Stealing personal information**: This includes unauthorized access to individuals' credentials, such as credit card numbers, passwords, and other sensitive data

- **Identity theft**: Personal identification details, such as Social Security numbers (in the US) or Aadhaar unique IDs (in India), names, bank information, biometric data, and driving licenses, can be targeted for identity theft

- **Discrimination and targeting individuals**: Certain types of data, such as medical records, can be exploited to discriminate against individuals based on their religious beliefs or can lead to consequences such as insurance rejection. Other examples include political posts, comments/opinions on social media, and more.

To address these concerns, privacy laws have been implemented worldwide. For instance, the **General Data Protection Regulation (GDPR)** in Europe and the **California Consumer Privacy Act (CCPA)** in the United States provide guidelines and regulations for the protection of sensitive data and address privacy-related breaches and issues.

It is crucial for organizations to be up to date with the practices of privacy in data analysis in order to be aware of privacy risks, comply with relevant privacy laws and regulations, and take necessary measures to protect individuals' data privacy.

Privacy in data analysis refers to the process of performing data analysis tasks while preserving the privacy and confidentiality of data. It involves applying various techniques and methodologies to extract meaningful insights from data without compromising the privacy rights of individuals whose data is being analyzed. The goal is to strike a balance between data utility and privacy, which is crucial in privacy data analysis. However, there are trade-offs associated with each approach:

- **High data privacy**: Emphasizing strong privacy measures may result in poor data utility. For instance, if sensitive personal data, such as Social Security numbers or medical records, is redacted or anonymized to safeguard privacy, it becomes challenging to link that data to other relevant data points. Consequently, the resulting insights may be less useful for analytical purposes.

- **High data utility**: Prioritizing high data utility may lead to weaker privacy protection. In this scenario, the data remains highly useful and valuable for analysis. However, there may be inadequate protection of sensitive information, increasing the risk of re-identification or the unintended disclosure of personal details.

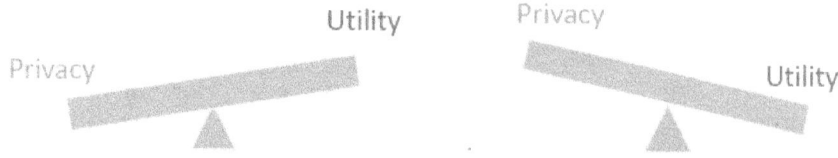

Figure 3.1 – Privacy versus utility trade-off

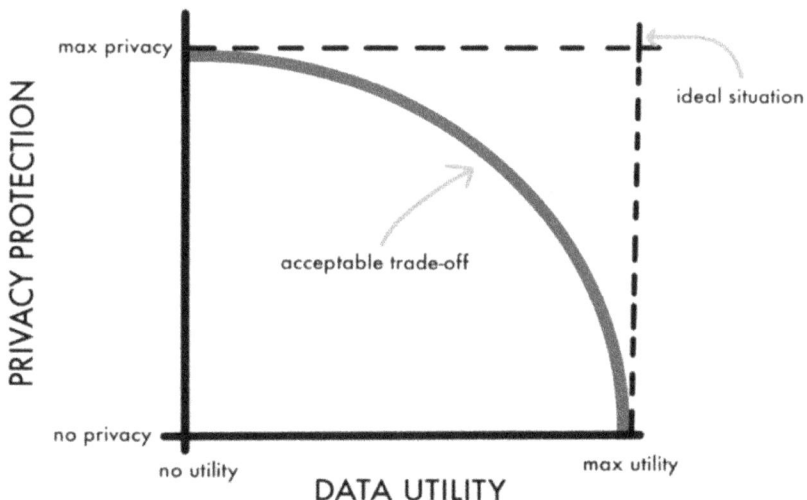

Figure 3.2 – Privacy versus utility trade-off

Image source: *Churi, Prathamesh & Pawar, Dr. Ambika & Moreno Guerrero, Antonio. (2021). A Comprehensive Survey on Data Utility and Privacy: Taking Indian Healthcare System as a Potential Case Study. Inventions. 6. 1-30. 10.3390/inventions6030045.*

Finding the right balance between privacy and utility is a critical challenge of privacy in data analysis. Organizations must implement privacy-preserving techniques, such as anonymization, aggregation, and noise addition, to protect sensitive information while still enabling meaningful analysis. By carefully considering both privacy and utility aspects, it is possible to derive valuable insights from data while safeguarding individuals' privacy rights.

Objectives of privacy in data analysis

The objectives of privacy in data analysis are as follows:

- **Privacy preservation**: The primary focus is on safeguarding the sensitive information contained within the dataset, ensuring that individuals' privacy is respected and protected. This involves preventing unauthorized access, disclosure, or the identification of individuals from the data.

- **Data utility**: While preserving privacy, the analysis should still provide valuable and accurate results that maintain the usefulness and quality of the data. The challenge lies in finding methods that allow for effective analysis while minimizing the impact on data utility.

- **Risk assessment**: Privacy in data analysis involves identifying and assessing potential privacy risks associated with the data being analyzed. This includes evaluating the likelihood of re-identification or the possibility of the unintended disclosure of sensitive information during the analysis process.

- **Anonymization techniques**: Various anonymization techniques are employed to remove or obfuscate identifying information from the dataset. This can include methods such as data generalization, suppression, noise addition, and data perturbation to protect individuals' identities.

- **Privacy-preserving algorithms/techniques**: Specialized algorithms and methodologies are utilized to perform analysis tasks while minimizing privacy risks. These algorithms ensure that the results reveal aggregated and anonymized insights rather than individual-level details.

- **Regulatory compliance and ethical considerations**: Privacy in data analysis adheres to relevant privacy laws and regulations to ensure compliance and ethical handling of sensitive data. Organizations must adhere to legal requirements and ethical guidelines to protect individuals' privacy rights.

By protecting privacy in data analysis techniques, organizations can gain valuable insights from sensitive data while respecting privacy constraints. This approach fosters trust among those individuals whose data is being analyzed and promotes responsible data-handling practices.

Privacy-preserving techniques

It's important to note that data privacy is a fundamental right, and individuals have the right to control how their personal information is collected, used, and shared. Moreover, as discussed above, a lack of data privacy protection can lead to serious consequences such as identity theft, financial fraud, and discrimination.

In the context of data analysis, an adversary refers to an entity or party that actively seeks to gain unauthorized access to, or exploit, sensitive data, disrupt the data analysis process, or manipulate the results for their own benefit or malicious intent. Adversaries in data analysis can include individuals, organizations, or automated systems that aim to compromise the integrity, confidentiality, or availability of the data or the analysis process itself.

Organizations can implement various privacy-preserving techniques that allow for the analysis of data while still protecting the privacy of individuals. We'll cover the following privacy-preserving techniques and associated privacy attacks:

- Data anonymization techniques/algorithms:

  - K-anonymity, ℓ-diversity, and t-closeness

- Data aggregation and privacy attacks associated with data aggregation

Data anonymization and algorithms for data anonymization

One of the techniques for safeguarding privacy and complying with data protection regulations is data anonymization. By anonymizing data, organizations can mitigate the risks associated with the unauthorized disclosure and misuse of personal information while still preserving the utility and analytical value of the data.

Typically, datasets consist of three types of attributes – namely, key identifiers, quasi-identifiers, and sensitive attributes:

- **Key identifiers**: These attributes uniquely identify individuals within the dataset.

- **Quasi-identifiers**: Quasi-identifiers are indirect identifiers that, when combined with other attributes, can be used to identify individuals or uncover personal information. They provide hints or partial identification.

- **Sensitive attributes**: These attributes contain sensitive personal data that needs to be protected to maintain privacy.

Let's consider an example dataset from a hospital or medical clinic that collects and maintains patient data. Assume that the hospital collects various data for each patient, including name, date of birth, gender, postal code/zip code, height, weight, blood pressure, SpO2 levels, current medical conditions, last visit date, and so on.

Here are some details about this dataset:

- The name serves as a key attribute since it uniquely identifies each person.

- Quasi-identifiers include attributes such as date of birth, gender, postal code, and height. Although knowing the date of birth alone may not be sufficient to identify an individual, combining it with other quasi-identifiers such as gender, postal code, and medical conditions might make it possible to identify the individual.

- Sensitive attributes encompass information related to the type of disease the patient is suffering from. Protecting this information is crucial to maintain the privacy of the patients.

By understanding the distinct types of attributes within a dataset, organizations can implement appropriate anonymization techniques to safeguard the privacy of individuals while still allowing for valuable analysis and research.

Now we know about the different types of attributes, let's try data anonymization on a sample toy dataset and analyze the results.

Let's consider an example dataset containing information about individuals' ages and medical conditions. The sensitive attribute in this case is the medical condition, which we want to protect while preserving the data's utility.

Key Attribute	Quasi-Identifiers				Sensitive Attribute	Other Data …
Name	Date of Birth	Gender	Zip Code	Height in cm	Disease	Other Data …
John	2000-09-15	M	90001	170	COVID	….
Rosy	2002-12-08	F	96162	165	COVID	
Robin	1945-07-24	M	92348	180	CANCER	
Hellen	1950-01-13	F	95411	156	CANCER	
Antonio	1970-01-13	M	95416	180	Fever	…..

Table 3.1 – Sample dataset

Obviously, the data collected contains personal information such as names, dates of birth, and so on, so in order to protect individuals, let's try the data anonymization technique and see whether it helps or not. Data anonymization means either removing or modifying the personally identifiable information so that others will not be able to detect individuals from the data.

Let us try a simple solution, that is, removing the personal information (in this case, the name of the person) from the given data. The data looks like the following after removing the names from the data.

Name	Date of Birth	Gender	Zip Code	Height in cm	Diagnosed Disease	Other Data …
	2000-09-15	M	90001	170	COVID	….
	2002-12-08	F	96162	165	COVID	
	1945-07-24	M	92348	180	CANCER	
	1950-03-13	F	95411	156	CANCER	
	1970-01-13	M	95416	180	Fever	…..

Table 3.2 – Dataset without names

Is the removal of the names sufficient to protect sensitive data? Can this data now be shared with others? Could anyone accurately identify the individual with cancer from this data?

Based on the preceding data table, it looks like people will not be able to find out because we have removed the name of the person from the data. This is the first impression from this data. Removing the names of individuals from a dataset is a step toward protecting sensitive data and preserving privacy.

However, it is important to note that simply removing the name does not guarantee complete privacy protection. Other quasi-identifiers and sensitive attributes within the dataset could still potentially lead to the re-identification of individuals.

Let's see how an adversary would be able to identify individuals using this so-called anonymized dataset.

Government organizations do a census data survey (population survey and voter list updates/additions, and so on) every 5 or 10 years, depending on a country's regulations, and share it with other departments as appropriate.

There is a possibility that the anonymized dataset (like the preceding data table) could be linked with publicly available datasets published by either government organizations or non-profit organizations and insights could be derived or individuals could be identified, that is, sensitive personal information could be found out.

The following is a sample dataset of voter ID data from the government:

Name	Date of Birth	Gender	Zip Code	Last Year Voted	Registration Date	Othe Data …
John	2000-09-15	M	90001	2020	2018	….
Rosy	2002-12-08	F	96162	-	2020	
Robin	1945-07-24	M	92348	2020	1963	

| Hellen | 1950-03-13 | F | 92411 | 2020 | 1968 | |
| Antonio | 1970-01-13 | M | 95416 | - | 1988 | |

Table 3.3 – Sample voter ID dataset

It is quite possible to link both these datasets, that is, patients' data with voters' data (refer to the LINDDUN framework's privacy threat linkage category, which was described in *Chapter 1*) and find out who has cancer from a given zip/postal code.

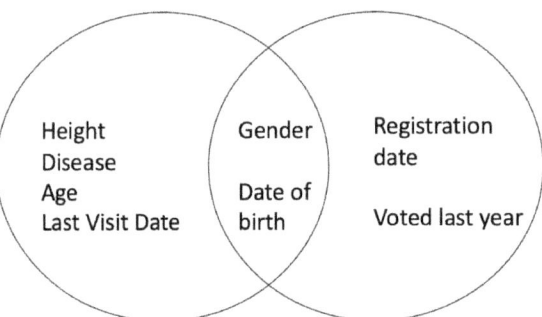

Figure 3.3 – Linking two datasets

It is very clear that Robin lives in postal code 92348, so an adversary could link voter ID and medical data records and identify that he suffers from cancer. Once the adversary finds the sensitive information in this case, that is, who has cancer, then possible discrimination as a side effect of this data privacy breach could be the rejection of medical insurance for Robin.

So, just removing one or two fields from the dataset is not enough to protect the sensitive information of individuals.

Let's try another data anonymization technique called k-anonymity.

K-anonymity – data anonymization

Latanya Sweeney and *Pierangela Samarati* introduced k-anonymity in 1998 and it is described as follows:

> *Given person-specific field-structured data, produce a release of the data with scientific guarantees that the individuals who are the subjects of the data cannot be re-identified while the data remain practically useful. A release of data is said to have the k-anonymity property if the information for each person contained in the release cannot be distinguished from at least k-1 individuals whose information also appears in the release.*

(Source: Wikipedia)

In order to apply k-anonymity anonymization, one needs to do the following:

1. Identify the quasi-identifier in the given dataset.

2. Generalize the attributes (instead of exact dates, use ranges) or suppress the attributes (instead of showing full zip codes show only partial ones) for at least k-1 records in the dataset.

Name	Date of Birth	Gender	Zip Code	Height in cm	Diagnosed Disease	Other Data...
	2000-09-15	M	90001	170	COVID	....
	2002-12-08	F	96162	165	COVID	
	1945-07-24	M	92348	180	CANCER	
	1950-03-13	F	92411	156	CANCER	
	1970-01-13	M	95416	180	Fever	

Table 3.4 – Dataset without names

In this example, the quasi-identifiers are zip code and date of birth. These are considered indirect identifiers as there could be individuals who have the same zip code and date of birth, but these two factors may not necessarily identify a specific individual. Once the quasi-identifiers are identified, either suppress them or generalize attributes. Let's try the last three digits of the zip code suppressed with *** and generalize the date of birth using a range rather than the exact date of birth.

The updated data (anonymized) looks as follows:

Name	Date of Birth	Gender	Zip Code	Height in cm	Diagnosed Disease	Other Data...
	20-30	M	90***	170	COVID	....
	20-30	F	96***	165	COVID	
	70-80	M	92***	180	CANCER	
	70-80	F	92***	156	CANCER	
	50-60	M	95***	180	FEVER	

Table 3.5 – Dataset after generalization and suppression

Go through these questions:

* Is this now sufficient to protect the sensitive data because we have anonymized the data (quasi-identifiers) using k-anonymity?

* Can the data now be shared with others?

* Could anyone now find out exactly who is the person with cancer from this data?

In this case, it is still possible to identify the individual because, within the same age range (70-80), every person has cancer. This vulnerability is known as a homogeneity attack, which can occur in k-anonymity when a set of k records have exact sensitive values. Despite the data being k-anonymized, the sensitive value for the set of k records can still be accurately predicted due to the identical date range for the sensitive value "cancer" in our example.

Let's now implement k-anonymity using Python for a sample dataset.

Source code: `K-Anonymity_Example.ipynb`

We'll create a synthetic dataset of medical records containing quasi-identifiers (age and gender) and a sensitive attribute (diagnosis):

```python
import pandas as pd
import random
# Step 1: Define the parameters
num_records = 1000  # Number of records in the dataset
k = 5  # Desired k-anonymity level
# Step 2: Generate the synthetic dataset
age_range = (20, 80)
gender_list = ["Male", "Female"]
diagnosis_list = ["Covid", "Cancer", "Fever", "Obesity"]
data = []
for _ in range(num_records):
    age = random.randint(age_range[0], age_range[1])
    gender = random.choice(gender_list)
    diagnosis = random.choice(diagnosis_list)
    data.append({"age": age, "gender": gender, "diagnosis":
diagnosis})
dataset = pd.DataFrame(data)
dataset.to_csv("dataset.csv", index=False)
# Step 3: Apply k-anonymity
groups = dataset.groupby(["age", "gender"])
anonymized_data = pd.DataFrame()
for _, group in groups:
    if len(group) < k:
        group["age"] = generalize(group["age"])
        group["gender"] = generalize(group["gender"])
    anonymized_data = anonymized_data.append(group)
# Save the anonymized dataset to a file
anonymized_data.to_csv("k_anonymized_dataset.csv", index=False)
def generalize(attribute):
```

```
if attribute.name == "age":
    # Generalize age into predefined ranges
    age_ranges = [(0, 20), (20, 40), (40, 60), (60, 80)]
    for start, end in age_ranges:
        if start <= attribute.iloc[0] < end:
            return f"{start}-{end}"
elif attribute.name == "gender":
    # Generalize gender to binary values
    return "Other"
else:
    # Handle other attributes if needed
    return attribute.
```

In this example, we first define the parameters, including the desired number of records (num_records) and the desired k-anonymity level (k). Then, we generate a synthetic dataset by randomly assigning values to the quasi-identifiers (age and gender) and the sensitive attribute (diagnosis). After generating the dataset, we apply k-anonymity by grouping the records based on the quasi-identifiers and checking whether the group size is less than k. If the group size is less than k, we generalize the quasi-identifiers by applying a suitable generalization technique.

The `generalize()` function takes an attribute (e.g., age or gender) as input and applies a specific generalization technique. Here, we implement the generalization for the age attribute by dividing it into predefined age ranges (e.g., 0-20, 20-40, 40-60, and 60-80). For the gender attribute, we generalize it as `"Other"`.

We can modify and extend this function based on the generalization requirements specific to the dataset. For instance, we may need to handle other attributes differently or define additional generalization rules for different quasi-identifiers.

Finally, the anonymized dataset is saved to a file named `k_anonymized_dataset.csv`.

Let's try another anonymization technique called ℓ-diversity and see whether it provides a solution to the privacy questions discussed in our examination of k-anonymity.

ℓ-diversity – data anonymization

ℓ-diversity is an approach that aims to mitigate the vulnerability of homogeneity attacks on anonymized data. It achieves this by ensuring that each sequence of quasi-identifiers is associated with a "well-represented" sensitive attribute. In other words, ℓ-diversity seeks to introduce diversity within a group of records that share the same quasi-identifiers, ensuring that multiple sensitive attribute values are present. By doing so, ℓ-diversity reduces the risk of accurately inferring sensitive information from the anonymized data.

Original dataset

Quasi-Identifiers		Sensitive Attribute
Date of Birth	ZIP Code	Disease
2000-09-15	90001	COVID
2002-12-08	96162	COVID
1945-07-24	92348	CANCER
1950-03-13	95411	CANCER
1970-01-13	95416	FEVER

Table 3.6 – Original dataset

K-anonymization dataset

Quasi-Identifiers		Sensitive Attribute
Date of Birth	ZIP Code	Disease
20-30	90***	COVID
20-30	96***	COVID
70-80	92***	CANCER
70-80	92***	CANCER
50-60	95***	FEVER

Table 3.7 – K-anonymization dataset

This data is not diverse when this data is shared with others.

To make this dataset ℓ-diverse (for $\ell=2$) we need to introduce a unique disease for at least one entry in each group with the same date of birth and zip code. Here is an example of how we could modify the dataset:

ℓ -Diversity Dataset Quasi-Identifiers		Sensitive Attribute
Date of Birth	Zip Code	Disease
20-30	90***	COVID
70-80	92***	CANCER
50-60	95***	FEVER

20-30	96***	~~COVID~~
		FLU
70-80	92***	~~CANCER~~
		HEART DISEASE

Table 3.8– ℓ-diversity dataset (somewhat diverse)

The 20-30 age group with zip code 96*** has `Disease` changed to `FLU`. The 70-80 age group with the zip code 92*** has `Disease` changed to `HEART DISEASE` for one record. Now, each set of similar dates of birth and zip codes has at least two unique diseases, which makes the dataset 2-diverse.

Achieving ℓ-diversity may seem straightforward in smaller datasets, as demonstrated in our example. However, there are limitations to its applicability, particularly when dealing with large datasets that contain millions or billions of records. In such cases, achieving ℓ-diversity becomes challenging and may not be necessary in all scenarios. It is important to note that ℓ-diversity, while helpful, is not a comprehensive solution for protecting sensitive information. While it introduces diversity within groups of records, it does not guarantee complete privacy. Additional privacy-preserving techniques and measures may be required to provide stronger protection for sensitive data.

Let's now implement ℓ-diversity using Python for a sample dataset.

Source code: L-Diversity_Example.ipynb

Let's implement ℓ-diversity on the synthetic dataset we generated earlier. We'll consider the diagnosis attribute as the sensitive attribute and ensure that each group has at least ℓ-distinctsensitive attribute values:

```
import pandas as pd
import random
# Step 1: Define the parameters
num_records = 1000  # Number of records in the dataset
l = 3  # Desired l-diversity level

# Step 2: Generate the synthetic dataset
age_range = (20, 70)
gender_list = ["Male", "Female"]
diagnosis_list = ["Covid", "Cancer", "Fever", "Obesity"]

data = []
for _ in range(num_records):
    age = random.randint(age_range[0], age_range[1])
    gender = random.choice(gender_list)
```

```
    diagnosis = random.choice(diagnosis_list)
    data.append({"age": age, "gender": gender, "diagnosis":
diagnosis})
dataset = pd.DataFrame(data)
dataset.to_csv("dataset_before_l_diversity.csv", index=False)


# Step 3: Apply l-diversity
groups = dataset.groupby(["age", "gender"])
anonymized_data = pd.DataFrame()


for _, group in groups:
    if len(group["diagnosis"].unique()) < l:
        # Generalize or suppress the sensitive attribute
        group["diagnosis"] = generalize(group["diagnosis"])
        anonymized_data = anonymized_data.append(group)
    else:
        anonymized_data = anonymized_data.append(group)


# Save the anonymized dataset to a file
anonymized_data.to_csv("l_diversity_dataset.csv", index=False)
```

We generate a synthetic dataset similar to before. Afterward, we group the records based on the quasi-identifiers (age and gender) and check whether the number of distinct sensitive attribute values (diagnosis) is less than "ℓ". If the condition is met, we apply generalization or suppression to the sensitive attribute to ensure there are at least "ℓ" distinct values within the group. Note that in this example, the `generalize()` function is called for the diagnosis attribute. Finally, the anonymized dataset is saved to a file named `l_diversity_dataset.csv`:

```
def generalize(attribute):
    if attribute.name == "diagnosis":
        # Generalize diagnosis to a higher-level category
        diagnosis_mapping = {
            "Covid": "Chronic Condition",
            "Cancer": "Chronic Condition",
            "Fever": "Non-Chronic Condition",
            "Obesity": "Non-Chronic Condition"
        }
        return attribute.map(diagnosis_mapping)
    else:
        # Handle other attributes if needed
        return attribute
```

In this example, the `generalize()` function takes an attribute (e.g., diagnosis) as input and applies a specific generalization technique. Here, we demonstrate the generalization for the diagnosis attribute by mapping specific diagnoses to higher-level categories. For instance, diagnoses such as `"Fever"` and `"Obesity"` are generalized to the category `"Non-Chronic Condition"`, while `"Cancer"` and `"Covid"` are generalized to `"Chronic Condition"`.

Now, let's delve into another approach, known as t-closeness.

t-closeness

The t-closeness approach ensures that the distribution of sensitive attributes within a group of individuals is not significantly different from the distribution of the same attributes in the overall population. It aims to maintain a balance between privacy and data utility by setting a threshold value, denoted by t, which represents the maximum allowed difference between the distributions.

To satisfy t-closeness, a dataset should exhibit similar probability distributions of sensitive attributes (such as age, income, and gender) in each group of individuals compared to the overall dataset. If the difference between the group and overall distributions exceeds the specified t-value, additional anonymization measures may be necessary.

Let's consider an example to illustrate t-closeness. Suppose we have a dataset containing information about individuals' ages and medical conditions. The sensitive attribute in this case is the medical condition, which we want to protect while preserving the data's utility.

We divide the dataset into groups based on a set of quasi-identifiers, such as age ranges and gender. For simplicity, let's focus on the age range. We have two groups: group A (age range 30-40) and group B (age range 40-50).

To satisfy t-closeness, we need to compare the distribution of medical conditions within each group to the overall distribution in the dataset. Let's assume the overall dataset has the following distribution of medical conditions: 60% cancer, 30% diabetes, and 10% asthma.

In group A, we find that the distribution of medical conditions is 70% cancer, 20% diabetes, and 10% asthma. In group B, the distribution is 50% cancer, 30% diabetes, and 20% asthma.

To determine whether the dataset satisfies t-closeness, we calculate the differences between the group distributions and the overall distribution. Let's assume we set the threshold value at t = 0.2 (20%).

In group A, the difference for cancer is 10% (70%–60%), which is within the t-value. For diabetes, the difference is 10% (20%–30%), also within the threshold. The same applies to asthma, with a difference of 0% (10%–10%).

In group B, the difference for cancer is 10% (50%–60%), within the t-value. For diabetes, the difference is 0% (30%–30%), and for asthma, it is 10% (20%–10%). Both differences are within the specified threshold.

Since all the differences between the group distributions and the overall distribution are within the t-value, we can conclude that the dataset satisfies t-closeness for the given threshold.

This example demonstrates how t-closeness ensures that the distribution of sensitive attributes within each group is not significantly different from the distribution in the overall dataset, providing a higher level of privacy while maintaining the data's utility.

Compared to k-anonymity, t-closeness provides a more stringent privacy requirement as it considers the actual distributions of sensitive attributes rather than just the group size. However, achieving t-closeness while maintaining the data's utility can be challenging. Various techniques and algorithms have been developed to strike a balance between privacy and utility when applying t-closeness in data anonymization processes.

Let's now implement t-closeness using Python for a sample dataset.

Source code:t_Closeness_Example.ipynb

Let's look at an example of implementing t-closeness on the synthetic dataset we generated earlier. We'll consider the diagnosis attribute as the sensitive attribute and ensure that the distribution of sensitive attribute values within each group does not deviate significantly from the overall distribution:

```python
import pandas as pd
import random
# Step 1: Define the parameters
num_records = 1000  # Number of records in the dataset
t = 0.2  # Desired t-closeness threshold
# Step 2: Generate the synthetic dataset
age_range = (20, 70)
gender_list = ["Male", "Female"]
diagnosis_list = ["Covid", "Cancer", "Fever", "Obesity"]
data = []
for _ in range(num_records):
    age = random.randint(age_range[0], age_range[1])
    gender = random.choice(gender_list)
    diagnosis = random.choice(diagnosis_list)
    data.append({"age": age, "gender": gender, "diagnosis": diagnosis})
dataset = pd.DataFrame(data)
dataset.to_csv("t-close_before_dataset.csv", index=False)
# Step 3: Calculate the overall distribution of the sensitive attribute
overall_distribution = dataset["diagnosis"].value_counts(normalize=True)
# Step 4: Apply t-closeness
groups = dataset.groupby(["age", "gender"])
anonymized_data = pd.DataFrame()
for _, group in groups:
    group_distribution = group["diagnosis"].value_counts(normalize=True)
```

```
        max_divergence = max(abs(group_distribution - overall_
distribution))
    if max_divergence > t:
        # Generalize or suppress the sensitive attribute
        group["diagnosis"] = generalize(group["diagnosis"])
    anonymized_data = anonymized_data.append(group)
# Save the anonymized dataset to a file
anonymized_data.to_csv("t_closeness_dataset.csv", index=False)
def generalize(attribute):
    if attribute.name == "diagnosis":
        # Generalize diagnosis to a higher-level category
        diagnosis_mapping = {
            "Covid": "Chronic Condition",
            "Cancer": "Chronic Condition",
            "Fever": "Non-Chronic Condition",
            "Obesity": "Non-Chronic Condition"
        }
        return attribute.map(diagnosis_mapping)
    else:
        # Handle other attributes if needed
        return attribute
```

In this example, we generate a synthetic dataset as before. We calculate the overall distribution of the sensitive attribute (diagnosis) in the dataset. Then, we group the records based on the quasi-identifiers (age and gender) and calculate the distribution of the sensitive attribute within each group. If the maximum divergence between the group's distribution and the overall distribution exceeds the specified threshold, t, we apply generalization or suppression to the sensitive attribute.

Comparison of k-anonymity, l-diversity, and t-closeness

Privacy Technique	Strengths	Weaknesses
k-anonymity	<ul><li>Simplicity and ease of implementation</li><li>Provides a strong guarantee of identity privacy</li><li>Effective in preventing attribute linkage attacks</li><li>Preserves data utility to a certain extent</li></ul>	<ul><li>Does not consider the distribution of sensitive attributes, which may lead to attribute disclosure</li><li>Can result in excessive generalization, reducing data utility</li><li>Sensitive attribute suppression may lead to information loss</li><li>Vulnerable to background knowledge attacks</li></ul>

ℓ -diversity	• Enhances k-anonymity by considering diversity within groups • Provides better protection against attribute disclosure • Allows for fine-grained control over the diversity requirement • Offers a trade-off between privacy and data utility	• Generalization and suppression may lead to information loss and reduced data utility • The choice of "ℓ" value may be subjective and challenging • Still vulnerable to background knowledge attacks • Cannot guarantee protection against all types of attribute disclosure
t-closeness	• Extends ℓ-diversity by considering the distribution of sensitive attributes • Provides statistical guarantees against attribute disclosure • Considers both diversity and distribution, striking a balance between privacy and data utility • Allows for fine-grained control over the t-closeness threshold	• Generalization and suppression may still result in information loss • The determination of an appropriate t-closeness threshold can be challenging • Statistical measures may require domain expertise • Vulnerable to background knowledge attacks and correlation attacks

Table 3.9– Comparison of k-anonymity, ℓ-diversity, and t-closeness

This data provides a high-level overview of the strengths and weaknesses of each technique. The actual effectiveness and limitations of each technique can vary depending on the specific implementation, dataset characteristics, and privacy requirements. It's important to carefully evaluate and select the most appropriate technique based on the specific use case.

Let's explore the data aggregation technique, which is another privacy-preserving technique. We will try implementing data aggregation and see whether aggregates solve the problem of predicting sensitive data.

Data aggregation

As we have seen, data cannot be fully anonymized and remain useful at the same time. Let's try aggregated datasets and see whether aggregated data will be useful for preserving privacy.

Data aggregation refers to the process of combining and summarizing data from multiple sources or individuals to obtain a consolidated view or analysis. It involves gathering data points or records and transforming them into a more concise representation, often in the form of statistical measures or aggregated values. The objective of data aggregation is to extract meaningful insights and patterns from large datasets while reducing complexity and maintaining data privacy.

Key aspects of data aggregation

Different aggregation methods are employed based on the nature of the data and the desired outcomes. Common techniques include summing, averaging, counting, maximum/minimum values, percentiles, or other statistical measures. These methods help condense the data while preserving essential characteristics or patterns.

Data aggregation can be performed at different levels of granularity, depending on the analysis requirements. It can involve aggregating data at the individual level, group level, geographic level, time intervals, or any other relevant grouping criteria. Adjusting the granularity allows for different perspectives and insights to be derived from the data.

Aggregated data provides a more manageable and concise representation of the original dataset, enabling effective analysis, visualization, and decision-making processes. It helps identify trends, patterns, correlations, or anomalies that may not be apparent when examining individual data points.

Privacy attacks associated with data aggregation

Data aggregation, while a useful technique for deriving insights from large datasets, can also introduce privacy risks. Here are some privacy attacks associated with data aggregation:

- **Re-identification attacks**: Aggregated data, even when individual identifiers are removed or anonymized, can still be susceptible to re-identification attacks. By combining multiple aggregated datasets or leveraging external information, an attacker may be able to identify individuals within the aggregated data, compromising their privacy.

- **Inference attacks**: Inference attacks occur when an adversary can infer sensitive information about individuals by analyzing the aggregated data. By identifying patterns, correlations, or statistical measures within the aggregated dataset, an attacker may be able to deduce private attributes or behaviors of individuals that were intended to remain confidential.

- **Membership inference attacks**: Membership inference attacks aim to determine whether a specific individual's data was included in the aggregated dataset. By leveraging statistical analysis techniques, an attacker can exploit patterns or characteristics in the aggregated data to infer the presence or absence of certain individuals, violating their privacy.

- **Attribute disclosure attacks**: Aggregated data may inadvertently disclose sensitive attributes or information about individuals, even if direct identifiers are removed. By analyzing the aggregated data and combining it with external knowledge, an attacker may be able to infer private attributes, preferences, or characteristics of individuals.

- **Composition attacks**: Composition attacks exploit the combination of multiple aggregated datasets to reveal sensitive information. Even if individual datasets do not pose privacy risks on their own, the aggregation of multiple datasets may enable attackers to identify individuals or extract private information by cross-referencing data from different sources.

It is important to strike a balance between data utility and privacy protection to mitigate the risks associated with privacy attacks during data aggregation.

Let's walk through key privacy attacks with aggregated datasets.

Differencing privacy attack with an aggregated dataset

A differencing privacy attack leverages the statistical differences between the outputs of different queries to infer sensitive information about individuals or their data points.

Example

An adversary attempts to infer the salary information of individual employees by exploiting the outputs or queries that provide aggregate statistics about salaries.

Consider a dataset containing the salary information of employees in a company. The dataset includes attributes such as employee name, age, and salary.

Toy dataset example (original dataset):

Employee Name	Age	Salary (USD K)
John	30	200
Rosy	35	300
Robin	40	250
Hellen	50	315

Table 3.10 – Toy dataset example

Let's assume that aggregated data protects the privacy of the individual by default.

If the HR department in this company publishes an aggregated dataset on a weekly basis instead of sharing the original dataset, it can help mitigate the privacy risks associated with individual salary information. By providing only aggregate statistics, the HR department can protect the sensitive details of individual salaries while still providing useful insights to stakeholders.

Aggregated dataset – week 1:

Average Age	Average Salary
38.75	266.25

Table 3.11 – Aggregated dataset – week 1

Suppose that, in the following week, an additional employee was hired. The raw data was then updated with the details of the new employee and the aggregated dataset was published.

Employee Name	Age	Salary (USD K)
John	30	200
Rosy	35	300
Robin	40	250
Hellen	50	315
Fedrick	55	335

Table 3.12 – Aggregated dataset – week 1

Aggregated dataset – week 2:

Average Age	Average Salary
42	280

Table 3.13 – Aggregated dataset – week 2

Since HR has not shared the actual dataset but the aggregates only, is this enough to protect the personal information? In this case, the salary of an individual is the personal information.

Anyone could guess the new employee's salary who joined in week 2 because we have the average salary before and after the new employee joined.

New employee salary = (total employees * average salary) $_{after\ they\ join}$ – (total employees * average salary) $_{before\ they\ joined}$ = 5* 280 – 4 * 266.25 = 335

This is called a differencing attack, that is, combining multiple aggregate queries to obtain precise information about specific individuals. In this way, aggregated datasets have problems with differencing attacks, so it is not completely possible to protect the personal information of individuals just by aggregating the data.

Reconstruction attacks

Let's consider a hypothetical dataset from which the following information can be inferred:

"Three groups with a higher occurrence of the specified disease are identified, and data is collected regarding the symptoms and treatments related to that particular illness."

Using this knowledge, the attacker narrows down the possibilities and identifies a potential candidate within the target age group. They compare the individual's characteristics, symptoms, and treatment history, available in public sources, with the aggregated data to find a match.

By leveraging statistical techniques, machine learning algorithms, or domain-specific knowledge, the attacker refines their assumptions and gradually reconstructs the medical history of the targeted individual. They can uncover sensitive information, such as the specific disease, related comorbidities, previous treatments, and potential health risks.

This reconstruction attack demonstrates how an adversary can exploit aggregated data, external information, and statistical analysis to infer detailed individual-level information. By combining publicly available data, patterns observed in the aggregated statistics, and auxiliary knowledge, the attacker can breach the privacy of individuals and unveil sensitive personal information.

Let's look at another reconstruction attack on public data to understand the problem in detail with aggregated datasets.

Database reconstruction attacks

As discussed earlier, most countries conduct a census survey (a survey of their population) every 5 or 10 years and share the aggregated data with orginazations as well as in various government portals.

Let's go through an example of a reconstruction attack with the following data:

Name	Age (years)	Gender	City	Education
A	15	M	Bangalore	Under Grad
B	80	M	Bangalore	Illiterate
C	25	M	Bangalore	Graduated
D	45	F	Bangalore	Graduated
E	40	F	Bangalore	Graduated
F	20	F	Bangalore	Graduated
G	15	F	Bangalore	Graduated

Table 3.14 – Reconstruction attack data

Assume that an intermediatory has access to the preceding original data but is only allowed to publish aggregated data publicly. The intermediatory publishes the following statistics dataset based on the original data:

Gender	Count	Average Age	Median Age
Male	3	40	25
Female	4	30	40
Total	7	34.28	25

Table 3.15 – Statistics dataset

Using the aggregated datasets only, is it possible to find out the approximate age of the individuals (consider the case of the male gender)?

Let's try to attempt to solve this using a SAT solver. A SAT solver is a program whose objective is to solve the Boolean satisfiability problem. I have used the Google open source OR-Tools package using Python to solve this.

Based on the preceding data, there are 3 men in the dataset with an average age of 40 and a median age of 25. Assuming x, y, and z are the ages of the 3 men, then y will be 25 (because the median age is 25). The average of x, y, and z is 40, so the sum of their ages will equal 120. Assume that the maximum age a person will live is 110 years:

```
from ortools.sat.python import cp_model

def AgeFindSATprogram():

    # Creates the model.
    model = cp_model.CpModel()

    # Creates 3 variable
    num_vals = 3
    x = model.NewIntVar(1,110, 'x')
    y = 25
    z = model.NewIntVar(1,110, 'z')

    # Creates the constraints.
    model.Add(x+y+z==120)

    # Creates a solver and solves the model.
    solver = cp_model.CpSolver()
    status = solver.Solve(model)
```

```
    if status == cp_model.OPTIMAL or status == cp_model.FEASIBLE:
        print('x = %i' % solver.Value(x))
        print('y = %i' % solver.Value(y))
        print('z = %i' % solver.Value(z))
    else:
        print('No solution found.')


AgeFindSATprogram()
```

This returns the following output:

```
x = 94
y = 25
z = 1
```

In this way, one can solve the approximate value of the original dataset by using the aggregate dataset. That is why governments in some countries (such as India) don't allow aggregated datasets to be published, publishing only raw counts.

Gender	Count
Male	3
Female	4
Total	7

Table 3.16 – Aggregate dataset

Mitigating database reconstruction attacks

To mitigate reconstruction attacks and protect the privacy of data, several techniques can be employed, including the following:

- **Strong anonymization**: Apply robust anonymization techniques, such as generalization, suppression, or aggregation, to the released data. By transforming the data in a way that preserves privacy, it becomes more challenging for attackers to reconstruct individual records accurately.

- **Noise addition**: Introduce controlled noise to the released data or queries to add an additional layer of privacy protection. By injecting random perturbations, the attacker's ability to uncover the original data or infer precise information is diminished.

- **Data perturbation**: Modify the data by perturbing or modifying certain attributes or values while maintaining the overall utility of the data. This can make it harder for attackers to reconstruct the original information accurately.

- **Query restriction**: Limit the types or number of queries that can be made to the data to reduce the exposure of sensitive information. By restricting access to certain queries or applying query-based privacy mechanisms, the risk of reconstruction attacks can be mitigated.

Tools/frameworks to protect privacy with SQL queries

Open Diffix is an open source implementation of Diffix that enables privacy-preserving data analysis using SQL. It provides a solution for privacy-conscious organizations or researchers who need to analyze sensitive data while safeguarding the privacy of individuals. Open Diffix employs strong anonymization techniques to achieve this goal, ensuring that the results of database queries do not reveal sensitive information about individuals.

Diffix acts as a SQL proxy between the client (adversary or benign) and the database. Diffix returns SQL query results and adds a minimal amount of noise. Open Diffix is designed to ensure that sensitive information about individuals cannot be inferred from the data released. The key idea behind Open Diffix is to introduce controlled noise into query responses or statistical aggregates to prevent the identification of individual records.

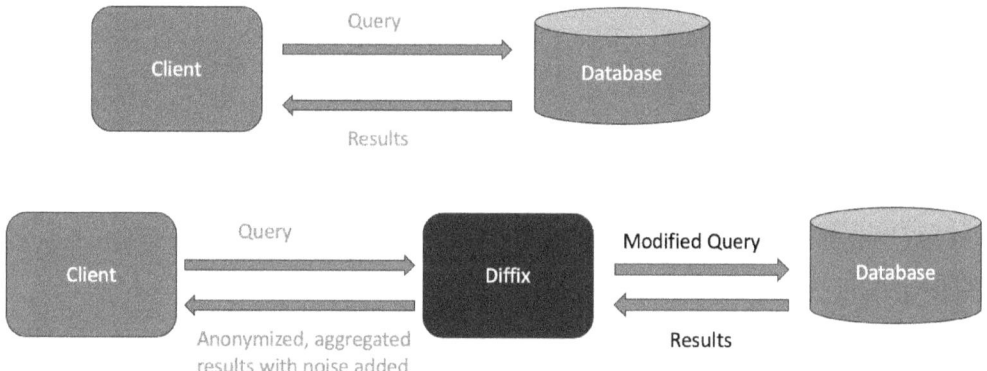

Figure 3.4– diffix system flow

Open Diffix has the following key features:

- Noise addition
- Suppression

Let's quickly review some more detail on these two features.

Noise addition

Open Diffix supports adding noise to query results. Noise can be added in different ways:

- **Fixed amount of noise**: In this case, a limited amount of noise is added to query results. This is called a privacy budget.

- **Sticky noise**: Each query, when executed the first time, will have random noise added, but repeated execution of the same query will give the same results, that is, the same amount of noise will be added to the query.

- **Proportional noise**: Based on the actual data and distribution of data (time series data), the tool will add the corresponding noise.

Suppression

Open Diffix supports the suppression of certain data (for example, extreme values in the case of numbers, and certain text data as well) and returns the results of the SQL query. In this way, apart from adding noise, suppression will also help to prevent revealing extreme values, and in turn, helps to protect privacy.

Sample use case using Open Diffix on Postgres DB

Let's go through a sample database using Postgres DB, execute the SQL queries using Open Diffix, and learn how Open Diffix anonymizes and aggregates the results by adding noise to protect the privacy of individuals.

These are the steps we'll work through in the following example:

1. Database creation
2. User role creation and grant access
3. Enabling Diffix on the DB tables to protect privacy

DB creation

In this case, we will create two tables (`Employee` and `Department`) and a sample database (`sample_db`) as well.

Database Name: `sample_db`

DB Users: `schema_owner, analyst_trusted, analyst_untrusted`

Tables: `Employee, Department`

```
CREATE TABLE department(
       deptno  INT GENERATED ALWAYS AS IDENTITY,
       deptname VARCHAR(255) NOT NULL,
       mgrno INT NOT NULL,
       mgrname VARCHAR(255)  NOT NULL,
       mgremailid VARCHAR(255) NOT NULL,
       location  VARCHAR(255) NOT NULL,
       PRIMARY KEY(deptno));
```

Column Name	Column Type	Description	PII Data	Data Masked
DEPTNO	Integer (primary key)	Department number or ID	No	No
DEPTNAME	Text	Department name	No	No
MGRNO	Integer	Department head employee ID	No	No
MGRNAME	Text	Department head employee Name	No	No
MGREMAILID	Text	Department head employee Email ID	Yes	Yes
LOCATION	Text	Location of the department	No	No

Table 3.17 - Creating a sample database

```
CREATE TABLE employee(
    empno  INT GENERATED ALWAYS AS IDENTITY,
    ename VARCHAR(255) NOT NULL,
    job VARCHAR(255) NOT NULL,
    emailid VARCHAR(255) NOT NULL,
    mgr INT NOT NULL,
    hiredate DATE NOT NULL,
    deptno INT NOT NULL,
    PRIMARY KEY(empno),
    CONSTRAINT fk_department
        FOREIGN KEY(deptno) REFERENCES department(deptno)
);
```

Column Name	Column Type	Description	PII Data	Data Masked
EMPNO	Integer (primary key)	Employee number or ID	No	No
ENAME	Text	Employee name	No	No
JOB	Text	Job title	No	No
EMAILID	Text	Employee email ID	Yes	Yes
MGR	Integer	Manager's employee ID	No	No
HIREDATE	Date	Join date	No	No
DEPTNO	integer	Department number or ID	No	No

Table 3.18 - Creating a sample database

Once the tables are created, insert the sample data as follows and run a sample query:

```
SELECT DEPTNO, DEPTNAME, MGRNO, MGRNAME, MGREMAILID,LOCATION FROM
DEPARTMENT;
```

DEPTNO	DEPTNAME	MGRNO	MGRNAME	MGREMAILID	LOCATION
1	Research and Development	45	Michael	o64jDLFOfpfCe1fENtwVSAAT	India
2	IT Services	47	Anna	i1ALN8ApLAaQrccuuGqAZgAP	India
3	Sales and Marketing	43	Srinivas	yvGkdh5dKQcN41nXWyYOjgAM	India
4	Accounts and Finance	44	Sankara	4jDkdh5LAN841nX9yYOjQrc2	India
5	Human Resources	43	Meena	2II8Wq_aJDy1hGuNdYMeKQAB	India
6	Infrastructures	42	Deepa	o2xkrwTjAcHghtiNWX5zSgAF	India

Table 3.19 - Inserting the sample data

```
SELECT EMPNO,ENAME,JOB,EMAILID,MGR,HIREDATE,DEPTNO FROM EMPLOYEE
LIMIT 5;
```

EMPNO	ENAME	JOB	EMAILID	MGR	HIREDATE	DEPTNO
1	Melissa	Tester	RGk3h5dKccN41nyYOj	45	2020-01-17	1
2	Lzenson	Cybersecurity Engineer	FJDy1hGuNdYMeKQAB	43	2020-06-11	2
3	Dyer	Program Manager	Hy1hGuNdYMeKQAthy	43	2020-11-13	4
4	Kesavan	Legal Assistant	HghtiNWX5zStsadgA1	42	2021-03-11	4
5	Jonathan	Data Engineer	pfC3e1fE3Nt4wVSAer	47	2021-10-06	6

Table 3.20 - Inserting the sample data

User creation

Users can have one of the following access levels to a database – three types of users are supported in Diffix:

- `direct`: Direct (non-anonymized) access to data. Restrictions listed in this document do not apply in direct mode.

- `anonymized_trusted`: Anonymized access to data. Prevents accidental release of personal data.

- `anonymized_untrusted`: Anonymized access to data. Prevents intentional release of personal data.

  ```
  CREATE USER schema_owner WITH PASSWORD 'schema_owner';

  CREATE USER analyst_trusted WITH PASSWORD 'analyst_trusted';

  CREATE USER analyst_untrusted WITH PASSWORD 'analyst_untrusted';
  ```

Granting access

Grant access to these three users so that they can select the data and execute functions:

```
GRANT SELECT ON ALL TABLES IN SCHEMA public TO schema_owner;
GRANT EXECUTE ON ALL FUNCTIONS IN SCHEMA public TO schema_owner;
GRANT SELECT ON ALL TABLES IN SCHEMA public TO analyst_trusted;
GRANT EXECUTE ON ALL FUNCTIONS IN SCHEMA public TO analyst_trusted;
GRANT SELECT ON ALL TABLES IN SCHEMA public TO analyst_untrusted;
GRANT EXECUTE ON ALL FUNCTIONS IN SCHEMA public TO analyst_untrusted;
```

Defining roles

Define a role for each user by calling Diffix's mark role function:

- Direct access:

  ```
  CALL diffix.mark_role('schema_owner', 'direct');
  ```

- Trusted access:

  ```
  CALL diffix.mark_role('analyst_trusted', 'anonymized_trusted');
  ```

- Untrusted access:

  ```
  CALL diffix.mark_role('analyst_untrusted', 'anonymized_untrusted');
  ```

Enabling Diffix on the tables

Enable Diffix on the tables by executing the following code:

```
CALL diffix.mark_personal('employee', 'empno');
CALL diffix.mark_personal('department', 'deptno');
```

Run SQL queries and see how privacy is protected.

Execute this query with different users (direct and anonymized trusted and untrusted users) and observe the results:

```
SELECT  deptname, count(*), diffix.count_noise(*)   FROM employee
GROUP BY deptname
```

For direct users:

Deptname	Count	Count_noise
Learning and Development	15959	0.0
Sales and Marketing	737	0.0
Accounts and Finance	573	0.0
Human Resources	435	0.0
Infrastructures	20448	0.0
IT Services	21021	0.0

Table 3.21 - Actual results for direct users

For `anonymized_trusted` and `anonymized_untrusted` users:

Deptname	Count	Count_noise
Learning and Development	15957	1.0
Sales and Marketing	736	1.0
Accounts and Finance	571	1.0
Human Resources	436	1.0
Infrastructures	20446	1.0
IT Services	21018	1.0

Table 3.22 - Protected results for anonymized users

In the case of direct users, Diffix will provide the actual results. For anonymized users, it provides the results by adding noise, so that an adversary will not be able to find the actual results.

Subqueries

Any SQL may be executed on the output of anonymizing subqueries as their output is no longer considered personal. The query results vary for direct users versus anonymized users:

```
SELECT avg(tab.deptwise_count) FROM
     (SELECT  deptname, count(*) AS deptwise_count
```

```
                FROM employee
                GROUP BY deptname) tab
```

- **For direct users:**

Average
8578.1428571428571429

Table 3.23 - Average for direct users

- **For** `anonymized_trusted` **and** `anonymized_untrusted` **users:**

Average
8576.8571428571428571

Table 3.24 - Average for anonymized users

Suppression

Let's consider a scenario where there are a total of 874 employees in an organization. Through queries on a specific table, it becomes apparent that only one employee was hired on that day. This observation raises privacy concerns, as without proper anonymization, if we possess knowledge of an individual's hiring date, it is highly likely that we can extract additional information about that person by querying other tables within the database.

The absence of multiple employees hired on the same day in the table implies a potential vulnerability in privacy protection. If an attacker or unauthorized entity gains access to the hiring date of an individual, they could exploit this knowledge by performing queries on other interconnected tables to infer additional personal information. This could compromise the privacy and confidentiality of employees' sensitive data.

To mitigate this risk and preserve privacy, it is crucial to implement proper privacy preservation techniques such as suppression.

SQL query to find out the exact dates employees joined

```
SELECT hiredate, count(*) FROM employee
where deptname='Research and Development'
GROUP BY hiredate ORDER BY count(*) DESC LIMIT 5
```

For direct users:

HireDate	Count
2021-12-10	723
2020-03-03	98
2020-03-01	36
2021-01-13	1
2020-03-19	1

Table 3.25 - Analysis of singular employee joining dates and privacy risks

From this result, there are only two dates on which a single employee joined the organization. This particular pattern poses a privacy risk, as individuals with knowledge of this information could potentially link it to other data sources, thereby uncovering additional personal details. To safeguard this sensitive information and prevent unauthorized inference, it is crucial to provide alternative results that differ from the actual ones for `anonymized_trusted` and `anonymized_untrusted` users.

Diffix, a privacy-preserving technique, addresses this concern by ensuring that untrusted and anonymized users do not receive the actual query results. Instead, Diffix suppresses data where the result count is too small to reveal meaningful information. By doing so, it effectively conceals the precise details while still providing statistical insights and analysis.

HireDate	Count
2021-12-10	723
2020-03-03	96
2020-03-01	37
*	18

Table 3.26 - Suppressed data counts for employee hire dates

In the preceding table, we see that almost all the data has been suppressed. Counts for only four dates are displayed. The * symbol represents the hire date with respect to individual employee and it is not displayed the exact hire data for the 18 employees because of the different hire dates which can revel the sensitive information.

Diffix Elm automatically recognizes this and suppresses the dates, merging them into a single bin with the value *, which we call the suppression bin. Along with suppression, noise is included in the count as well.

By suppressing data with low result counts, Diffix limits the possibility of discovering sensitive information through queries. This approach safeguards the privacy of individuals and prevents potential linkage attacks, where attackers try to correlate multiple datasets to reveal personal details. Diffix's mechanism of altering the query results without compromising statistical accuracy adds an extra layer of protection to the data.

In summary, we have explored different anonymization techniques and aggregated datasets, understanding the advantages and limitations of each approach. However, it is important to note that these techniques may not provide complete privacy protection for individuals. This realization has led to the development of a new technique known as differential privacy, which we will learn about in the upcoming section in detail.

Privacy-enhancing technologies

Privacy-enhancing technologies (**PETs**) are a set of technologies and techniques that help protect sensitive information while still allowing useful analysis and processing of the data. Here is a high-level introduction to some of the commonly used PETs.

Differential privacy

This is a technique that adds a certain amount of noise to a dataset to protect the privacy of individual records while still allowing for statistical analysis. Differential privacy ensures that any queries made on a dataset do not reveal information about specific individuals, making it a powerful tool for protecting privacy in large datasets. We will go through differential privacy in this chapter and the rest of the PETs in other, subsequent chapters.

Federated learning

This is a technique for training machine learning models on data that is distributed across multiple devices or servers, without the need to centralize the data. In federated learning, the model is trained locally on each device, and then the updated model is sent back to a central server for aggregation. This approach helps to protect the privacy of individual data while still allowing for useful analysis and processing. We will learn more about federated learning in *Chapter 7*.

Secure multi-party computation (SMC)

This is a technique that enables multiple parties to compute a function on their respective inputs without revealing their inputs to each other. SMC can be used for a range of privacy-preserving computations, including secure voting, privacy-preserving data mining, and secure auctions. We will learn more about SMC in *Chapter 9*.

Homomorphic encryption

This is a technique that allows computation on encrypted data without decrypting it first. This means that data can be processed while still being protected, making homomorphic encryption a powerful tool for protecting privacy in data analysis. We will learn more about homomorphic encryption in *Chapter 9*.

Anonymization

This is the process of removing identifying information from a dataset, such as names, addresses, and other personally identifiable information. Anonymization helps to protect the privacy of individuals in the dataset while still allowing for useful analysis. We have learned about data anonymization already, in the previous section.

De-identification

This is a technique for removing identifying information from a dataset, while still preserving the usefulness of the data for analysis. De-identification involves transforming data in a way that makes it difficult to link it back to an individual, but still maintains its statistical properties. One-way hashing is one of the techniques to implement the de-identification of data to preserve privacy.

Overall, PETs are an important set of tools for protecting privacy in data analysis, allowing organizations to balance the need for useful data analysis with the need to protect the privacy of individuals.

Let's now learn more about differential privacy.

Differential privacy

Differential privacy is a concept in data privacy that provides a rigorous framework for quantifying and controlling the privacy guarantees of data analysis algorithms. It ensures that the presence or absence of an individual's data does not significantly affect the outcome of the analysis.

Mathematically, differential privacy is defined using two key concepts: **sensitivity** and the **privacy budget**.

Sensitivity (Δf)

Sensitivity measures the maximum amount that the output of a function f can change when a single data point is added to or removed from the dataset. It quantifies the impact of an individual's data on the analysis outcome. The sensitivity is typically defined using a metric called the L1 or L2 norm.

Privacy budget (ε)

The privacy budget, also known as the privacy parameter, controls the amount of privacy protection provided by a differentially private algorithm. A smaller value of ε indicates stronger privacy guarantees. It represents the maximum allowable probability that the algorithm's output will change significantly when a single individual's data is included or excluded.

Using these concepts, we can define differential privacy as follows:

A mechanism or algorithm L, which is randomized, provides ε-differential privacy if there are any two adjacent datasets D1 and D2 differing by the data of just one person, and for any subset of outputs defined as S.

In mathematical terms, differential privacy is defined as follows:

$$\Pr(\ L[D1] \in S\ ) \leq \exp(\varepsilon)\ .\ \Pr.(\ L[D2]\ [[\text{OMML-EQ-4}]]\ S\ ) + \delta$$

Where Pr is the probability and:

- D1 and D2 are the two datasets that differ by a single record.
- S is all subsets of the randomized algorithm/mechanism L, which is applied.
- ε (epsilon) is a positive real number that controls the privacy loss or referred privacy budget.

ε determins much the algorithm can differ between the two databases and captures how much privacy loss there is when the algorithm runs on the database.

When ε is set to zero, the answers to both queries will closely resemble each other, leading to a reduction in privacy.

This inequality states that the probability of obtaining an output in the set S when the algorithm operates on dataset D1 is at most $\exp(\varepsilon)$ times the probability of obtaining the same output when the algorithm operates on a neighboring dataset D2.

The $\exp(\varepsilon)$ factor quantifies the privacy guarantee, where smaller values of ε provide stronger privacy protection.

Note that there are variations and extensions of the differential privacy definition that incorporate additional parameters, such as the number of queries or the privacy loss over multiple analyses.

Differential privacy provides a rigorous and quantifiable approach to balancing privacy and utility, allowing for the analysis of datasets while minimizing the risk of privacy breaches.

Differential privacy has the following properties, which are quite useful in overall privacy:

- Privacy guarantee
- Composability
- Group privacy
- Robustness to auxiliary information

Privacy guarantee

Differential privacy provides a formal guarantee that the output of a query does not reveal sensitive information about any individual in the dataset. This means that even if an attacker has access to all but one individual's data, they will not be able to determine that individual's information with high probability.

The following is an example to illustrate privacy guarantees in differential privacy.

Let's consider a scenario where a company wants to release aggregate information about the average salary of its employees. However, they also want to protect the privacy of individual employees. To achieve this, they decide to use differential privacy.

Using differential privacy, the company adds random noise to the data before releasing the aggregated result. Suppose the actual average salary of the employees is $50,000. By applying differential privacy, the released result might be slightly perturbed, such as $50,150 or $49,850, to ensure privacy protection.

Now, let's assume an external attacker tries to identify the salary of a specific employee by analyzing the released aggregate information. Due to the added noise, the attacker cannot accurately determine the salary of any individual employee. The randomness introduced by differential privacy ensures that the released data does not reveal precise information about any individual's salary.

By controlling the level of noise added (based on the epsilon parameter), the company can adjust the privacy guarantee. For example, if a smaller epsilon value is chosen, the added noise will be higher, providing a stronger privacy guarantee but potentially sacrificing some accuracy in the aggregated result.

In this example, differential privacy guarantees that the individual salary information remains protected, even when aggregate statistics are released. It ensures that no external party can precisely identify the salary of any specific employee, thus preserving privacy while still allowing useful insights to be derived from the data.

Randomized response

Differential privacy achieves its privacy guarantee by adding random noise to the output of a query. The amount of noise is calibrated to balance the privacy guarantee with the accuracy of the output.

Randomized response is a method that introduces randomness into the data collection process to protect individual privacy. Let's explain this technique with another example.

Suppose a survey is conducted to gather sensitive information about a population's illegal activities. The goal is to estimate the percentage of people engaged in illegal activities without revealing specific individuals' responses.

Instead of directly asking individuals whether they participate in illegal activities, which may lead to reluctance or dishonesty due to privacy concerns, the surveyors employ the randomized response technique.

Here's how it works:

1. Each participant is given a fair coin (equally likely to land on heads or tails).

2. The participant flips the coin in private without revealing the result.

3. If the coin lands on heads, the participant truthfully answers the question about illegal activities.

4. If the coin lands on tails, the participant provides a random response, independent of their actual behavior. For example, they might say "yes" regardless of their involvement in illegal activities.

By using this randomized response approach, the true response of any individual is masked by randomness. The probability of providing a true response or a random response is equal (50% in this example), making it difficult to infer an individual's true behavior based on their response.

Now, when the survey results are analyzed, the researchers take into account the randomized responses and the overall statistics to estimate the percentage of people engaged in illegal activities. This estimation incorporates the randomness introduced by the randomized response technique, providing privacy protection for individuals while still allowing valuable insights to be derived from the aggregated data.

The randomized response technique is just one of the many methods used in differential privacy to ensure privacy guarantees. It effectively adds noise and uncertainty to individual responses, making it challenging to identify specific individuals and their sensitive information.

Composability

Composability is a key property of differential privacy, which refers to the ability to combine multiple differentially private mechanisms in a way that preserves the overall privacy guarantees.

In differential privacy, a mechanism is said to be ε-differentially private if the probability of any two neighboring datasets producing the same output is at most e^{ε} times the probability of the output being produced by one of the datasets alone.

The composability property ensures that if we apply multiple ε-differentially private mechanisms sequentially, the overall privacy guarantee remains ε-differential private. That is, the total privacy loss due to the composition of multiple mechanisms is at most the sum of the individual privacy losses.

This property is important because it allows us to design complex systems that use differentially private mechanisms for different tasks and still provide strong privacy guarantees.

For example, we can use differentially private mechanisms to collect sensitive data from multiple sources and then aggregate the data to obtain useful insights while preserving privacy. Let's now illustrate this with an example.

Consider a company that collects customer data, including their purchase history. The company wants to provide personalized recommendations to customers while ensuring their privacy using differential privacy. They decide to apply differential privacy mechanisms to protect the data. First, the company uses differential privacy to generate personalized recommendations for each customer based on their purchase history. This analysis introduces random noise to the recommendation algorithm, ensuring that no individual's purchase history is precisely revealed in the recommendations. Next, the company wants to perform a separate analysis to determine the average amount spent by customers in different age groups. They utilize the same differentially private mechanism to compute the average spend, incorporating random noise to protect individual spending information.

The key aspect of composability in differential privacy is that the privacy guarantees hold even when these analyses are combined. In this example, the privacy guarantees are preserved when both the personalized recommendation and the average spend analysis are performed on the same dataset.

Suppose an external attacker attempts to identify the purchase history or spending details of a specific customer. Due to the added noise and the composability property of differential privacy, the attacker cannot accurately distinguish the true individual data from the randomized data generated during the analysis. The privacy guarantees provided by differential privacy extend across multiple analyses, protecting individual privacy even when multiple queries are performed on the same dataset.

Composability ensures that the privacy guarantees of differential privacy remain intact, allowing organizations to perform various analyses on sensitive data while maintaining a consistent level of privacy protection.

However, the composability property assumes that the differentially private mechanisms are independent and do not share any information. If the mechanisms share information or if there are correlations between them, then the overall privacy guarantee may be weaker than the sum of the individual privacy guarantees. Therefore, it is important to carefully design and analyze the composition of differentially private mechanisms to ensure that privacy is preserved.

Group privacy

One of the key concepts in differential privacy is group privacy, which refers to the protection of privacy for groups of individuals, as opposed to individual privacy.

Group privacy in differential privacy is achieved by ensuring that the results of data analysis do not reveal information about any specific individual or subgroup of individuals.

This is typically accomplished by adding noise to the data in such a way that the statistical properties of the data remain largely unchanged, but the specific details about any individual or subgroup are obscured.

For example, suppose a researcher wants to analyze the average income of a certain demographic group. In a differentially private framework, the researcher would first add random noise to the income data of each individual in the group, such that the overall statistical properties of the data are preserved, but the income of any individual cannot be determined. The researcher can then calculate the average income of the group without revealing any individual's income.

Group privacy in differential privacy is important because it ensures that even if an attacker has some additional information about a subset of individuals, they cannot use that information to learn anything about other individuals in the group.

This makes differential privacy a powerful tool for protecting the privacy of individuals and groups in data analysis.

In dataset terminology, privacy can be protected on not only one row but X number of rows in the dataset:

$$Pr \, (\, L[D1] \in S \,) \leq \, exp(\, X * \varepsilon) \, . \, \, Pr(\, L[D2] \in S \,) + \delta$$

X defines the number of persons (rows) in the dataset that need to be protected using differential privacy.

Robustness to auxiliary information

Robustness to auxiliary information is a key property of differential privacy that ensures that the privacy guarantees of the framework remain intact even if an attacker (adversary) has some additional information about the individuals whose data is being analyzed.

In the privacy frameworks that we have studied, such as k-anonymity and ℓ-diversity, the privacy of individuals is protected by ensuring that each individual's data is indistinguishable from that of at least k or ℓ-1 other individuals, respectively. However, these frameworks do not take into account any additional information that an attacker might have about the individuals. For example, an attacker might be able to infer an individual's sensitive information based on their zip code or occupation, even if their data is anonymized according to these frameworks.

Differential privacy, on the other hand, provides strong privacy guarantees even in the face of such auxiliary information. This is achieved by adding random noise to the data in such a way that the noise makes it impossible to determine with high confidence whether a particular individual's data is included in the dataset or not, regardless of any additional information an attacker might have.

Even if an attacker knows the exact income of one individual in the group, the added noise makes it impossible to determine with high confidence whether that individual's data is included in the dataset or not. Robustness to auxiliary information is an important property of differential privacy because it ensures that the privacy guarantees of the framework remain intact even in the face of sophisticated attacks by adversaries who might have access to additional information about the individuals in the dataset.

Local and global differential privacy

Differential privacy is a concept that helps to protect the privacy of individuals while still allowing useful analysis to be performed on data.

Local differential privacy (**LDP**) and **global differential privacy** (**GDP**) are two different approaches to achieving differential privacy. LDP is a method of differential privacy that involves adding noise to individual data points before they are shared. This means that each data point is perturbed before it is released, which helps to protect the privacy of individuals in the dataset. LDP is often used in situations where individuals are providing their own data, such as in mobile apps, and it ensures that the data is kept private even from the server.

GDP, on the other hand, involves adding noise to the aggregate data. This means that privacy protection is applied to the overall results of an analysis rather than to individual data points. GDP is often used in situations where the data is collected centrally, such as in a hospital or government agency, and the privacy of the individuals in the dataset must be protected.

Both LDP and GDP have their own advantages and disadvantages. LDP is more effective at protecting the privacy of individual data points, but it can be less accurate when it comes to analyzing the data. GDP, on the other hand, is more accurate but may not provide as strong privacy protection for individual data points. Ultimately, the choice between LDP and GDP will depend on the specific situation and the trade-offs between accuracy and privacy that are required.

Summary

In conclusion, we have explored the realm of privacy in data analysis and the importance of privacy-preserving techniques. We discussed concepts such as anonymization, k-anonymity, t-closeness, and ℓ-diversity, which play a crucial role in safeguarding privacy during data analysis. However, we also acknowledged the limitations of these techniques. Furthermore, we delved into a high-level overview of privacy-enhancing technologies, with a specific focus on differential privacy. We explored essential concepts such as privacy loss, privacy budgets, and differential privacy mechanisms, understanding how they contribute to privacy preservation. We also examined the implementation of differential privacy properties and their significance in ensuring privacy guarantees.

In the next chapter, our focus will shift toward differential privacy algorithms. We will get an overview of these algorithms and explore the importance of concepts such as sensitivity and clipping in the context of differential privacy. Additionally, we will delve into the generation of aggregates using differential privacy, understanding how this approach provides statistical insights while preserving privacy. By delving deeper into these topics, we will enhance our understanding of the practical applications of differential privacy and how it can be effectively implemented to balance privacy and data utility.

4

Overview of Differential Privacy Algorithms and Applications of Differential Privacy

The concept of differential privacy holds great significance in the realm of data privacy and its importance continues to grow as more and more data is collected and analyzed. Differential privacy algorithms offer a means to safeguard individual privacy while still allowing for valuable insights to be derived from this data.

In this chapter, we will gain an overview of differential privacy algorithms, along with a comprehension of crucial concepts such as sensitivity and clipping in the context of differential privacy. Additionally, we will explore how aggregates are generated through the use of differential privacy, including in real-world applications.

The following main topics will be covered in this chapter:

- Differential privacy algorithms:

    - The Laplace algorithm for differential privacy
    - The Gaussian algorithm for differential privacy
    - Generating aggregates using differential privacy

- Sensitivity and its role in aggregate generation algorithms using differential privacy:

  - Queries using differential privacy: Count, sum, and mean Count, sum, and mean

- Clipping:

  - Understanding its significance within the realm of differential privacy and examples

- Overview of real-world applications that make use of differential privacy

Differential privacy algorithms

Differential privacy is a fundamental concept that's designed to safeguard individual privacy while enabling the statistical analysis of sensitive data. It establishes a mathematical framework that guarantees the preservation of individual privacy during data analysis and sharing processes. Differential privacy algorithms play a vital role by introducing random noise into the data, making it challenging to identify specific records.

In the preceding chapter, we learned that the core idea behind differential privacy is to introduce random noise into the data analysis process. This noise makes it difficult for an attacker to determine whether a particular individual's data was included in the analysis, thus preserving privacy. The fundamental concept is that the inclusion or exclusion of any individual's data should not significantly impact the results of the analysis:

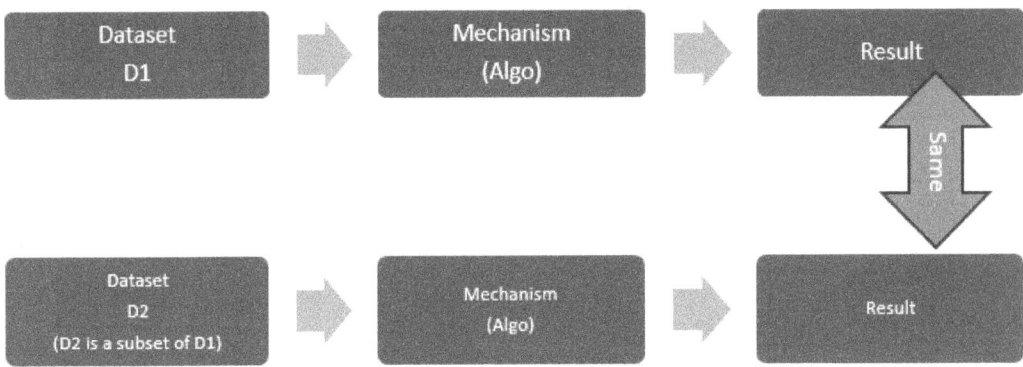

Figure 4.1 – Illustrating the primary concept of differential privacy

Differential privacy can be defined mathematically. For two datasets, *D1* and *D2*, which differ by a single record, the probability that the randomized algorithm or mechanism, *L* (such as count, sum, average, and so on), applied to *D1* yields a result in subset *S* is bounded by the exponential of epsilon (ε) multiplied by the probability that *L* applied to *D2* yields a result in S, plus a delta (δ) term.

Mathematically, this is given as follows:

$$P(L[D1] \in S) \leq \exp(\varepsilon) * P(L[D2] \in S) + \delta$$

In this equation, the following applies:

- *P* denotes probability.

- *D1* and *D2* represent the two datasets that differ by a single record.

- *S* encompasses all subsets of the randomized algorithm or mechanism, *L*.

- *ε* (epsilon) is a positive real number that controls the privacy loss and is referred to as the privacy budget. It determines the extent to which the algorithm can differ between the two databases and quantifies the privacy loss incurred when the algorithm is applied to the database. If ε is zero, the queries will yield similar answers, which compromises privacy to a lesser extent.

To ensure differential privacy, random noise (in the form of a random number) is added to the results. This noise prevents adversaries from determining whether the returned results are genuine or have noise added, and if so, the magnitude of the added noise. The added noise guarantees that the data cannot be utilized to identify individuals while still allowing useful statistical properties to be calculated.

The choice of the random number distribution depends on how to select the right random number. Next, we'll explore two well-known distributions, the Laplace distribution and the Gaussian distribution, to assess their suitability for the privacy parameter, ε, and to guarantee differential privacy.

Laplace distribution

The Laplace distribution, also referred to as the double exponential distribution, is a continuous probability distribution named after the renowned mathematician Pierre-Simon Laplace. It arises from merging two exponential distributions, one positive and one negative.

To better comprehend the Laplace distribution, let's understand the exponential distribution. The exponential distribution is defined by this function:

$$f(x) = e^{-x}$$

When x equals 0, f(x) is equal to 1, and as x increases, f(x) gradually approaches 0.

To observe this behavior, you can try a simple Python program to analyze the exponential distribution.

Try out a simple Python program and test this behavior:

Source code: Exponential_Laplace_Guassian_Clipping.ipynb

```
%matplotlib inline

import numpy as np
```

```
import matplotlib.pyplot as plt

x = np.arange(0,10, 0.25)

# define f(x) = e power (-x)

plt.plot(x, np.exp(-x));

plt.xlabel("x")

plt.ylabel("f(x)")
```

Here's the output:

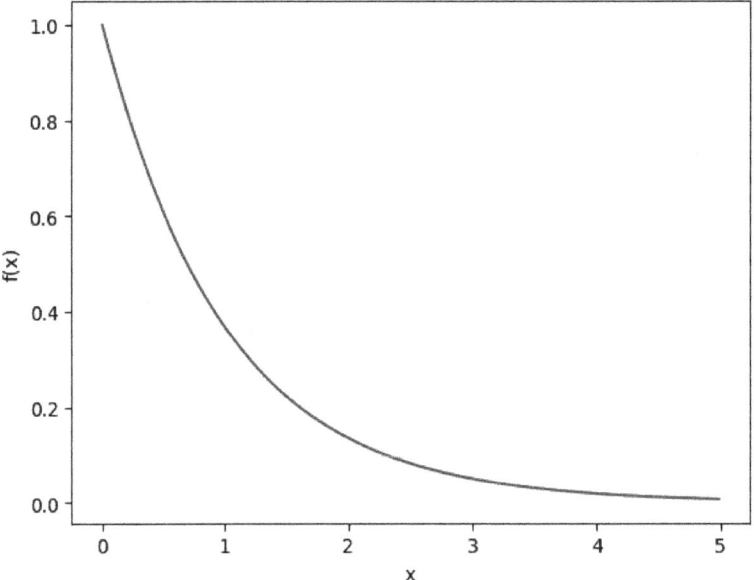

Figure 4.2 – Exponential distribution

Laplace distribution – mathematical definition

Laplace distribution's probabilistic density function is defined as follows:

$$f(x \mid \mu, b) = \frac{1}{2b} \exp\left(-\frac{|x - \mu|}{b} \right)$$

Here, x is a random variable, µ is the location parameter that determines the center of the distribution, and b > 0 is the scale parameter that controls the spread or variability of the distribution.

The Laplace distribution has several important properties. It is symmetric around its mean (μ), with a peak at μ. The tails of the distribution are heavy, meaning that it has a higher probability of observing extreme values compared to a Gaussian (normal) distribution. This property makes the Laplace distribution suitable for modeling data with outliers or heavy-tailed behavior. The Laplace distribution finds applications in various fields, including statistics, signal processing, image processing, and machine learning. It is particularly useful in scenarios where robustness to outliers is desired or when modeling data with a Laplacian noise assumption.

If μ equals 0 and b equals 1, then the equation is as follows:

$$f(x) = \frac{1}{2}\ e\text{-}x$$

This is nothing but half of the exponential distribution. Let's implement Laplace in simple Python code and understand it further:

```
def laplace_dist(x, μ, b):
    return 1 / (2 * b) * np.exp(-np.abs(x - μ) / b)


def plot_laplace_dist(x, μ, b):

    plt.plot(x, laplace_dist(x, μ, b),
    label="μ={}, b={}".format(μ, b))


x = np.arange(-10, 10, 0.1)

plot_laplace_dist(x, -2, 4)

plt.axvline(x=-2, linestyle='dotted', linewidth=3, label='mean')

plot_laplace_dist(x, 4, 8)

plt.axvline(x=4, color='orange', linestyle='dotted', linewidth=3,
label='mean')

plt.legend();
```

Here's the output:

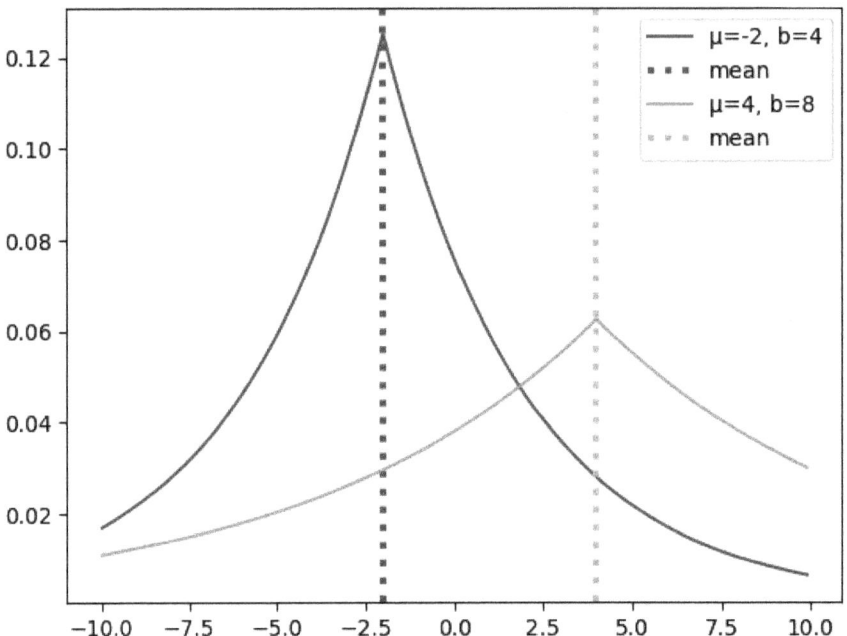

Figure 4.3 – Laplace distribution with different means and scaling parameters

From this graph, we can see that as the scaling parameter (b) increases in the Laplace distribution, the tails contribute a greater proportion compared to the central point. This characteristic can be leveraged to enhance privacy by utilizing larger scaling parameters.

By increasing the scaling parameter, the Laplace distribution's tails become heavier, resulting in a higher likelihood of extreme values occurring. This increased variability makes it more challenging for adversaries to extract sensitive information or identify individual data points accurately. Therefore, larger scaling parameters can be employed as a privacy-enhancing measure as they contribute to a broader spread of the distribution and provide a greater level of privacy protection.

Why is the Laplace distribution one of the algorithms of choice for differential privacy?

Take a look at the following example, which shows that Laplace distribution is one of the right algorithms to implement for differential privacy:

```
ε = 0.5
x = np.arange(90.2, 110, 0.20)
query_result=np.average(x)
dist1 = laplace_dist(x, query_result, 1 / ε)
dist2 = laplace_dist(x, query_result + 1, 1 / ε)
plt.plot(x, dist1, label="distribution 1")
plt.plot(x, dist2, label="distribution 2")
```

```
plt.axvline(x=query_result, c="black", linestyle='dotted',
label="query result")
plt.legend()
```

Here, we have taken a simple dataset that consists of numbers from 90 to 110 in increments of 0.2. The average of these numbers is close to 100, which is the query result – that is, the query finds the average of these numbers.

We find the Laplace distribution for this dataset with the mean as the query result and the scaling factor (b) set to 2, which is **1 / ε** in this case:

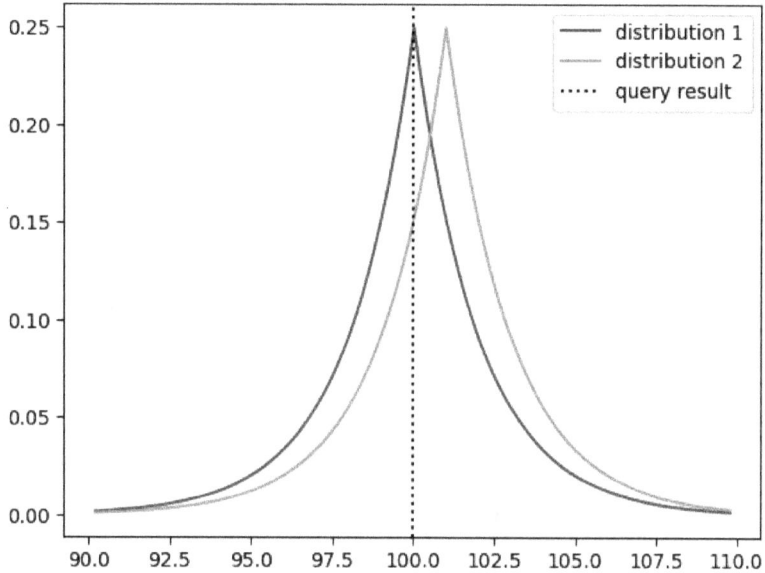

Figure 4.4 – Distributions using Laplace

The query result can come from any of these possible distributions. As per the differential privacy formula (DP), query results from one distribution should be less than or equal to $e^ε$ times the second distribution when they are compared.

In this particular instance, the resultant value of 100 from the query is derived from distribution 1. However, it's not possible to definitively exclude the possibility that it originates from the second distribution, particularly if the probability of the outcome is less than or equal to $e^ε$ times:

```
p1 = laplace_dist(query_result, query_result, 1 / ε)

p2 = laplace_dist(query_result, query_result+1, 1 / ε)

print(p1,p2)
```

The result is *0.25 0.15163266492815836*.

```
p1, p2 * np.exp(ε)
```

The result is 0.25 0.25.

This proves that the Laplace distribution can be used as one of the mechanisms for differential privacy and that the b scaling parameter can be set to 1/ε, which is the privacy budget.

Gaussian distribution

The Gaussian distribution, also known as normal distribution, is a continuous probability distribution that is widely used in statistical analysis to model real-world phenomena. It is characterized by its bell-shaped curve, which is symmetrical around the mean value.

Gaussian distribution – mathematical definition

Gaussian distribution is also called normal distribution and it is a continuous probability distribution.

The formula for Gaussian distribution is defined as follows:

$$f(x) = \frac{1}{\sigma\sqrt{2\pi}} e^{-\frac{1}{2}\left(\frac{x-\mu}{\sigma}\right)^2}$$

Here, x is the random variable in the Gaussian distribution; μ is the location parameter, as in, it will be the mean of the data; σ is the standard deviation; and $\sigma 2$ is the variance of the data.

The following code is an implementation of a Gaussian function:

```
def guissan dist(x, μ, σ):

    return (1/(σ * np.sqrt(2 * np.pi)) * np.exp(-(x - μ)**2 / (2 *
σ**2)))

def plot_guissan(x, μ, σ ):
    plt.plot(x, guissan_dist(x, μ, σ),
    label="μ={}, σ ={}".format(μ, σ ))

x = np.arange(-10, 10, 0.1)

stdv=np.std(x)

plot_guissan(x, -2, stdv)
```

```
plot_laplace_dist(x,-2,4)
```

```
plt.legend()
```

Let's generate plots for both the Gaussian distribution and the Laplace distribution using the random dataset we've created:

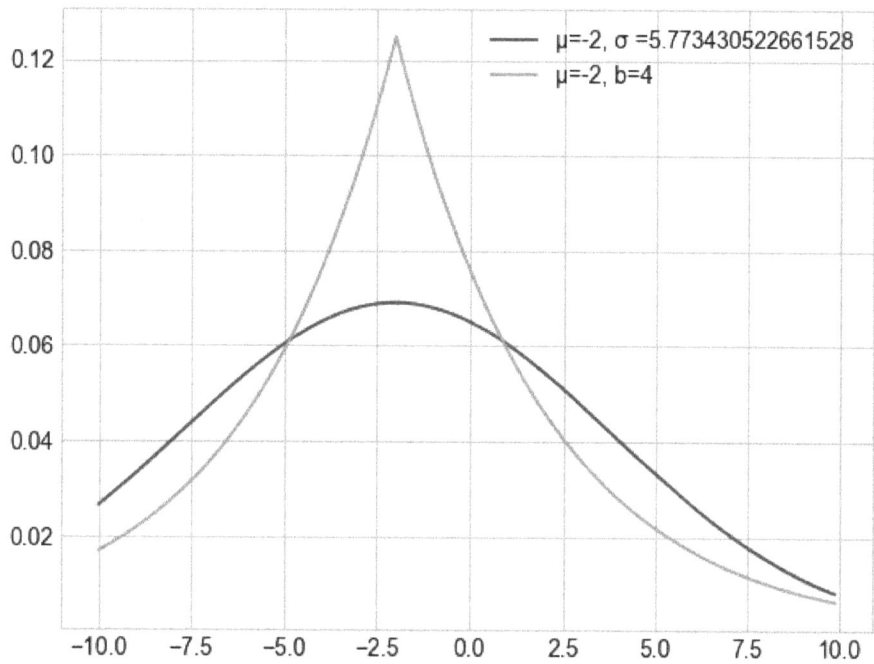

Figure 4.5 – Gaussian distribution with different means and standard deviation

Based on this graph, we can see that the Laplace distribution has a sharper peak compared to the Gaussian distribution.

Now, let's determine if the Gaussian mechanism complies with ε differential privacy:

```
query_result=100
```

```
p1 = guissan_dist(query_result, query_result, 1 / ε)
```

```
p2 = guissan_dist(query_result, query_result+1, 1 / ε)
```

```
print(p1,p2)
```

```
0.19947114020071635  0.17603266338214976
```

```
p1, p2 * np.exp(ε)
```

```
0.19947114020071635, 0.29022879645614585
```

The Gaussian mechanism does not satisfy ε differential privacy. Let's check whether it satisfies (ε, δ) differential privacy by adding a δ value in this case:

```
p1, p2 * np.exp(ε)-0.09075
```

```
0.19947114020071635, 0.19947879645614586
```

This shows that the Gaussian mechanism supports (ε, δ) differential privacy, which is sometimes called approximate differential privacy due to the addition of delta (δ).

Comparison of noise-adding algorithms to apply differential privacy

The following table provides a high-level comparison of three noise-adding mechanisms:

	Exponential Mechanism	**Laplace Mechanism**	**Gaussian Mechanism**
Usage	Mostly used to select a specific record or choose the best element, specifically with non-numeric or categorical data	Widely used in numerical aggregates due to its balance between privacy and utility and its ease of implementation	Used when dealing with a smaller scale of data and where differential privacy needs to be applied repeatedly
Noise type	Exponential noise	Laplace (double exponential) noise	Gaussian noise
Privacy	High privacy level due to the exponential distribution of output probabilities	Random noise might significantly change outcomes, compromising data utility	Achieves either delta privacy or epsilon-delta privacy, which is a trade-off for better utility
Implementation	Complex to implement due to the nature of exponential distribution	Relatively easy to implement due to the feature of the Laplace distribution	Complex to implement but generally suitable when data sensitivity is low
Adaptability	Mainly used for categorical data	Used mostly with numerical data	Suitable for low-sensitivity and multi-query data

Table 4.1 - High-level comparison of three noise-adding mechanisms

Generating aggregates using differential privacy

Differential privacy is a technique that's used to protect the privacy of individuals when they're collecting and sharing data. One of the ways that differential privacy can be used is in the generation of aggregates, which are statistical summaries of data that can provide useful insights while also protecting the privacy of individuals.

Once the noise has been added to the data, statistical aggregates can be calculated and released without compromising individual privacy. Examples of statistical aggregates that can be generated using differential privacy include means, medians, and histograms. One of the challenges of using differential privacy to generate aggregates is balancing privacy and accuracy. The amount of noise that is added to the data will affect the accuracy of the statistical aggregates that are generated. Therefore, there is often a trade-off between the amount of privacy protection and the accuracy of the results. Overall, the use of differential privacy to generate aggregates is a powerful tool for protecting individual privacy while still allowing useful statistical properties to be calculated. However, it requires careful consideration of the specific use case and the trade-offs between privacy and accuracy.

Sensitivity

Sensitivity plays a crucial role in the realm of differential privacy. It refers to the maximum amount by which the output of a function or computation can change when a single individual's data point is added or removed from a dataset. Sensitivity provides a measure of the privacy risk associated with performing computations on sensitive data.

Let's look at an example of a real-life scenario that illustrates the need to measure the impact of changing a dataset using sensitivity analysis.

Scenario – financial risk assessment model

Suppose a financial institution develops a machine learning model to assess the credit risk of loan applicants. The model takes various features, such as income, credit history, employment status, and outstanding debt, into account to predict the likelihood of loan default. The institution wants to ensure that the model is robust and not overly sensitive to the presence or absence of any individual in the dataset. In this case, sensitivity analysis can be employed to evaluate the impact of changing the dataset, specifically by removing or modifying the data of certain individuals. The institution wants to determine whether the exclusion or modification of data for specific individuals significantly alters the model's predictions or introduces bias, and measure the privacy risk associated with this change.

Let's break this down in mathematical terms.

For example, if we have a function, f, that takes a dataset, D, as input and returns a numerical output, then the sensitivity of f is defined as the maximum absolute difference between the outputs of f on two datasets, D and D', that differ by the presence or absence of a single individual's data point.

So, here, the sensitivity of f is $sensitivity(f) = max_{\{D, D'\}} ||f(D) - f(D')||$.

The sensitivity of a function is a fundamental concept in differential privacy because it determines the amount of noise that needs to be added to the output of the function to guarantee privacy. Naturally, functions with higher sensitivity are more prone to leaking information about individuals in the dataset and therefore require more noise to be added to their outputs to maintain privacy.

To understand sensitivity, let's consider the following dataset:

$$X = [\ 0,\ 1,2,3,4,5,6,7,8,9,10,11,.....N]$$

Two functions/queries are defined in this dataset: $f1(x) = x$ and $f2(x) = x^3$.

Let's define another function/query that calculates the following difference:

$$\Delta f(xa,\ xb) = |\ f(xa) - f(xb)\ |$$

Let's calculate f1, f2, $\Delta f1$, $\Delta f2$ and observe the data:

X→	X_0	X_1	X_2	X_3	X_4	X_5	X_6	X_7	....
	0	1	2	3	4	5	6	7	...
$f_1(x_i)=x$	0	1	2	3	4	5	6	7	...
$f_2(x_i)=x3$	0	1	8	27	64	125	216	343	...
$\Delta f_1(xi,\ x_{i+1})$	1	1	1	1	1	1	1	1	....
$\Delta f_2(xi,\ x_{i+1})$	1	7	21	37	61	91	127	...	...

Table 4.2 - Sensitivity analysis of functions/queries on datasets

Sensitivity refers to the impact a change in the underlying dataset can have on the result of a query.

Let xA, xB be any dataset from all possible datasets of X differing by at most one element.

So, the sensitivity is calculated with the following equation:

$$\text{Sensitivity} = \max |\ f(xa) - f(xb)\ ||,\ \text{where Xa,Xb} \subseteq X$$

Based on this definition, it is very clear that the sensitivity for $f1(x) = x$ is 1 because the maximum difference is constant. With the second query, $f2(x)=x^3$, the difference is not constant and it is unbounded, growing based on the number of data elements in the dataset. To calculate the sensitivity of the second function/query, we need to specify the lower bounds and upper bounds – for example, from x_5 to x_{10}, x_0 to x_5, and so on.

Let's define another query that is bounded within the input values of 5 to 10 or 0 to 5 using the same function:

$$f3(x) = x^3, \text{ where } 0<=x<=5$$

$$\Delta f(xa, xb) = |\, f(xa) - f(xb)\,|$$

Here's the table:

X→	X_0	X_1	X_2	X_3	X_4	X_5
	0	1	2	3	4	5
$fx(x_i)=x^3$	0	1	8	27	64	125
$\Delta f_2(xi, x_{i+1})$	1	7	21	37	61	91

Table 4.3 - Sensitivity analysis of query bounded within input values

In this case, the sensitivity will be max $|\,1+7+21+37+61+91\,| = 216$:

```
def f(x):
    return x ** 3
def sensitivity(f, xa, xb):
    sensitivity = abs(f(xa) - f(xb))
    return sensitivity
# Sensitivity calculation for the input range 0 to 6 and summing the
sensitivities
total_sensitivity = 0
for x in range(6):
    xa = x
    xb = x + 1
    sensitivity_value = sensitivity(f, xa, xb)
    print(sensitivity_value)
    total_sensitivity += sensitivity_value
print("Total Sensitivity within the range 0 to 6:", total_sensitivity)
```

Output:

```
1
7
19
37
```

```
61
91
Total sensitivity within the range 0 to 6: 216
```

The sensitivity of queries can be further classified into global sensitivity and local sensitivity, depending on whether the sensitivity is calculated using universal datasets or local datasets.

Global sensitivity

Global sensitivity refers to the maximum possible change in query results when considering all possible datasets. It is calculated based on the entire population or a universal dataset. Global sensitivity provides an upper bound on the potential privacy risk and captures the worst-case scenario.

Local sensitivity

Local sensitivity, on the other hand, focuses on the sensitivity of a query based on a specific local dataset. It measures the maximum change in query results when a single data point is added or removed from this local dataset. Local sensitivity provides a more fine-grained and context-specific measure of privacy risk.

Based on this, we can formulate the following:

Query/Function	Global Sensitivity	Comments
$f(x) = x$	1	Changing x by 1 will result in changes of $f(x)$ by 1
$f(x) = x^2$	Unbounded	A change in $f(x)$ depends on the value of x
$f(x) = x + x$	2	Changing x by 1 changes $f(x)$ by 2
$f(x) = 5 * x$	5	Changing x by 1 changes $f(x)$ by 5

Table 4.4 - Local sensitivity analysis of query results

Differential privacy is achieved by adding noise to the query result so that privacy is guaranteed.

The amount of noise that needs to be added in differential privacy depends on four key parameters:

- **The privacy parameter (ε):** The privacy parameter, ε, controls the level of privacy protection provided by differential privacy. A smaller value of ε implies stronger privacy guarantees but may result in higher noise being added to the query result.

- **The privacy budget ($\Delta\varepsilon$):** The privacy budget represents the total amount of privacy that can be expended over multiple queries or operations. It is typically divided equally between the individual queries to ensure that the overall privacy guarantees are maintained.

- **Sensitivity (Δf)**: Sensitivity refers to the maximum amount by which the query result can change when a single data point is added or removed. It quantifies the impact of individual data on the query result and helps determine the amount of noise to be added.

- **Data size (N)**: The size of the dataset being analyzed also influences the amount of noise added in differential privacy. Larger datasets generally allow for less noise to be added, resulting in improved accuracy, while smaller datasets may require more noise to preserve privacy.

By carefully choosing appropriate values for these parameters and considering the trade-off between privacy and accuracy, the noise can be tailored to provide effective privacy protection while still yielding useful and reliable query results:

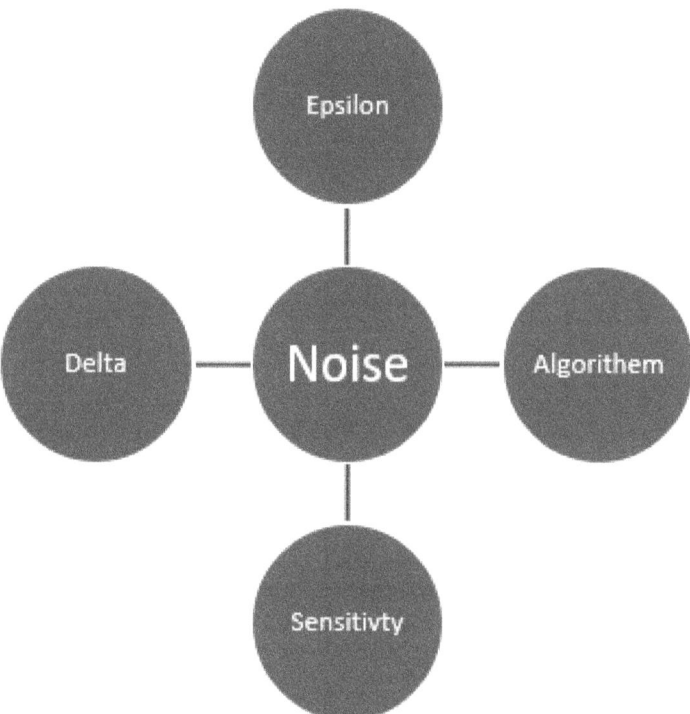

Figure 4.6 – Parameters that influence the addition of noise

Queries that use differential privacy

Queries that use differential privacy allow analysts and data scientists to retrieve aggregated information from a dataset while ensuring that the privacy of individual data points is protected. These queries are designed to add a controlled amount of noise to the query results, making it difficult to discern the contribution of any particular individual in the dataset.

Various types of queries can be performed using differential privacy. Some commonly used ones include the following:

- **Count queries**: These queries aim to determine the number of records that satisfy certain conditions in a dataset while preserving privacy. The query result is perturbed by adding noise to the true count, ensuring that individual contributions cannot be accurately determined.

- **Sum queries**: Sum queries involve calculating the sum of specific values in a dataset while maintaining privacy – for example, computing the total income of a group of individuals while protecting the confidentiality of individual income values.

- **Average queries**: Average queries compute the average value of a specific attribute in a dataset while ensuring privacy. Noise is added to the computed average to protect individual data points.

- **Top-K queries**: Top-K queries aim to identify the K largest values or records in a dataset while preserving privacy. Noise is added to the query result to hide the specific contributions of individuals.

- **Range queries**: Range queries involve retrieving records or values within a specific range from a dataset while maintaining privacy. Noise is added to the query result to prevent individual data points that have been identified falling within the range.

To achieve differential privacy, various mechanisms are used to add noise to the query results. These mechanisms include the Laplace mechanism, the Gaussian mechanism, and randomized response, among others. The choice of mechanism depends on the specific requirements of the query and the desired privacy guarantees.

It is important to note that the level of privacy protection provided by differential privacy is quantifiable through a privacy parameter called epsilon (ε). A smaller value of ε provides stronger privacy guarantees but may introduce more noise into the query results, potentially reducing their accuracy.

Queries that use differential privacy offer a balance between data utility and privacy protection. They allow for meaningful analysis and insights to be derived from sensitive datasets while safeguarding the confidentiality of individuals' information. By incorporating differential privacy techniques into data analysis processes, organizations can ensure compliance with privacy regulations and build trust with their users or customers.

Let's deep dive into these queries one by one with the following dataset:

Name	Age	Salary	Gender	Title
A	32	200K	M	Staff Engineer
B	44	300K	F	Manager
C	55	400K	F	Director
D	66	500K	M	VP

Table 4.6 - Differential privacy queries analysis with the provided dataset

Count queries using differential privacy

Analyze the results of the following queries and figure out how sensitivity operates:

- How many total employees are there in this dataset? The answer is 4.

- How many people have below 300K salary? The answer is 1.

- How many male employees are working? The answer is 2.

To ensure the answer includes noise and guarantees differential privacy, it is essential to determine the sensitivity of count queries and the appropriate value to use for sensitivity.

Count queries have a sensitivity of 1, meaning that if one record is added, the query result will increase by 1, or if one row is removed, the query result will decrease by 1. By understanding the sensitivity of count queries, we can accurately incorporate noise while preserving privacy.

We will utilize the NumPy library to generate differentially private results for the sample count queries.

NumPy provides Laplace noise generation based on the given sensitivity, mean, and epsilon values:

```
import numpy as np
Total_employees = 4
sensitivity = 1
print("sensitivity:",sensitivity )
# epsilon =privacy loss or budget
epsilon= 0.8
noise = np.random.laplace(loc=0, scale=sensitivity/epsilon)
print("noise:",noise)
count_employee_dp = Total_employees + noise
print ("count with DP:", count_employee_dp)
```

Ouput:

```
sensitivity: 1
noise: 2.247932787995729
count with DP: 6.24793278799573
```

Sum queries using differential privacy

Let's analyze the following queries and understand how sensitivity works in this case:

- What is the total salaries of all employees in this dataset? The total sum of salaries is reported as 1,400K (without applying differential privacy).

- What is the total age of employees? The total sum of ages is reported as 197 (without applying differential privacy).

To ensure the answers include the noise that was added by guaranteed differential privacy, it is important to determine the sensitivity for the sum queries and identify the appropriate value to use for sensitivity.

Unlike count queries, sensitivity for sum queries is not straightforward because the addition of a new row will increase the output by the added value.

Therefore, sensitivity in this case is unbounded since it depends on the data of the record being added.

For the query regarding the age of a person, we can estimate the upper bound by considering the historical maximum age to which a person has lived, which is around 126. Thus, we can assume an upper bound of 126 for this query.

However, determining the real upper bound for the salary query is challenging. It could be 1M, 10M, 100M, or even higher. As a result, in sum queries, the sensitivity depends on the lower and upper bounds provided to the queries.

To automatically calculate the lower and upper bounds, a technique called clipping can be employed, which we will explore in the next section:

```
Total_salary = 1400
Lower_bound=100
Upper_bound=9999
sensitivity = Upper_bound - Lower_bound
print("sensitivity:",sensitivity )
# epsilon =privacy loss or budget
epsilon= 0.9
noise = np.random.laplace(loc=0, scale=sensitivity/epsilon)
print("noise:",noise)
Total_salary_dp = Total_salary + noise
print ("total salary with DP:",Total_salary_dp)
```

Output:

```
sensitivity: 9899
noise: 4724.329480136737
total salary with DP: 6124.329480136737
```

Guaranteeing accuracy and privacy in the results of differential privacy are heavily influenced by their sensitivity, which is determined by the lower and upper bounds specified for sum queries.

Average queries using differential privacy

Analyze the following queries and figure out how sensitivity works in this case:

- What is the average salary? The answer is 350 K.

- What is the average age? The answer is 49.26.

To ensure that the answers for average queries include the noise added by guaranteed differential privacy, it is important to determine the sensitivity for these queries and identify the appropriate value to use for the sensitivity parameter in the differential privacy algorithm.

One approach to considering average queries is to treat them as two separate queries: a summation query and a count query. The sensitivity of count queries can be calculated using known methods; the sensitivity of sum queries can be determined similarly.

By understanding the sensitivity of both the count and sum queries, we can then provide the differentially private result for average queries. This involves incorporating noise in a way that preserves privacy while ensuring the accuracy of the average calculation:

```
Average_salary = 350
Total_salary_dp = 6124.329480136737
count_dp = 6.24793278799573
dp_average = (Total_salary_dp)/count_dp
dp_average
```

```
980.2169274137402
```

The noisy average salary with differential privacy will be distorted due to the added noise, while the average salary without differential privacy remains unchanged.

Now, let's implement the same use case using differential privacy techniques such as sensitivity, clipping, and noise. We'll use the Laplace mechanism to inject noise into the average calculation.

While the current discussion aims to provide an understanding of the base concepts, it's important to note that production-grade frameworks that implement differential privacy take into account various intricacies in more detail. These frameworks address factors such as handling overflows of numbers during calculations and employing appropriate algorithms to ensure accurate and privacy-preserving computations. These frameworks prioritize robustness and reliability to meet the demands of real-world scenarios.

Clipping

As discussed earlier, unbounded queries have an infinite sensitivity value, which cannot be directly utilized to provide results with differential privacy. One approach to addressing this issue is to transform unbounded queries into bounded ones by specifying their lower and upper bounds.

In differential privacy, clipping is a technique that's used to bind the sensitivity of a function by constraining its output within a specific range. The fundamental concept is to clip or limit the output of a function to fall into a predetermined range, such as [-c, c], where c is a positive constant. Afterward, noise is introduced to the clipped output to ensure privacy guarantees.

The clipping procedure involves two steps:

1. **Scaling the function's output**: The output of the function is scaled by dividing it by a scaling factor, denoted as s. This scaling ensures that the absolute value of the scaled output is less than or equal to the clipping threshold, c.

   Mathematically, this can be represented as $f'(D) = f(D) / s$, where $|f'(D)| <= c$.

2. **Adding noise**: Once the output has been clipped, noise is incorporated to uphold differential privacy. This noise can be generated from a Laplace or Gaussian distribution, with appropriate parameters depending on the desired level of privacy.

The clipping technique serves to restrict the influence of individual data points on the function's output and facilitates the addition of sufficient noise to preserve privacy effectively. By bounding the sensitivity using clipping, differential privacy mechanisms can maintain data confidentiality while providing accurate results.

Let's revisit the simple dataset that we used earlier:

Name	Age	Salary	Gender	Title
A	32	200K	M	Staff Engineer
B	44	300K	F	Manager
C	55	400K	F	Director
D	66	500K	M	VP

Table 4.7 - Sensitivity bounding analysis using clipping technique

Given the actual query, "How much is the total salary of all employees in this dataset?" and the actual answer, which is 1,400K, providing the total salary with differential privacy requires the addition of noise using a noise mechanism such as a Laplace or Gaussian mechanism. This noise mechanism relies on the sensitivity of the query.

To proceed, we must establish the lower and upper bounds as arbitrary fixed values. By calculating the difference between the lower and upper bounds, we obtain the sensitivity of the query. However, to delve deeper into this process, we will apply the clip function to the input data.

By applying the clip function, we constrain the input data within a specific range, which aids in managing the sensitivity of the query. This technique helps limit the influence of individual data points and facilitates the addition of noise in a controlled manner. By employing clipping, we can strike a balance between data privacy and utility, ensuring accurate results while preserving the confidentiality of sensitive information.

Let's play with the clip function on a dataset and understand this further.

Clipping example 1

Let's consider the following example use case: a social media platform wants to determine the average number of likes per user for a given post. The platform wants to protect the privacy of its users while still obtaining an accurate estimate.

Here's an implementation in Python without differential privacy:

Source code: DP_End_to_End.ipynb

```
import random
# Generate a list of simulated likes per user
likes_per_user = [random.randint(0, 100) for _ in range(1000)]
print(likes_per_user)
# Calculate the average number of likes per user
average_likes = sum(likes_per_user) / len(likes_per_user)
print("Average likes per user (without differential
privacy):", average_likes)
```

Average likes per user (without differential privacy): 51.146

Now, let's implement the same use case using differential privacy techniques such as sensitivity, clipping, and noise. We'll use the Laplace mechanism to inject noise into the average calculation:

```
import random
import numpy as np
# Generate a list of simulated likes per user
likes_per_user = [random.randint(0, 100) for _ in range(1000)]
average_likes = sum(likes_per_user) / len(likes_per_user)
print("Average likes per user (without differential
privacy):", average_likes)
# Set the sensitivity
sensitivity = 1
# Clip the likes per user to a specified range
clipped_likes_per_user = np.clip(likes_per_user, 0, 100)
# Calculate the average number of likes per user
clipped_average_likes = np.mean(clipped_likes_per_user)
# Define the privacy budget and epsilon value
privacy_budget = 1.0
epsilon = 0.1
# Calculate the scale parameter for the Laplace distribution
scale = sensitivity / (privacy_budget * epsilon)
# Inject noise using the Laplace mechanism
noisy_average_likes = clipped_average_likes + np.random.laplace(0,
scale)
```

```
print("Average likes per user (with differential privacy):", noisy_
average_likes)
```

Average likes per user (without differential privacy): 51.252
Average likes per user (with differential privacy): 41.61714462702061

In this example, we set the sensitivity to 1 since the average calculation has a sensitivity of 1 (changing one user's likes can affect the average by, at most, 1). Then, we clipped the likes per user to ensure they fall within a specific range (in this case, 0 to 100).

Next, we calculated the clipped average likes and defined the privacy budget and epsilon value. The privacy budget represents the total amount of privacy protection available and the epsilon value controls the level of privacy.

Finally, we calculated the scale parameter for the Laplace distribution based on the sensitivity, privacy budget, and epsilon. We added noise to the clipped average likes using the Laplace mechanism, considering the scale parameter.

The resulting noisy_average_likes value provides an estimate of the average number of likes per user while incorporating differential privacy techniques to protect users' privacy.

Clipping example 2

The following is another example of clipping:

Source code : DP_End_to_End.ipynb

```
import pandas as pd
data = {'age': [32, 44, 55, 66],
    'salary': [200, 300, 400, 500]}
df = pd.DataFrame(data)
df['salary'].clip(lower=0, upper=999)
```

Output:

```
0     200
1     300
2     400
3     500
Name: salary, dtype: int64
```

In this case, if we apply the clip function provided by Pandas with a range of 0 and 9,999, then the salaries don't change because of the lower and upper bound values in the clip function.

Clipping example 3

The following is another example of clipping:

Source code: DP_End_to_End.ipynb

```
import pandas as pd
data = {'age': [32, 44, 55, 66], 'salary': [200, 300, 400, 500]}
df = pd.DataFrame(data)
df['salary'].clip(lower=250, upper=400)
```

Output:

```
0     250
1     300
2     400
3     400
Name: salary, dtype: int64
```

In this case, when we set the upper bound to 400 and the lower bound to 250, the clip function changed salaries above 400 to 400 and below 250 to 250 (for example, 200 becomes 250 and 500 becomes 400).

pandas' `clip` function will enforce that the input data is within the specified clipping bounds. If one of the values is too small, it will be converted into the lower clipping bound, and if a value is too large, it will be converted into the upper clipping bound.

The following schema illustrates this operation:

Clip (250, 200, 300, 400, 500, **400**)

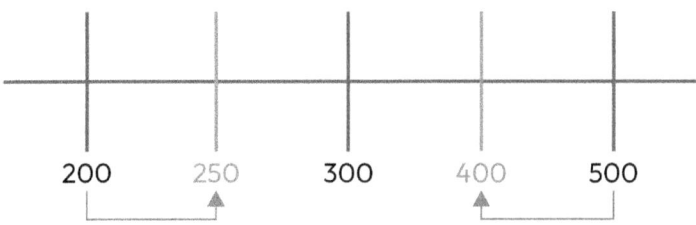

Figure 4.7 – Illustrating clipping functionality

Since it is a toy dataset, we can look at the data and come up with proper lower and upper bounds for clipping the data.

Let's add noise to the clipped inputs' sum with sensitivity (the clipped range difference) and epsilon:

```
def laplace_mech(v, sensitivity, epsilon):
    return v + np.random.laplace(loc=0, scale=sensitivity/epsilon)
epsilon = 0.9
```

```
sensitivity=400-100
print(laplace_mech(df['salary'].clip(lower=100, upper=400).
sum(),sensitivity,epsilon))
print(laplace_mech(df['salary'].sum(),sensitivity,epsilon))
```

Output:

```
1693.8114539575836.   (With clipped)
1793.7575175454226.   (without clipping)
```

By employing appropriate lower and upper bounds, the results that were obtained using differential privacy are closer to the actual values, resulting in increased accuracy compared to using random bounds. In this example, the actual total salary is 1,400K, but with added noise, the returned value is 1,693K.

However, one potential drawback of using specific lower and upper bounds is that they may inadvertently reveal the actual salaries of individuals present in the dataset, thereby compromising privacy.

An alternative approach is to initialize the lower bound with a value of zero or a suitably low value and gradually increase the upper bound. The process involves observing the query result and monitoring when it stops changing or becomes stable. At this point, the value of the upper bound is stopped, ensuring that the query result does not reveal sensitive information and preserving privacy.

This iterative approach allows for a balance between accuracy and privacy as it dynamically determines the bounds based on the characteristics of the data. It helps prevent the disclosure of specific salary information while still providing reasonably accurate results within the scope of differential privacy.

Let's explore this approach with the toy dataset:

Source code: DP_End_to_End.ipynb

```
import pandas as pd
import numpy as np
import matplotlib.pyplot as plt
plt.style.use('seaborn-whitegrid')
data = {'age': [32, 44, 55,66], 'salary': [200, 300, 400, 500]}
df = pd.DataFrame(data)
def laplace_mech(queryresult, sensitivity, epsilon):
    return queryresult + np.random.laplace(loc=0, scale=sensitivity/
epsilon)
epsilon = 0.5
plt.plot([laplace_mech(df['salary'].clip(lower=0, upper=i).sum(), i,
epsilon) for i in range(100,800,100)])
plt.xlabel('Clipping ranges for salary')
plt.ylabel('Total salary');
```

Here's the output:

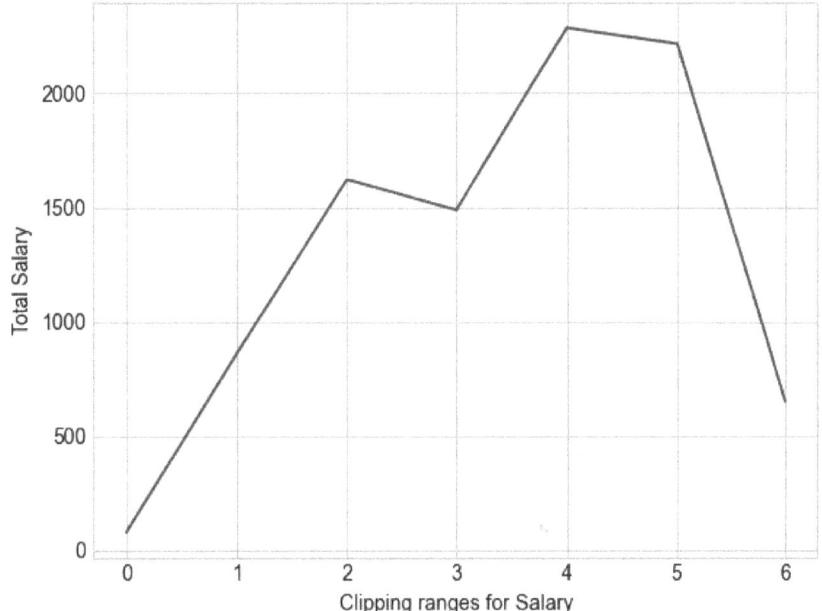

Figure 4.8 –Total Salary vs Clipping ranges for Salary

Given the small size of the dataset, it's relatively straightforward to identify the optimal clipping value for the query results – in this scenario, the sum of the salaries doesn't increase beyond 500. However, with larger datasets, the noise can sometimes become too dominant to analyze for all different sensitivity or clipping values.

Overview of real-life applications of differential privacy

In this section, we will go through a high-level summary of a few real-world applications that make use of differential privacy.

Differential privacy usage at Uber

Uber Technologies, Inc. provides mobility as a service – that is, ride-hailing services. It has to protect the privacy of the customers who make use of its services but at the same time analyze its data to provide a better service. It has developed an open source framework to achieve differential privacy with SQL queries, which it made to analyze and derive insights. All SQL queries are transformed using Uber's differential privacy framework, which executes SQL queries and provides the query results with differential privacy guaranteed.

It can be found on GitHub at `https://github.com/uber-archive/sql-differential-privacy`.

Differential privacy usage at Apple

Apple, Inc. has incorporated local differential privacy on a large scale in its products with the aim of understanding and enhancing the user experience. They've employed a private count mean sketch algorithm, which utilizes count, mean, sketch, and a privacy budget (epsilon). For each key feature, they've established a privacy budget and the maximum number of records (differentially private records) that can be transmitted to a remote server after sensitive features such as IP addresses have been removed:

Key Feature	Epsilon (ε)	Number of Records per Day for Each Device
Emoji suggestions	4	1
Lookup hints	4	2
QuickType	8	2
....		

Table 4.8 - Implementation of local differential privacy by Apple, Inc.

It injects noise after encoding the input vector using a hash function and flips each vector element with a probability of $1/(1 + e^{\varepsilon/2})$, where ε is the privacy budget.

More details about this algorithm can be found at `https://machinelearning.apple.com/research/learning-with-privacy-at-scale`.

Jun Tang, Aleksandra Korolova, Xiaolong Bai, Xueqiang Wang, Xiaofeng Wang, et al. have undertaken a detailed analysis of Apple's differential privacy implementation by debugging the client-side code and published a research article in Arxiv. I strongly recommend that you read this paper to understand Apple's implementation better and the additional details that were considered: *Privacy Loss in Apple's Implementation of Differential Privacy on MacOS 10.12* (`https://arxiv.org/abs/1709.02753`).

Differential privacy usage in the US Census

The US Census Bureau used differential privacy to apply noise to the 2020 census data to protect respondents' confidentiality in the collected and shared census data.

More details can be found at `https://www.census.gov/data/academy/webinars/2021/disclosure-avoidance-series/differential-privacy-101.html`.

Differential privacy at Google

Google uses differential privacy in its Chrome browser to find out about frequently visited pages; Google Maps and Google Assistant also have this functionality. The differential privacy system that it has developed is called **Randomized Aggregatable Privacy-Preserving Ordinal Response (RAPPOR)** and provides an epsilon value of 2 as the lower limit and 8 to 9 as the upper limit.

Google has open sourced the implementation of RAPPOR. You can learn more about this at `https://github.com/google/rappor`.

Summary

In this chapter, we covered the concept of differential privacy and explored how the Laplace and Gaussian mechanisms can generate noise to ensure privacy while generating aggregate query results. We discussed the significance of parameters such as epsilon, delta, and sensitivity, and how they are used to calculate noise using Laplace or Gaussian distributions. Additionally, we learned about the process of determining upper and lower bounds using the clipping technique. Finally, we provided a summary of how differential privacy is used in real-world applications at Apple and Uber and by the US Census Bureau.

In the next chapter, we will delve into open source frameworks for differential privacy. We will explore how to develop applications using these frameworks and dive into the realm of machine learning with differential privacy in detail. This will provide a comprehensive understanding of how to implement differential privacy in practical scenarios and how to leverage its benefits in the context of machine learning applications.

5

Developing Applications with Differential Privacy Using Open Source Frameworks

In this chapter, we will explore open source frameworks (**PyDP**, **PipelineDP**, **tmlt-analytics**, **PySpark**, **diffprivlib**, **PyTorch**, and **Opacus**) used to develop machine learning, deep learning, and large-scale applications with the power of differential privacy.

We will cover the following main topics:

- Open source frameworks for implementing differential privacy:

  - Introduction to the PyDP framework and its key features

  - Examples and demonstrations of PyDP in action

  - Developing a sample banking application with PyDP to showcase differential privacy techniques

- Protecting against membership inference attacks:

  - Understanding membership inference attacks and their potential risks

  - Techniques and strategies to safeguard against membership inference attacks when applying differential privacy

- Applying differential privacy on large datasets to protect sensitive data:

  - Leveraging the open source PipelineDP framework to apply differential privacy on large-scale datasets

  - Leveraging the open source Tumult Analytics and PySpark frameworks to apply differential privacy on large datasets

- Machine learning with differential privacy:

  - Running a fraud detection classification model on synthetic data using differential privacy

  - A clustering example, applying differential privacy using IBM's open source `diffprivlib` framework

- Deep learning with differential privacy:

  - Implementing a fraud detection model using the PyTorch deep learning framework

  - Utilizing the open source PyTorch and Opacus frameworks to develop deep learning models for fraud detection with differential privacy

- Differential privacy machine learning frameworks

- Limitations of differential privacy and some strategies to overcome them

Open source frameworks to implement differential privacy

There are several open source frameworks available to implement differential privacy. We will go through the PyDP framework in detail in this section.

Introduction to the PyDP framework and its key features

Google has released an open source framework called differential privacy that facilitates the implementation of differential privacy. This framework offers support for both ε- and (ε, δ)-differentially private statistics. It includes various features such as the ability to introduce noise using Laplace and Gaussian mechanisms. Additionally, the framework provides support for aggregated differential privacy algorithms including sum, count, mean, variance, and standard deviation. The libraries within this framework are implemented in the C++, Java, and Go languages, and it also offers a **command-line interface** (**CLI**) to execute differential privacy SQL queries. For further information, you can visit the GitHub repository at `https://github.com/google/differential-privacy`.

PyDP, developed by OpenMined in 2020, is another framework that implements Python wrapper functions for Google's differential privacy tools. While PyDP is not an exhaustive implementation of Google's differential privacy toolkit, it supports a subset of ε-differentially private algorithms from Google's toolkit. These algorithms enable the generation of aggregate statistics over numeric datasets that contain private or sensitive information. You can find the PyDP framework at `https://github.com/OpenMined/PyDP`.

Examples and demonstrations of PyDP in action

Installation of PyDP

PyDP installation is done the same way as installing any other Python package.

For Python 3.x, use the following:

```
pip3 install python-dp
```

For earlier versions of Python, use this line:

```
pip install python-dp
```

Sample program to calculate the mean using PyDP

PyDP supports the Laplacian noise mechanism and offers a range of aggregate functions such as sum, average, count, and more. When calculating the mean, PyDP requires the provision of bounds in the form of lower and upper values. To facilitate this, PyDP provides a class called BoundedMean, which offers the following constructors:

```
BoundedMean(epsilon: float, delta: float = 0, lower_bound: int, upper_
bound)
BoundedMean(epsilon: float)
```

The BoundedMean class in PyDP is used to calculate the differentially private mean of bounded data. It utilizes the epsilon and delta parameters to provide privacy guarantees. This class supports the Laplacian noise mechanism to add privacy-preserving noise to the mean calculation.

Here is example code to demonstrate the bounded mean:

Source code: Sample_Mean_Using_PyDP.ipynb

```
import pydp as dp
import numpy as np
from pydp.algorithms.laplacian import BoundedMean

#Generate simple data number from 1 to 10 in an array
data = np.arange(1,10,1)

# privacy budget as 0.6, delta as 0 and lower and upper bounds as 1
and 10

x = BoundedMean(0.6,0,1,10)
privacy_mean = x.quick_result(data)
mean = np.mean(data)
print("Data", data)
print("Normal Mean", mean)
print("Mean with differential privacy",privacy_mean )
```

This results in the following output:

```
Data [1 2 3 4 5 6 7 8 9]
Normal Mean 5.0
Mean with differential privacy 6.960764656372703
```

Developing a sample banking application with PyDP to showcase differential privacy techniques

In this application scenario, we take the example of a financial bank that aims to collaborate with merchants, companies, and other banks to launch a marketing campaign while safeguarding the sensitive personal information of its customers. Customers make purchases using their credit or debit cards in two different scenarios: offline transactions at merchant outlets, where the cards are physically swiped (known as card-present transactions), and online transactions, where card details such as the card number, name, expiry date, and CVV are entered (known as card-not-present transactions). The objective is to provide credit card transaction data to support the campaign without revealing individual customer details.

To successfully launch marketing campaigns or loyalty programs, the bank needs to share certain transaction details such as the type of purchases, transaction volumes, and transaction values associated with specific locations. This information allows the bank to partner with interested banks/merchants, enabling the promotion of large-scale product sales and the delivery of benefits to customers.

In this context, we will synthetically generate a significant number of transactions using predefined business rules and calculate statistics such as count, average, and sum for each location. The intention is to share these aggregated statistics with the partnering companies without compromising customer privacy.

There are two approaches to sharing these statistics. The first approach involves sharing the aggregates as they are, which may potentially reveal private information about individual customers. The second approach utilizes differential privacy to generate the aggregates, ensuring that they do not leak any sensitive customer information while preserving privacy.

By applying differential privacy techniques, the bank can protect customer privacy by introducing carefully calibrated noise to the aggregated statistics. This noise ensures that the shared aggregates do not disclose specific individuals' details while still providing valuable insights for the marketing campaign and loyalty program planning.

By adopting differential privacy, the bank can strike a balance between data utility and privacy protection, allowing it to collaborate with companies and merchants while maintaining the confidentiality of customer information.

Let's generate the synthetic datasets needed to develop this banking application.

The following is the customer data:

```
CUSTOMER_ID   latitude   longitude   mean_amount   tx_per_day
          0  54.881350   71.518937            62            4
          1  84.725174   62.356370            41            8
          2  89.177300   96.366276            41            7
          3  81.216873   47.997717            42            9
          4   8.712930    2.021840            84            1
          5  97.861834   79.915856            48            4
          6  67.887953   72.063265            60            1
          7  52.184832   41.466194            30            3
          8  45.615033   56.843395             6            6
          9  14.967487   22.232139            41            5
```

Figure 5.1 – Customer Dataset

The dataset used in this context comprises various attributes, including the customer ID, the location represented as latitude and longitude coordinates, the number of transactions per day, and the average transaction amount. It is important to note that the actual credit card number, **Card Verification Value (CVV)** code, and the expiry date associated with each customer's card are maintained in a separate table and are not shown in the provided dataset. Furthermore, these parameters are not utilized in generating the statistical information. In this particular example, the transactions generated pertain to card-present scenarios occurring at various merchant locations, using **point-of-sale (POS)** terminals to swipe the cards.

The following is the sample POS terminal data:

```
TERMINAL_ID   latitude   longitude
          0  54.881350   71.518937
          1  60.276338   54.488318
          2  42.365480   64.589411
          3  43.758721   89.177300
          4  96.366276   38.344152
          5  79.172504   52.889492
          6  56.804456   92.559664
          7   7.103606    8.712930
          8   2.021840   83.261985
          9  77.815675   87.001215
```

Figure 5.2 – Terminal Dataset

In this scenario, it is assumed that customers predominantly visit nearby merchant outlets for their day-to-day purchases, with the distance between the customer's location and the merchant locations typically falling within a range of around 5 miles.

To generate the transactions, the Euclidean distance between the customer's location and the available merchant locations is calculated. Using this distance information, merchants are randomly selected from the nearby options. This approach ensures that the generated transactions reflect the realistic behavior of customers going to nearby merchants within a specific distance radius for their purchases.

The following is the transaction data:

TX_DATETIME	CUSTOMER_ID	TERMINAL_ID	TX_AMOUNT
2023-02-01 16:54:00	0	9	62
2023-02-01 14:43:07	0	3	64
2023-02-01 17:11:15	0	9	61
2023-02-01 14:38:20	0	3	61
2023-02-02 11:42:47	0	0	62
2023-02-02 12:24:00	0	2	63
2023-02-02 14:06:50	0	5	62
2023-02-02 13:13:58	0	3	62
2023-02-03 16:09:00	0	3	61
2023-02-03 12:52:10	0	9	61
2023-02-03 04:54:30	0	9	62
2023-02-03 14:24:04	0	2	61
2023-02-04 18:18:17	0	3	60
2023-02-04 12:07:37	0	2	61
2023-02-04 16:15:27	0	5	63
2023-02-04 12:25:49	0	1	62
2023-02-05 09:32:02	0	5	60
2023-02-05 11:02:00	0	1	62
2023-02-05 15:25:02	0	2	63
2023-02-05 10:55:26	0	1	61

Figure 5.3 – Transaction Dataset

Let's generate aggregates using differential privacy on this dataset so that it can then be shared with merchants/banks in order to design marketing campaigns and loyalty programs.

Source code: Sample_Finance_App_DP.ipynb

 # unzip the transactions.csv.zip file (provided in this book's GitHub repo as transactions.csv).

```
https://github.com/PacktPublishing/Privacy-Preserving-Machine-
Learning/blob/main/Chapter%205/transactions.csv.zip
```

Data loading:

```
import pydp as dp
from pydp.algorithms.laplacian import BoundedSum, BoundedMean, Count,
Max
import pandas as pd

url = "transactions.csv"
df_actual = pd.read_csv(url, sep=",")
df = df_actual[['TRANSACTION_ID',
'TX_DATETIME','CUSTOMER_ID','TERMINAL_ID','TX_AMOUNT']]
df.head()
```

	TRANSACTION_ID	TX_DATETIME	CUSTOMER_ID	TERMINAL_ID	TX_AMOUNT
0	0	2023-02-01 00:43:37	901	8047	82
1	1	2023-02-01 01:20:13	2611	7777	15
2	2	2023-02-01 01:22:52	4212	3336	53
3	3	2023-02-01 01:26:40	1293	7432	59
4	4	2023-02-01 01:52:23	2499	1024	25

Figure 5.4 – Sample transaction data

In this section, we will generate differentially private aggregates for various scenarios. The focus will be on comparing the results obtained using traditional statistical methods with those achieved through differential privacy techniques. The following scenarios will be explored:

- Mean transaction amount for a given terminal

- Mean transaction amount for terminal IDs 1 to 100

- Number of customers who make purchases worth $25 or more via terminals 1 to 100

- Maximum transaction amount for a given terminal

- Sum of the transaction amounts for each terminal on a given day or month

Mean transaction amount for a given terminal

In order to calculate the average transaction amount for a given day or month via a specific POS terminal, we can define the following methods. These methods will be used to compare the results obtained using traditional statistical methods with the aggregates generated through differential privacy techniques:

- **Traditional average calculation**: We will implement a method that calculates the average transaction amount for a given day or month on a particular POS terminal using traditional statistical methods. This method will take as input the relevant transaction data, such as the transaction amounts and the date of the day or month of interest. The average will be computed by summing all the transaction amounts and dividing the sum by the total number of transactions on the specified day or month on the chosen POS terminal. The traditional average will serve as a baseline for comparison.

- **Differentially private average calculation**: We will develop a method that leverages differential privacy techniques to calculate the differentially private average transaction amount for a given day or month on the selected POS terminal. This method will take the same input as the traditional average calculation method. It will utilize differential privacy mechanisms, such as adding noise to the aggregated statistics, to protect the privacy of individual transactions while generating the average. The differentially private average will provide a privacy-preserving alternative to the traditional average. By utilizing these methods and comparing the results, we can assess the differences between the traditional average calculation and the average generated through differential privacy techniques.

This analysis will demonstrate the impact of differential privacy on aggregate calculations and highlight the trade-off between accuracy and privacy preservation in generating average transaction amounts on a given day or month via a specific POS terminal:

```
def mean_tx_amount(tid:int) -> float:
    dft = df[df["TERMINAL_ID"] == tid]
    return statistics.mean(list(dft["TX_AMOUNT"]))
mean_tx_amount(1)
```

56.22097378277154

```
# calculates mean applying differential privacy

def private_mean_tx_amount(privacy_budget: float, tid:int) -> float:
    x = BoundedMean(privacy_budget,1,1,300)
    dft = df[df["TERMINAL_ID"] == id]
    return x.quick_result(list(dft["TX_AMOUNT"]))
private_mean_tx_amount(0.6,1)
```

220.98103940917645

In the preceding calculation, we can see the average generated using the traditional method returns 56.22 whereas the differentially private version produces an average of 220.98 for POS terminal ID 1. In this way, the private average helps not to disclose the actual average.

Mean transaction amount for terminal IDs 1 to 100

Let's generate the mean transaction amount for the terminals 1 to 100 and make use of the private mean function defined earlier:

```
terminal_mean_vs_privacy_means=[]
for i in range (1, 100):
    mean = mean_tx_amount(i)
    privacy_mean = private_mean_tx_amount(0.9,i)
    terminal_mean_vs_privacy_means.append([i, mean,privacy_mean])

terminal_mean_vs_privacy_means_df =
pd.DataFrame(terminal_mean_vs_privacy_means,
columns=['Terminal Id','Mean','privacy_mean'])

terminal_mean_vs_privacy_means_df.head(10)
```

	Terminal Id	Mean	privacy_mean
0	1	56.220974	96.594534
1	2	52.366505	300.000000
2	3	40.870763	1.000000
3	4	52.102000	300.000000
4	5	47.911175	118.168421
5	6	52.302128	1.000000
6	7	51.335979	186.967339
7	8	50.100806	163.890127
8	9	52.631893	1.000000
9	10	48.978903	1.000000

Figure 5.5 – Actual Mean vs Privacy Mean

In the following sections, we will also generate the count and sum of transaction amounts for a given day or month on a specific POS terminal and compare the results obtained using traditional statistical methods with those generated through differential privacy techniques.

Number of customers who make purchases worth $25 or more via terminals 1 to 100

Next, we will implement a method that calculates the number of customers who made purchases of $25 or more on terminals 1 to 100.

This method will take the transaction data as input, including the terminal number and the corresponding transaction amounts. It will iterate through the transactions for each terminal ID from 1 to 100 and count the number of customers whose transaction amounts exceed $25. The count will provide an indication of the customer base that made higher-value purchases, helping to analyze the impact of differential privacy on identifying such customers. By utilizing this method, we can compare the results obtained using traditional statistical methods with the counts generated through differential privacy techniques.

This analysis will shed light on the differences between the approaches and demonstrate the effectiveness of differential privacy in identifying customers who made purchases worth more than $25 for terminals 1 to 100 while preserving individual privacy:

```
def count_tx_amount_above(limit: float,tid) -> int:
    dft = df[df["TERMINAL_ID"] == tid]
    return dft[dft.TX_AMOUNT > limit].count()[0]
count_tx_amount_above(25.0,1)
```

232

```
def private_tx_amount_above(privacy_budget: float, limit:
float,tid:int) -> float:
    dft = df[df["TERMINAL_ID"] == tid]
    x = Count(privacy_budget, dtype="float")
    return x.quick_result(list(dft[dft.TX_AMOUNT > limit]["TX_
AMOUNT"]))
private_tx_amount_above(0.1,25.0,1)
```

257

```
terminal_amount_vs_privacy_amont=[]
for i in range (1, 100):
    count = count_tx_amount_above(25.0,i)
    privacy_count = private_tx_amount_above(0.1,25.0,i)
    terminal_amount_vs_privacy_amont.append([i, count,privacy_count])

terminal_amount_vs_privacy_amont_df =
pd.DataFrame(terminal_amount_vs_privacy_amont, columns=['Terminal
Id','Count','privacy_count'])
terminal_amount_vs_privacy_amont_df.head(10)
```

	Terminal Id	Count	privacy_count
0	1	232	224
1	2	285	274
2	3	381	368
3	4	338	315
4	5	279	278
5	6	337	329
6	7	267	270
7	8	318	323
8	9	514	502
9	10	370	364

Figure 5.6 – Actual transaction counts vs privacy added counts

By adopting this approach, we ensure that the actual count of customers who made purchases worth more than $25 is not revealed or shared with banks/merchants. Instead, we provide those parties with differentially private counts, thus preserving individual privacy while still offering valuable insights. This allows banks/merchants to launch loyalty programs based on the differentially private counts, tailoring their initiatives based on the available data. Thus, by employing differential privacy techniques, institutions can strike a balance between providing useful information for loyalty programs and safeguarding the sensitive details of individual customers.

Maximum transaction amount for a given terminal

Let's define the functions to calculate the maximum transaction amount and the differentially private amount:

```
def max_tx_amount(tid:int) -> int:

    dft = df[df["TERMINAL_ID"] == tid]

    return dft.max()["TX_AMOUNT"]

max_tx_amount(1)
```

```
87
```

```
def private_max_tx_amount(privacy_budget: float,tid:int) -> int:
    dft = df[df["TERMINAL_ID"] == tid]
    x = Max(epsilon = privacy_budget, lower_bound = 100.0, upper_bound
= 50000.0, dtype="float"
    return x.quick_result(list(dft["TX_AMOUNT"]))
private_max_tx_amount(0.5,1)
```

```
167.51941105013407
```

As we can see in the preceding code, by employing differential privacy techniques, we can calculate an approximate maximum transaction amount for a given terminal while preserving the privacy of individual transactions. These values depend on the privacy budget. This example used 0.5 as the privacy budget. This allows us to share valuable aggregated information with banks/merchants without compromising the sensitive details of individual customers. In this example, the actual maximum transaction amount is 87. With added noise based on the privacy budget (i.e., 0.5) the value becomes 167. Thus, it may not be very useful in terms of utility. This illustrates the trade-off between privacy and utility. One needs to experiment with different privacy budgets to decide the best fit for the use case/application, deciding whether they want prioritize more privacy and less utility or more utility with less privacy.

Sum of the transaction amounts for each terminal on a given day or month

Let's define the functions to calculate the sum of the transaction amounts and the differentially private amounts:

```
def sum_tx_amount(tid:int) -> float:
    dft = df[df["TERMINAL_ID"] == tid]
    return dft.sum()["TX_AMOUNT"]
sum_tx_amount(1)
```

```
15011
```

```
def private_sum_tx_amount(privacy_budget: float, tid:int) -> float:
    dft = df[df["TERMINAL_ID"] == tid]
    x = BoundedSum(epsilon = privacy_budget, delta = 0,
lower_bound= 100.0, upper_bound = 50000.0, dtype="float")
    return x.quick_result(list(dft["TX_AMOUNT"]))
private_sum_tx_amount(0.6,1)
```

```
27759.46144104004
```

As an exercise, you can implement the count and sum functions for all POS terminals and compare the results obtained using traditional statistical methods with those generated through differential privacy techniques.

Protecting against membership inference attacks

Membership inference attacks pose a significant threat to the privacy of individuals in machine learning systems. These attacks aim to determine whether a specific data point was part of the training dataset used to create a machine learning model, potentially exposing sensitive information about individuals. To mitigate the risk of such attacks, differential privacy techniques can be employed.

To protect against membership inference attacks using differential privacy, several approaches can be adopted:

- **Noise addition**: During the training process, noise is added to the computations to introduce randomness and mask individual data points. This makes it challenging for attackers to identify whether a specific data point was used in the training.

- **Privacy budget management**: Differential privacy operates under a privacy budget that determines the maximum amount of privacy loss allowed. By carefully managing and allocating the privacy budget, the risk of membership inference attacks can be minimized.

- **Generalization and aggregation**: Applying generalization and aggregation techniques helps in obfuscating individual data points. By grouping similar data points together, the information about any specific individual becomes less distinguishable.

- **Perturbation mechanisms**: Utilizing perturbation mechanisms, such as adding noise to the model's outputs or gradients, enhances privacy protection. These mechanisms make it more challenging for attackers to infer membership status accurately.

- **Adversarial training**: Incorporating adversarial training techniques helps in training models that are robust against membership inference attacks. This involves training the model against a sophisticated attacker who tries to distinguish the presence of specific data points.

By combining these strategies and adopting a privacy-by-design approach, machine learning systems can better protect against membership inference attacks. It is important to note that while differential privacy provides strong privacy guarantees, there might still be cases where additional privacy-preserving techniques or post-processing is necessary to address specific attack scenarios.

Following is an example to demonstrate this:

Let's generate two datasets that differ by exactly one record. We'll create a copy of the original dataset and refer to it as the **redacted dataset**. In the redacted dataset, we'll remove one record to create the difference.

Here's how to proceed:

1. Start with the original dataset containing the desired records. This dataset represents the baseline or complete set of records.

2. Create a copy of the original dataset and label it as `redact_dataset`. This dataset will closely resemble the original dataset but with one record removed. Choose any record from the redacted dataset and remove it to create the difference. In the example, the first record is removed.

By creating the redacted dataset as a modified version of the original dataset, specifically by removing one record, we establish a distinct dataset that differs from the original by only that single record.

Use the following source code to create a redacted one.

Source code: Sample_Finance_App_DP.ipynb

```
import pandas as pd
url = "2023-07-08.csv"
df = pd.read_csv(url, sep=",")
redact_dataset = df.copy()
redact_dataset = redact_dataset[1:]
df.head()
```

	TRANSACTION_ID	TX_DATETIME	CUSTOMER_ID	TERMINAL_ID	TX_AMOUNT
0	3909694	2023-07-08 00:42:26	2079	5549	36
1	3909695	2023-07-08 01:12:58	478	8310	43
2	3909696	2023-07-08 01:43:53	2105	6875	11
3	3909697	2023-07-08 01:51:21	1094	4877	87
4	3909698	2023-07-08 01:54:43	3968	6327	31

Figure 5.7 - Sample transaction dataset

```
redact_dataset.head()
```

	TRANSACTION_ID	TX_DATETIME	CUSTOMER_ID	TERMINAL_ID	TX_AMOUNT
1	3909695	2023-07-08 01:12:58	478	8310	43
2	3909696	2023-07-08 01:43:53	2105	6875	11
3	3909697	2023-07-08 01:51:21	1094	4877	87
4	3909698	2023-07-08 01:54:43	3968	6327	31
5	3909699	2023-07-08 02:10:16	1217	3667	47

Figure 5.8 – Redact dataset

We have removed just one record (customer ID 2079) from the original dataset, who made a transaction of $36. This was done to form the redacted dataset.

Let's calculate the sum of transaction amounts from the original dataset and the redacted dataset to determine the difference. This difference will correspond to the exact transaction amount made by the customer with the ID 2079:

```
sum_original_dataset = round(sum(df["TX_AMOUNT"].to_list()), 2)
sum_redact_dataset =
round(sum(redact_dataset["TX_AMOUNT"].to_list()), 2)
tx_amount_2079 = round((sum_original_dataset - sum_redact_dataset), 2)
tx_amount_2079
```

36

Let's apply differential privacy techniques to calculate the sum of transaction amounts from the original dataset and the redacted dataset. The difference between these two sums should not reveal the exact transaction amount made by the customer with ID 2079. Here's how you can approach this:

Sum using differential privacy on the original dataset:

```
dp_sum_original_dataset = BoundedSum(
    epsilon=1, lower_bound=1.0, upper_bound=500.0, dtype="float"
)

dp_sum_original_dataset.reset()
dp_sum_original_dataset.add_entries(
    df["TX_AMOUNT"].to_list()
)

dp_sum_og = round(dp_sum_original_dataset.result(), 2)
print(dp_sum_og)
```

```
1300958.19
```

Sum using differential privacy on the redacted dataset:

```
dp_redact_dataset = BoundedSum(epsilon=1, lower_bound=1.0, upper_
bound=500.0, dtype="float")
dp_redact_dataset.add_entries(redact_dataset["TX_AMOUNT"].to_list())
dp_sum_redact = round(dp_redact_dataset.result(), 2)
print(dp_sum_redact)
```

```
1302153.33
```

Difference between the two datasets using differential privacy:

```
round(dp_sum_og - dp_sum_redact, 2)
```

```
-1195.14
```

In this example, when calculating the sum of transaction amounts using differential privacy for both the original dataset and the redacted dataset, the difference between these two sums resulted in negative numbers. However, it is important to note that these negative values do not represent the actual transaction amount made by customer ID 2079.

The negative values in the difference arise due to the inherent noise added during the differential privacy calculations. Differential privacy techniques introduce randomization to protect individual privacy, and this random noise can sometimes lead to negative perturbations in the aggregated results.

Therefore, it is crucial to interpret these negative values correctly. They should not be considered as the actual transaction amount made by customer ID 2079, but rather as an indication that the differential privacy mechanisms have successfully introduced noise to protect individual privacy while providing approximate aggregate results.

It is essential to understand that differential privacy focuses on preserving privacy rather than exactness in the calculated results. The negative difference serves as a reminder of the privacy guarantees provided by differential privacy, ensuring that individual transaction details are safeguarded even in the presence of aggregate computations.

Applying differential privacy to large datasets

In the previous examples, we focused on calculating differentially private aggregates (such as count, sum, and average) on smaller datasets, involving a single terminal or a limited number of terminals. However, in this section, we will explore how to generate differentially private aggregates on large datasets, including millions or even billions of records. Specifically, we will consider a use case involving a dataset of approximately 5 million credit card transactions across 1,000 point-of-sale terminals and 5,000 customers.

Use case – generating differentially private aggregates on a large dataset

Let's generate the dataset comprising credit card transactions recorded on a daily basis across 1,000 POS terminals. These transactions involve a total of 5,000 customers, resulting in an extensive collection of approximately 5 million records.

To calculate differentially private aggregates on such a large dataset, specialized techniques and frameworks are employed to handle the scale and complexity of the data. These techniques ensure that privacy is preserved while providing meaningful aggregate statistics.

By leveraging differential privacy on large datasets, organizations can extract valuable insights without compromising the privacy of individual customers. The generated differentially private aggregates enable data-driven decision-making and analysis while protecting sensitive information. It is worth noting that the methods and frameworks used to apply differential privacy on large datasets may vary depending on the specific requirements and available resources. Techniques such as data partitioning, parallel processing, and optimized algorithms play a crucial role in efficiently computing differentially private aggregates on such vast datasets.

By understanding how to generate differentially private aggregates on large datasets, organizations can derive actionable insights from their data while upholding the privacy of individuals involved in the transactions. The dataset is organized in a specific format, with each day's transactions stored in separate files that include the date in the filename. To illustrate, let's consider an example where the dataset corresponds to transactions on February 1, 2022. The file for this day uses the following format:

Filename: `2022-02-01.csv`

The filename consists of the specific date in the format YYYY-MM-DD. In this case, 2022-02-01 represents the date of the transactions contained in the file.

The actual content of the file will be the transaction data for that specific day, including details such as customer ID, transaction amount, POS terminal ID, and any other relevant information.

This file format, where each day's transactions are stored in separate files with the date included in the filename, helps in organizing and managing the dataset chronologically.

It enables easy retrieval and analysis of transaction data from specific dates, facilitating time-based analysis and reporting tasks:

TRANSACTION_ID	TX_DATETIME	CUSTOMER_ID	TERMINAL_ID	TX_AMOUNT
0	2023-02-01 00:43:37	901	8047	82
1	2023-02-01 01:20:13	2611	7777	15
2	2023-02-01 01:22:52	4212	3336	53
3	2023-02-01 01:26:40	1293	7432	59
4	2023-02-01 01:52:23	2499	1024	25
5	2023-02-01 02:11:03	2718	168	68
6	2023-02-01 02:11:56	2998	5513	80

Table 5.1 – First few rows of transactions data

TRANSACTION_ID	TX_DATETIME	CUSTOMER_ID	TERMINAL_ID	TX_AMOUNT
24901	2023-02-02 01:34:52	4999	4536	43
24902	2023-02-02 01:44:39	580	3511	29
24903	2023-02-02 01:48:04	3309	7661	50
24904	2023-02-02 01:58:12	2919	5322	94
24905	2023-02-02 02:07:07	3868	3217	97
24906	2023-02-02 02:08:43	1822	489	15

Table 5.2 – Last few rows of transactions data

For our use case scenario, where the core exercise is to createdifferentially private aggregates , let's assume that the system receives data covering several months, and the objective is to generate differentially private aggregates of transactions for each POS terminal.

POS Terminal	Differentially private aggregates		
	Count	Sum	Average
1			
2			

POS Terminal	Differentially private aggregates		
3			
4			
..			

Table 5.3 – Differentially private aggregates

While the use of frameworks such as Pytorch, pandas, and PyDP can be effective for generating differentially private aggregates, it is true that processing large datasets using these methods alone may be time-consuming and not scalable. However, there are alternative approaches and tools available to address these challenges.

These approaches include the following:

- **Parallel processing**: We could utilize parallel processing techniques to distribute the computation across multiple processors or machines. This can significantly reduce the processing time and enable scalability when dealing with large datasets.

- **Distributed computing**: We could employ distributed computing frameworks such as Apache Spark or Hadoop to handle big data processing. These frameworks provide distributed data processing capabilities, allowing for efficient processing of large-scale datasets.

- **Cloud computing**: We could leverage cloud computing platforms such as **Amazon Web Services** (**AWS**) or **Google Cloud Platform** (**GCP**) to harness the power of scalable infrastructure. These platforms offer services such as Amazon EMR, Google Dataproc, or Azure HDInsight, which can handle large-scale data processing in a cost-effective and scalable manner.

- **Optimized differential privacy libraries**: We could explore specialized differential privacy libraries, such as PipelineDP, TensorFlow Privacy, or Opacus, that are designed to provide efficient and scalable implementations of differential privacy algorithms. These libraries offer optimizations specific to privacy-preserving computations, enabling faster, more scalable processing.

- **Data partitioning and pre-aggregation**: We could divide the data into manageable partitions and perform pre-aggregation to reduce the overall computational load. This approach can improve performance by minimizing the amount of data processed at each step.

By incorporating these approaches and tools, it is possible to overcome the challenges of processing large datasets when generating differentially private aggregates. These methods can significantly reduce the processing time and enhance scalability, enabling organizations to efficiently analyze and derive insights from their data while preserving privacy.

Here are some questions/designs to consider:

- How can you solve or generate DP aggregates on large datasets?

- How can you partition the data? (i.e., partitioning based on a date or on a POS terminal)

- How can you apply DP within a given partition?

  - How do you find out the maximum and minimum value bounds to cacluate the (clip) function

  - Should we apply DP (bounds/sensitivity) for each data value in a partition to generate the DP aggregates? OR

  - Should we generate statistical aggregates first and then apply DP within the partition?

PipelineDP high-level architecture

The PipelineDP framework is designed to address the questions and considerations mentioned earlier when generating differentially private aggregates for large datasets.

PipelineDP (`https://pipelinedp.io`) is an open source framework that supports generating differentially private aggregates on large datasets using open source frameworks such as Apache Spark and Apache Beam. The PipelineDP framework was developed by Google in collaboration with OpenMined. As of writing this book, the PipelineDP team has a disclaimer that this framework isn't recommended for production deployments but is considered good enough for development mode and demonstration purposes to enhance your understanding.

Figure 5.9 – PipelineDP architecture

PipelineDP provides APIs in three modes (Apache Spark, Apache Beam, and local mode) and access to their corresponding implementations through the DP engine.

Here are the key concepts used in PipelineDP:

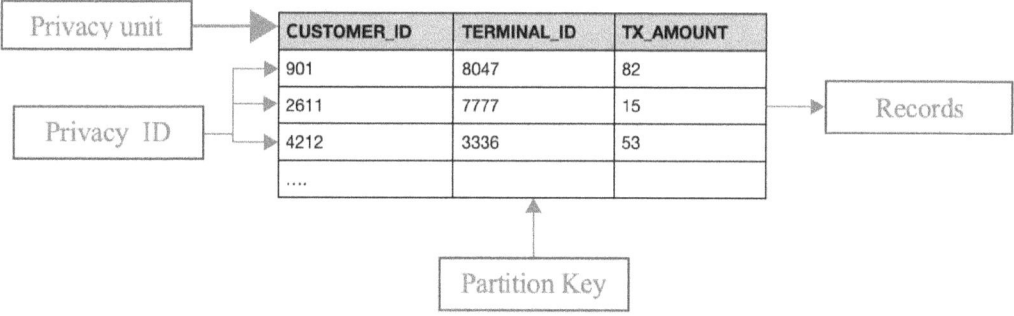

Figure 5.10 – Example to show privacy unit, privacy-id, and partition keys for sample data

- **Record** is an element in the input dataset in PipelineDP.

- **Partition** is a subset of the data corresponding to a given value of the aggregation criterion. In our case, we want the results per POS terminal, so the partition will be TERMINAL_D.

- **Partition key** is the aggregation key corresponding to a partition. TERMINAL_D is the partition key in this example.

- **Privacy unit** is the entity that we want to protect with differential privacy.

- **Privacy ID** is an identifier of a privacy unit. In our example, CUSTOMER_IDis the privacy ID.

Now that we understand the high-level architecture and key concepts, let's implement differential privacy on a large dataset using PipelineDP.

The following line of code installs the PipelineDP framework.

```
pip install PipelineDP
```

Source code : Large_Data_Sets-DP_PipelineDP.ipynb

```
Import pipeline_dp
Import pandas as pd
Import numpy as np
url ="transactions.csv"
df_actual = pd.read_csv(url, sep=",")
df_transactions = df_actual[['TRANSACTION_ID', 'TX_
DATETIME','CUSTOMER_ID','TERMINAL_ID','TX_AMOUNT']]
df_transactions
```

	TRANSACTION_ID	TX_DATETIME	CUSTOMER_ID	TERMINAL_ID	TX_AMOUNT
0	0	2023-02-01 00:43:37	901	8047	82
1	1	2023-02-01 01:20:13	2611	7777	15
2	2	2023-02-01 01:22:52	4212	3336	53
3	3	2023-02-01 01:26:40	1293	7432	59
4	4	2023-02-01 01:52:23	2499	1024	25
...	...	...	...	...	...
4557161	4557161	2023-08-02 21:46:12	1465	7455	92
4557162	4557162	2023-08-02 21:47:08	4009	3429	36
4557163	4557163	2023-08-02 21:54:43	1336	3116	50
4557164	4557164	2023-08-02 22:02:05	1611	3314	81
4557165	4557165	2023-08-02 22:02:13	4367	2968	21

4557166 rows × 5 columns

Figure 5.11 – Transactions data

```
rows =
[index_row[1] for index_row in transactions_df.iterrows()]
```

The following code will produce the differentially private counts (total number of customers) for each terminal:

```
#In this example we use a local backend, but Spark and Apache
#backends also can be tried in a similar way by making use of the
provided classes.

backend = pipeline_dp.LocalBackend()

# Define the total budget.

budget_accountant = pipeline_dp.NaiveBudgetAccountant(total_epsilon=1,
total_delta=1e-6)

# Create DPEngine which will execute the logic to generate the
aggregates

dp_engine = pipeline_dp.DPEngine(budget_accountant, backend)

# Define privacy ID, partition key, and aggregated value extractors.

# The aggregated value extractor isn't used for Count aggregates, but
this is required for SUM, AVERAGE aggregates

data_extractors = pipeline_dp.DataExtractors(
    partition_extractor=lambda row: row.TERMINAL_ID,
    privacy_id_extractor=lambda row: row.CUSTOMER_ID,
    value_extractor=lambda row: 1)

# Configure the aggregation parameters. Number of partitions is 10000
because the number of terminals is 10,000

params = pipeline_dp.AggregateParams(
    noise_kind=pipeline_dp.NoiseKind.LAPLACE,
    metrics=[pipeline_dp.Metrics.COUNT],
    max_partitions_contributed=100,
    max_contributions_per_partition=10)
public_partitions=list(range(1, 10000))

#Create a computational graph for the aggregation.

dp_result = dp_engine.aggregate(rows, params, data_extractors, public_
```

```
partitions)

#Compute budget per each DP operation.

budget_accountant.compute_budgets()
dp_result = list(dp_result)
dp_dict=dict(dp_result)
myKeys = list(dp_dict.keys())
myKeys.sort()
sorted_dict = {i: dp_dict[i] for I in myKeys}
print(sorted_dict)
dp_count = [0] * 100
for count_sum_per_day in dp_result:
  index =  count_sum_per_day[0] - 1
  dp_count[index] = count_sum_per_day[1][0]
  print(dp_count[index])
```

Now generate the actual counts and compare them with differential privacy ones:

```
df_counts = df_transactions.groupby(by='TERMINAL_ID').agg('count')
df_counts
```

	TRANSACTION_ID
TERMINAL_ID	
0	375
1	267
2	412
3	472
4	500
...	...
9995	348
9996	335
9997	499
9998	493
9999	361

10000 rows × 4 columns

Figure 5.12 – Transactions aggregates

By utilizing the PipelineDP framework, we can address the challenges and considerations involved in generating differentially private aggregates for large datasets. It provides a comprehensive solution that combines scalability, privacy preservation, and accurate aggregation, allowing us to effectively leverage differential privacy techniques for large-scale data analysis.

Tumult Analytics

Tumult Analytics is a robust and feature-rich Python library designed for performing aggregate queries on tabular data while ensuring the principles of differential privacy. The library offers an intuitive interface, making it accessible to users familiar with SQL or PySpark. It provides a wide range of aggregation functions, data transformation operators, and privacy definitions, ensuring versatility in analytical tasks. Developed and maintained by a team of experts in differential privacy, Tumult Analytics guarantees reliability and is even utilized in production environments by reputable institutions such as the U.S. Census Bureau. Powered by Spark, the library demonstrates excellent scalability, enabling efficient processing of large datasets. With its comprehensive functionality and emphasis on privacy preservation, Tumult Analytics is a valuable tool for data analysis with a focus on maintaining data privacy.

The following is the citation for the Tumult Analytics open source framework:

```
@software{tumultanalyticssoftware,
    author = {Tumult Labs},
    title = {Tumult {{Analytics}}},
    month = dec,
    year = 2022,
    version = {latest},
    url = {https://tmlt.dev}
}
```

Installation of Tumult Analytics

To utilize Tumult Analytics, it is essential to have Python installed, as the library is built using Python. It is compatible with Python versions 3.7 to 3.10. Additionally, since Tumult Analytics leverages PySpark for its computations, it is necessary to have Java 8 or 11 installed as well.

The following code installs the Tumult Analytics framework.

```
pip3 install tmlt.analytics
```

Key features of Tumult Analytics

Tumult Analytics provides classes and methods to build aggregates on large datasets using differential privacy. Some of the high-level classes are as follows:

Class	Methods
Session (tmlt.analytics.session)	The Session module offers a convenient interface for managing data sources and conducting differentially private queries on them. Creating a session is straightforward, with Session.from_dataframe() for a simple session involving a single private data source or Session.Builder for more complex scenarios involving multiple data sources. Once the session is set up, queries can be executed on the data using Session.evaluate(). When initializing instance type of Session , PrivcyBudget is specified to ensure that the queries performed on the private data do not exceed this allocated budget. By default, the Session instance enforces privacy protection at the row level. This means that the queries prevent any potential attacker from deducing whether individual rows have been added or removed from the private tables. However, this privacy guarantee applies only to the queries themselves and assumes that the private data is not utilized elsewhere in the computation process.
PureDPBudget (tmlt.analytics.privacy_budget)	A privacy budget that provides pure differential privacy.
ApproxDPBudget tmlt.analytics.privacy_budget)	A privacy budget that provides approximate differential privacy.
QueryBuilder (tmlt.analytics.query_builder)	A high-level interface for specifying DP queries. The QueryBuilder class can apply transformations, such as joins or filters, as well as compute aggregations including counts, sums, and standard deviation.
KeySet (tmlt.analytics.keyset)	A KeySet specifies a list of values for one or more columns. Currently, KeySets are used as a simpler way to specify domains for groupby transformations.

Table 5.4 – tmlt.analytics.session description

Example application with Tumult Analytics

Let's generate the aggregates using tmlt-analytics on the large dataset used in the previous section (i.e., our transaction data).

Source code: DP_Large_Data_Sets_TMLT.ipynb

Let's import the required Python packages as follows:

```
import os
from pyspark import SparkFiles
from pyspark.sql import SparkSession
from tmlt.analytics.privacy_budget import PureDPBudget
from tmlt.analytics.protected_change import AddOneRow
from tmlt.analytics.query_builder import QueryBuilder
from tmlt.analytics.session import Session
```

Next, initialize the Spark session:

```
spark = SparkSession.builder.getOrCreate()
```

Let's load the dataset that contains information about credit card transactions. We get the data from the local directory and load it into a Spark DataFrame:

#Create a downloads directory and copy the transactions.csv file by unzipping the transactions.csv.zip file:

```
https://github.com/PacktPublishing/Privacy-Preserving-Machine-
Learning/blob/main/Chapter%205/transactions.csv.zip
```

```
spark.sparkContext.addFile(
    "/downloads/transactions.csv")
trans_df = spark.read.csv(
    SparkFiles.get("/downloads/transactions.csv"), header=True,
inferSchema=True
)
trans_df.head(5)
```

```
[Row(_c0=0, TRANSACTION_ID=0, TX_DATETIME=datetime.datetime(2023, 2,
1, 0, 43, 37), CUSTOMER_ID=901, TERMINAL_ID=8047, TX_AMOUNT=82, TX_
TIME_SECONDS=2617, TX_TIME_DAYS=0),
 Row(_c0=1, TRANSACTION_ID=1, TX_DATETIME=datetime.datetime(2023, 2,
1, 1, 20, 13), CUSTOMER_ID=2611, TERMINAL_ID=7777, TX_AMOUNT=15, TX_
TIME_SECONDS=4813, TX_TIME_DAYS=0),
 Row(_c0=2, TRANSACTION_ID=2, TX_DATETIME=datetime.datetime(2023, 2,
1, 1, 22, 52), CUSTOMER_ID=4212, TERMINAL_ID=3336, TX_AMOUNT=53, TX_
TIME_SECONDS=4972, TX_TIME_DAYS=0),
 Row(_c0=3, TRANSACTION_ID=3, TX_DATETIME=datetime.datetime(2023, 2,
1, 1, 26, 40), CUSTOMER_ID=1293, TERMINAL_ID=7432, TX_AMOUNT=59, TX_
TIME_SECONDS=5200, TX_TIME_DAYS=0),
 Row(_c0=4, TRANSACTION_ID=4, TX_DATETIME=datetime.datetime(2023, 2,
1, 1, 52, 23), CUSTOMER_ID=2499, TERMINAL_ID=1024, TX_AMOUNT=25, TX_
TIME_SECONDS=6743, TX_TIME_DAYS=0)]
```

Initiating a tmlt-analytics session

To perform queries using Tumult Analytics, the data needs to be encapsulated within a `Session` object, which facilitates query tracking and management. The following code snippet demonstrates how to create a `Session` by wrapping a DataFrame containing private data using the `from_dataframe()` method:

```
session = Session.from_dataframe(
    privacy_budget=PureDPBudget(3.5),
    source_id="transactions",
    dataframe=trans_df,
    protected_change=AddOneRow(),
)
```

When a session is initialized with a finite privacy budget, it offers a straightforward interface promise: all queries executed on this session, collectively, will yield differentially private outcomes with a maximum epsilon value of 3.5. Epsilon serves as a metric for quantifying potential privacy loss, where a lower epsilon implies a more stringent constraint on privacy loss and, consequently, a higher level of protection. In this context, the interface promise corresponds to a privacy guarantee, ensuring a minimum level of safeguarding for private data.

In addition to the data itself, several additional pieces of information are required:

- The `privacy_budget` parameter specifies the privacy guarantee that the session will provide.

- The `source_id` parameter serves as an identifier for the DataFrame. It will be used to reference this specific DataFrame when constructing queries.

- The `protected_change` parameter defines the unit of data for which the differential privacy guarantee is applied. In this example, `AddOneRow()` is used to protect individual rows within the dataset.

Exeucting DP queries using Session

Our first query finds the number of total transactions in the data with DP enabled:

```
count_query = QueryBuilder("transactions").count()
```

To execute the query on the desired private data, we utilize `QueryBuilder("transactions")` in the first step, which indicates the specific source data (data source) we want to query, corresponding to the `source_id` parameter specified earlier. In the following line, the `count()` statement is used to retrieve the total number of records in the dataset. Once the query is constructed, we proceed to run it on the data by employing the `evaluate` method of our session. To accomplish this, we allocate a privacy budget to the evaluation process. In this case, we evaluate the query with differential privacy, setting ε=1 as the privacy parameter:

```
total_count = session.evaluate(
    count_query,
  privacy_budget=PureDPBudget(epsilon=1)
)
```

The results of the query are returned as a Spark DataFrame. We can see them using the `show()` method of this DataFrame. We have utilized 1 out of 3.5 of our allocated privacy budget for this query, so the remaining privacy budget will be 2.5:

```
total_count.show()

+-------+
|  count|
+-------+
|4557168|
+-------+
```

If you are following along with the example and executing the code, you may observe varying values. This variation is a fundamental aspect of differential privacy, as it introduces randomization (referred to as noise) during query execution. To showcase this characteristic, let's proceed to evaluate the same query one more time and see the result.

The amount of noise added to the query computation can vary based on the privacy parameters, the type of aggregation, and the underlying data. However, in many instances, the query result still provides reliable insights into the original data. In this particular scenario, we can confirm this by executing a count query directly on the original DataFrame, which will yield the true and accurate result:

```
total_count = trans_df.count()
print(total_count)
```

```
4557166
```

Playing with privacy budgets

Describe the session object to know the attributes and the remaining privacy budget.

```
session.describe()
```

```
The session has a remaining privacy budget of
PureDPBudget(epsilon=2.5).
```

```
The following private tables are available:
Table 'transactions' (no constraints):
  Columns:
     - '_c0'               INTEGER
     - 'TRANSACTION_ID'    INTEGER
     - 'TX_DATETIME'       TIMESTAMP
     - 'CUSTOMER_ID'       INTEGER
     - 'TERMINAL_ID'       INTEGER
     - 'TX_AMOUNT'         INTEGER
     - 'TX_TIME_SECONDS'   INTEGER
     - 'TX_TIME_DAYS'      INTEGER
```

Utilizing privacy budgets with privacy queries

Let's find out the number of customers whose purchases are worth less than $25.

This query consumes epsilon=1 out of our total budget:

```
low_purchagers= QueryBuilder("transactions").filter("TX_AMOUNT < 25").
count()
low_purchagers_count = session.evaluate(
    low_purchagers,
    privacy_budget=PureDPBudget(epsilon=1),
)
low_purchagers_count.show()
```

```
+-------+
|  count|
+-------+
|1024844|
+-------+
```

```
print(session.remaining_privacy_budget)
```

PureDPBudget(epsilon=1.5)

We have utilized 1 unit of our remaining total privacy budget of 2.5, so there is 1.5 left. Let's try another query to consume another 1 unit of the privacy budget:

Let's find out the number of customers whose purchases are greater than $25 but less than $50.

```
med_purchagers= QueryBuilder("transactions").filter("TX_AMOUNT >25 AND
TX_AMOUNT <50").count()
med_purchagers_count = session.evaluate(
```

```
    med_purchagers,
    privacy_budget=PureDPBudget(epsilon=1),
)
med_purchagers_count.show()
```

```
+-------+
|  count|
+-------+
|1165384|
+-------+
```

```
print(session.remaining_privacy_budget)
```

PureDPBudget(epsilon=0.5)

We have utilized a budget of 3 out of a total of 3.5, so there is 0.5 left. Let's try another query to consume the privacy budget of 1 (run the query more than the available budget and observe the results).

```
high_purchagers= QueryBuilder("transactions").filter("TX_AMOUNT >
50").count()
high_purchagers_count = session.evaluate(
    high_purchagers,
    privacy_budget=PureDPBudget(epsilon=1),
)
high_purchagers_count.show()
```

It will throw a runtime error as follows:

```
-----------------------------------------------------------------------
------
InsufficientBudgetError                         Traceback (most recent call
last)
/usr/local/lib/python3.10/dist-packages/tmlt/analytics/session.py in
evaluate(self, query_expr, privacy_budget)
    1283               try:
->  1284                     answers = self._accountant.measure(
    1285                         measurement, d_out=adjusted_budget.value

/usr/local/lib/python3.10/dist-packages/tmlt/core/measurements/
interactive_measurements.py in measure(self, measurement, d_out)
    1343            if not self._privacy_budget.can_spend_budget(d_out):
->  1344                 raise InsufficientBudgetError(
    1345                     self.privacy_budget,
```

```
InsufficientBudgetError: PrivacyAccountant's remaining privacy budget
is 1/2, which is insufficient for this operation that requires privacy
loss 1.
RuntimeError: Cannot answer query without exceeding the Session
privacy budget.
Requested: ε=1.000
Remaining: ε=0.500
Difference: ε=0.500
```

Let's display the remaining privacy budget for use in the subsequent queries.

```
print(session.remaining_privacy_budget)
```

```
PureDPBudget(epsilon=0.5)
```

Rewrite the last query so that it will make use of the remaining available budget so that the privacy budget is used completely, instead of wasting the unused budget. In this case, we will not specify the privacy budget as 1 or 2, but will make use of the remaining privacy budget from the `Session` class itself:

```
high_purchagers= QueryBuilder("transactions").filter("TX_AMOUNT >
50").count()
high_purchagers_count = session.evaluate(
    high_purchagers,
    privacy_budget=session.remaining_privacy_budget,
)
high_purchagers_count.show()
```

```
+-------+
| count|
+-------+
|2271804|
+-------+
```

Total number of high purchase counts by applying differential privacy.

Groupby queries

In Tumult Analytics, the KeySet class is utilized to define the list of groupby keys. It allows us to specify both the columns by which we intend to group the data and the potential values associated with those columns. The KeySet class serves as a convenient means of specifying the grouping criteria in Tumult Analytics.

Let's now write a query to find the average transaction amount on each terminal (taking just the first 10 terminals for this example).

Create a session with the privacy budget set to 2.5:

```
budget = PureDPBudget(epsilon=2.5) # maximum budget consumed in the
Session
session = Session.from_dataframe(
    privacy_budget=budget,
    source_id="transactions",
    dataframe=trans_df,
    protected_change=AddOneRow(),
)
```

Make use of the KeySet class and define the groupby column (TERMINAL_ID) or columns and values as well, to filter:

```
from tmlt.analytics.keyset import KeySet
terminal_ids = KeySet.from_dict({
    "TERMINAL_ID": [
        1,2,3,4,5,6,7,8,9,10
    ]
})
```

Execute the query and provide the lower and upper bounds to clip TX_AMOUNT:

```
average_purchase_query = (
    QueryBuilder("transactions")
    .groupby(terminal_ids)
    .average("TX_AMOUNT", low=5, high=100)
)
average_purchages= session.evaluate(
    average_purchase_query,
    privacy_budget=PureDPBudget(1),
)
average_purchages.show(truncate=False)
```

```
+-----------+------------------+
|TERMINAL_ID|TX_AMOUNT_average |
+-----------+------------------+
|1          |55.93609022556391 |
|2          |52.93446601941748 |
|3          |40.95974576271186 |
|4          |52.02414486921529 |
|5          |47.511428571428574|
|6          |52.276595744680854|
|7          |51.566233766233765|
```

```
|8               |50.12273641851107 |
|9               |52.88358208955224 |
|10              |48.98945147679325 |
+----------+------------------+
```

In this way, we can generate the aggregates using differential privacy, ensuring they are not the same as the actual aggregates.

Queries using privacy IDs

Previously, we focused on working with tables where each individual in the dataset was linked to a single row. However, this is not always the case. In certain datasets, it is possible for the same individual to appear in multiple rows. In such cases, it is typical to assign a unique identifier to each person (i.e., across different rows). The objective then shifts toward concealing whether all the rows associated with a particular identifier are present in the data.

Tumult Analytics refers to these identifiers as privacy IDs. Each privacy ID corresponds to a one-to-one mapping with a person or any other entity that requires protection. The aim is to safeguard the privacy of individuals or entities by preserving the anonymity of their presence within the dataset. This can be achieved by making use of the `AddRowsWithID` protected change. This protected change will prevent arbitrarily adding and removing many rows all sharing the same ID.

Initializing a session with the privacy ID:

```python
from tmlt.analytics.protected_change import AddRowsWithID
budget = PureDPBudget(epsilon=2.5) # maximum budget consumed in the
Session
session = Session.from_dataframe(
    privacy_budget=budget,
    source_id="transactions",
    dataframe=trans_df,
    protected_change=AddRowsWithID(id_column="CUSTOMER_ID"),
)
```

Execute the query with the privacy ID:

```python
keyset = KeySet.from_dataframe(
    trans_df.select("TERMINAL_ID", "TX_AMOUNT")
)
count_query = (
    QueryBuilder("transactions")
    .groupby(keyset)
    .count()
)
result = session.evaluate(count_query, PureDPBudget(1))
```

Let's execute this code and observe the output:

```
RuntimeError: A constraint on the number of rows contributed by each
ID is needed to perform this query (e.g. MaxRowsPerID).
```

This error arises due to the absence of a constraint on the number of rows a single individual can contribute to the dataset. It is possible for a single customer to do many transactions, even exceeding 1,000 or more. However, differential privacy necessitates concealing the influence of an individual's data through the introduction of statistical noise.

In order to address this issue, it is necessary to establish a restriction on the maximum influence that a single customer can exert on the computed statistic before conducting aggregations. This constraint is enforced on the data to mitigate any potential impact. The most straightforward constraint, known as MaxRowsPerID, restricts the total number of rows contributed by each privacy ID. To enforce this constraint, we can simply pass it as a parameter to the enforce() operation. For the specific query at hand, we will set the maximum number of contributed rows per library member to 100:

```
from tmlt.analytics.constraints import (
    MaxGroupsPerID,
    MaxRowsPerGroupPerID,
    MaxRowsPerID,
)
keyset = KeySet.from_dataframe(
    trans_df.select("TERMINAL_ID", "TX_AMOUNT")
)
count_query = (
    QueryBuilder("transactions")
    .enforce(MaxRowsPerID(100))
    .groupby(keyset)
    .count()
)
result = session.evaluate(count_query, PureDPBudget(1))
top_five = result.sort("count", ascending=False).limit(5)
top_five.show()
```

```
+-----------+----------+-----+
|TERMINAL_ID|TX_AMOUNT|count|
+-----------+----------+-----+
|       3001|       98| 1240|
|       3536|       42| 1217|
|       4359|       71| 1212|
|       9137|       97| 1145|
|       7179|       76| 1143|
+-----------+----------+-----+
```

```
result.show()
```

```
+----------+---------+-----+
|TERMINAL_ID|TX_AMOUNT|count|
+----------+---------+-----+
|         0|        4|  401|
|         0|        7|  224|
|         0|       11|   -7|
|         0|       12|  131|
|         0|       16|  -35|
|         0|       18|  -68|
|         0|       20| -126|
|         0|       24|  -46|
|         0|       26| -162|
|         0|       28|  -30|
|         0|       31|  447|
|         0|       33|   23|
|         0|       35|   20|
|         0|       44|   96|
|         0|       49|  -56|
|         0|       51|  211|
|         0|       58|  -88|
|         0|       59|  -27|
|         0|       60| -254|
|         0|       61|  525|
+----------+---------+-----+
```

The preceding output only showing the top 20 rows.

Tumult Analytics provides filters, joins, and transformations of the data to execute complex queries, apply differential privacy to large datasets, and make use of Spark distributed processing. We have covered the basic key features, but there are many more features to explore based on the use cases in the application/system.

Machine learning using differential privacy

In this section, our objective is to develop a machine learning classification model that can accurately distinguish between fraudulent and genuine credit card transactions. To ensure privacy protection, we will also apply differential privacy techniques to the model. The classification model will be trained on a labeled dataset consisting of historical credit card transactions, where each transaction is labeled as either fraudulent or genuine. Popular machine learning algorithms such as logistic regression, decision trees, or neural networks can be applied to build the classification model and will make use of neural networks in our case.

To incorporate differential privacy, we will leverage techniques such as the addition of noise to the training process and the use of privacy-preserving algorithms. These techniques ensure that the model's training process and subsequent predictions do not compromise the privacy of individual transactions or sensitive customer information.

By integrating differential privacy into the classification model, we can provide robust privacy guarantees while maintaining high accuracy in identifying fraudulent transactions. This ensures that the model can effectively protect customers' privacy and prevent unauthorized access to sensitive financial data.

Throughout this section, we will explore the steps involved in training the classification model, evaluating its performance, and applying differential privacy techniques to enhance privacy protection. By the end, we will have a powerful model capable of accurately classifying credit card transactions as either fraudulent or genuine while ensuring the privacy of the individuals involved.

Synthetic Dataset Generation: Introducing Fraudulent Transactions

We will make use of the same transaction data used earlier, add another column to the dataset called TX_FRAUD, and mark any transaction greater than $75 as fraudulent. This obviously doesn't reflect the real world, but we will generate our example synthetic data using this rule. In this dataset, roughly 25% of the data is marked as fraudulent transactions while 75% of the data is genuine. In a real-world scenario, fraudulent transactions will likely be less than 1% in most datasets, which is highly imbalanced.

Source code: Fraud_Transactions_Generator.ipynb

```
import pandas as pd
import numpy as np
url="transactions.csv"

df_actual = pd.read_csv(url, sep=",")
df_actual.head()
df_transactions = df_actual[['TRANSACTION_ID', 'TX_
DATETIME','CUSTOMER_ID','TERMINAL_ID','TX_AMOUNT']]
df_transactions
df_transactions.insert(5, 'TX_FRAUD', 0, True)
df_transactions
```

TRANSACTION_ID	TX_DATETIME	CUSTOMER_ID	TERMINAL_ID	TX_AMOUNT	TX_FRAUD
0	2023-02-01 00:43:37	901	8047	82	0
1	2023-02-01 01:20:13	2611	7777	15	0
2	2023-02-01 01:22:52	4212	3336	53	0
3	2023-02-01 01:26:40	1293	7432	59	0
4	2023-02-01 01:52:23	2499	1024	25	0
...	...	...	...	...	...
4557161	2023-08-02 21:46:12	1465	7455	92	0
4557162	2023-08-02 21:47:08	4009	3429	36	0
4557163	2023-08-02 21:54:43	1336	3116	50	0
4557164	2023-08-02 22:02:05	1611	3314	81	0
4557165	2023-08-02 22:02:13	4367	2968	21	0

Figure 5.13 – Transactions Data

4557166 rows × 6 columns

```
df_transactions.loc[df_transactions.TX_AMOUNT>75, 'TX_FRAUD']=1
nb_frauds=df_transactions.TX_FRAUD.sum()
print("Number of fraud transaction",nb_frauds)
```

Number of fraud transaction 1106783

```
df_transactions.head()
```

TRANSACTION_ID	TX_DATETIME	CUSTOMER_ID	TERMINAL_ID	TX_AMOUNT	TX_FRAUD
0	2023-02-01 00:43:37	901	8047	82	1
1	2023-02-01 01:20:13	2611	7777	15	0
2	2023-02-01 01:22:52	4212	3336	53	0
3	2023-02-01 01:26:40	1293	7432	59	0
4	2023-02-01 01:52:23	2499	1024	25	0

Figure 5.14 – First few rows of transactions data

```
df_transactions.to_csv("fraud_transactions.csv")
```

Develop a classification model using scikit-learn

The following are the high-level steps for developing a classification model:

1.   Load the fraud_transactions dataset.

2. Split the dataset into train and test in the ratio of (70:30).

3. Initialize the classifier (logistic regression from sci-kit learn).

4. Train the classifier with the training data (70% of the transactions).

5. Find out the accuracy of the classifier with the test dataset (30% of the transactions).

6. Find out the calculated weights/coefficient used in the `decision` function and intercept from the logistic regression model.

Source code: Noise_Gradient_Final.ipynb

```
from sklearn.linear_model import LogisticRegression
from sklearn.model_selection import train_test_split
from sklearn.model_selection import cross_val_score
import numpy as np
import pandas as pd
url = "fraud_transactions.csv"

df_actual = pd.read_csv(url, sep=",")
df_transactions =
df_actual[['CUSTOMER_ID','TERMINAL_ID','TX_AMOUNT','TX_FRAUD']]
df_transactions
```

	CUSTOMER_ID	TERMINAL_ID	TX_AMOUNT	TX_FRAUD
0	901	8047	82	1
1	2611	7777	15	0
2	4212	3336	53	0
3	1293	7432	59	0
4	2499	1024	25	0
...	...	...	...	...
4557161	1465	7455	92	1
4557162	4009	3429	36	0
4557163	1336	3116	50	0
4557164	1611	3314	81	1
4557165	4367	2968	21	0

4557166 rows × 4 columns

Figure 5.15 – Fraud transactions dataset

```
print('No Frauds', round(df_transactions['TX_FRAUD'].value_counts()
[0]/len(df_transactions) * 100,2), '% of the dataset')
```

```
print('Frauds', round(df_transactions['TX_FRAUD'].value_counts()[1]/
len(df_transactions) * 100,2), '% of the dataset')
```

No Frauds 75.71 % of the dataset
Frauds 24.29 % of the dataset

```
X = df_transactions.drop('TX_FRAUD', axis=1)
y = df_transactions['TX_FRAUD']
X_train, X_test, y_train, y_test = train_test_split(X, y, test_
size=0.3, random_state=42)

# Turn the values into an array for feeding the classification
algorithms.
X_train = X_train.values
X_test = X_test.values
y_train = y_train.values
y_test = y_test.values
logreg = LogisticRegression(random_state=0)
logreg.fit(X_train, y_train)

training_score = cross_val_score(logreg, X_train, y_train, cv=2)
print('Logistic Regression Cross Validation Score: ',
round(training_score.mean() * 100, 2).astype(str) + '%')
```

Logistic Regression Cross Validation Score: 100.0%

```
np.sum(logreg.predict(X_test) == y_test)/X_test.shape[0]
```

1.0

```
logreg.intercept_[0], logreg.coef_[0]
```

(-1168.308115256604,
 array([-2.47724513e-05, 3.17749573e-06, 1.54748556e+01]))

Once we know the gradients/coefficients of the classifier and intercept as well, then it will be easy to calculate the predictions. In our case, we have three features, CUSTOMER_ID, TERMINAL_ID, and TRANSACTON_AMOUNT. The linear equation will come with three features:

$$y = w1 * x1 + w2 * x2 + w3 * x3 + b$$

Once we know the feature values (x1, x2, x3…. xn), weights (w1, w2, w3, …, wn), and b value (bias /intercept), then we can calculate the y-hat value (predictions). Logistic regression uses a logistic function to estimate/predict the probabilities. In our case, it will be the following:

$$\text{y-hat} = 1.0 \; / \; 1.0 + e^{\,-(w1\,*\,x1\; +\; w2\,*\,x2\; +\; w3\,*\,x3\; +\; b)}$$

We will take one actual transaction and calculate the prediction value using the weights obtained from the model:

CUSTOMER_ID	TERMINAL_ID	TX_AMOUNT	TX_FRAUD
79	3115	78	1

Table 5.16 – Prediction for CUSTOMER_ID

```
data=[79,3115,78]
weights = [-2.47724513e-05, 3.17749573e-06, 1.54748556e+01]
intercept = -1168.308115256604

def predict(data,coefficients,intercept):
    yhat = intercept
    for i in range(len(data)):
        yhat += coefficients[i] * data[i]
    return 1.0 / (1.0 + np.exp(-yhat))
yhat = predict(data,weights,intercept)
yhat
```

```
1.0
```

Logistic regression uses the stochastic gradient descent algorithm internally in order to calculate gradients/coefficients.

Let's implement our own stochastic gradient descent algorithm to calculate the weights instead of using the scikit-learn-provided one in the logistic regression model.

High-level implementation of the SGD algorithm

The steps to implement the SGD algorithm are as follows:

1. Initialize the initial weights with all zeros (one zero for each feature and the bias/intercept also as zero): Initial weights = [0,0,0] and intercept=0

2. Do the following for each row in the training data:

    A. Calculate the predictions based on the initial weights and intercept.

    B. Find out the error between the actual value and the predicted value:

    Error = Actual Value – Predicted Value

    C. Update the intercept based on the error and the learning rate value:

    intercept= intercept + l_rate * error * yhat * (1.0 - yhat)

    D. Update the weights for all training data in the training set.

3. Based on the number of epochs, repeat the preceding steps.

```
def predict(data,coefficients,intercept):
    yhat = intercept
    for i in range(len(data)):
        yhat += coefficients[i] * data[i]
    return 1.0 / (1.0 + np.exp(-yhat))


def final_gradients(gradients):
    length_grads=len(grads)
    avg_grads=[0,0,0]
    for i in range(0,length_grads):
        avg_grads[0]+=grads[i][0]
        avg_grads[1]+=grads[i][1]
        avg_grads[2]+=grads[i][2]
    avg_grads=[i/length_grads for i in avg_grads]

    return avg_grads

def sgd(train,y_train,l_rate, n_epoch):
    coef = [0,0,0]
    final_grads = [0,0,0]
    intercept = 0
    for epoch in range(n_epoch):
        predictions=[]
        gradients=[]
        sum_error = 0.0
        for i in range(len(train)):
            yhat = predict(train[i], coef,intercept)
            predictions.append(yhat)
            error = y_train[i] - yhat
            sum_error += error**2
            intercept= intercept + l_rate * error * yhat * (1.0 -
```

```
yhat)
## intercept
            temp=train[i]
            for j in range(3):
                coef[j] = coef[j] + l_rate * error * yhat * (1.0 -
yhat) * temp[j]
            gradients.append(coef)
        final_grads = final_gradients(gradients)
        print('>epoch=%d, lrate=%.3f, error=%.3f, intercept=%.3f '%
(epoch, l_rate, sum_error,intercept))
    return final_grads


l_rate = 0.24
n_epoch = 4
coef = sgd(X_train[:10],y_train[:10],l_rate, n_epoch)
print(coef)



>epoch=0, lrate=0.240, error=2.250, intercept=-0.030
>epoch=1, lrate=0.240, error=2.000, intercept=-0.030
>epoch=2, lrate=0.240, error=2.000, intercept=-0.030
>epoch=3, lrate=0.240, error=2.000, intercept=-0.030
[-136.44000000000003, -263.88000000000005, -0.5099999999999999]
```

In this way, we can calculate the coefficients. The final coefficients are calculated as the average of all coefficients/gradients.

Applying differential privacy options using machine learning

Applying differential privacy to the preceding algorithm means generating gradients with differential privacy so that the model doesn't reveal the details of the training examples. When applying differential privacy to SGD, the objective is to incorporate privacy protection into the training process of the machine learning model. This involves adding noise to the gradients computed during the SGD optimization steps to ensure that the trained model does not reveal specific details about any individual data point.

The addition of noise in SGD with differential privacy helps prevent potential privacy breaches by making it difficult to distinguish the impact of any particular training example on the model's updates. It ensures that the model's parameters do not memorize or overfit specific training samples, thereby offering privacy guarantees for the individuals whose data was used in the training process.

Generating gradients using differential privacy

There are two approaches for generating gradients using differential privacy in the context of machine learning.

Approach 1:

1. Generate the final gradients using the normal SGD algorithm on the training data.

2. Calculate the sum of the gradients obtained from the preceding SGD step.

3. Add noise to the sum of the gradients using the Laplace mechanism, taking into account the desired sensitivity and privacy budget.

Approach 2:

1. Apply the clipping method to each training example or the overall training data.

2. Generate gradients using the clipped training data inputs.

3. Compute the sum of the gradients obtained in the previous step.

4. Add noise to the sum of the gradients using either the Gaussian or Laplace mechanism, considering the desired sensitivity and privacy budget.

5. Calculate the count of the training examples using differential privacy, treating it as a count query with sensitivity set to 1 and utilizing the required privacy budget.

6. Compute the average of the noisy gradients sum and the noisy count to obtain the differentially private gradients average.

Both approaches aim to incorporate differential privacy into the gradient calculation process, thereby protecting the privacy of individual training examples while training the machine learning model. The choice between the two approaches depends on the specific requirements of the application and the desired level of privacy guarantees.

By adding appropriate noise and applying privacy-preserving mechanisms, these approaches ensure that the gradients used for updating the model parameters do not reveal sensitive information about individual training examples. This enables the training process to provide privacy guarantees while still achieving accurate and reliable model performance.

Let's now implement *Approach 2* for our fraud detection example.

As we know, clipping is the process of setting the lower and upper bounds for the data, so implement the following `clip` function:

```
def clip(iv, b):
    norm = np.linalg.norm(iv, ord=2)
    if norm > b:
        return b * (iv / norm)
    else:
        return iv
print( clip([[4548, 8796,   17]],5.0) )
```

```
[[2.29645183 4.44142267 0.00858392]]
```

```
clip(X_train[:5], 5)
```

```
array([[1.35772596e+00, 2.62589215e+00, 5.07505304e-03],
       [3.40625619e-01, 2.83605905e-02, 1.55236917e-02],
       [1.41504420e+00, 1.98583840e+00, 1.52251591e-02],
       [1.06964206e+00, 1.67446897e+00, 2.08972772e-02],
       [6.95282267e-01, 2.50737474e+00, 1.19413013e-03]])
```

```
def dp_final_gradients(gradients):

    length_grads=len(grads)
    sensitivity = 1
    epsilon= 0.8
    noise = np.random.laplace(loc=0, scale=sensitivity/epsilon)
    noise_lenth = length_grads + noise
    avg_grads=[0,0,0]
    for i in range(0,length_grads):
        avg_grads[0]+=grads[i][0]
        avg_grads[1]+=grads[i][1]
        avg_grads[2]+=grads[i][2]
        avg_grads=[i/noise_lenth for i in avg_grads]
    return avg_grads

def dp_sgd(train,y_train,l_rate, n_epoch):

    train = clip(train, 5)
    coef = [0,0,0]
    final_grads = [0,0,0]
    intercept = 0
    for epoch in range(n_epoch):
        predictions=[]
        gradients=[]
        sum_error = 0.0
        for i in range(len(train)):
            yhat = predict(train[i], coef,intercept)
            predictions.append(yhat)
            error = y_train[i] - yhat
            sum_error += error**2
            intercept= intercept + l_rate * error * yhat * (1.0 -
```

```
yhat)
## intercept
            temp=train[i]
            for j in range(3):
                coef[j] = coef[j] + l_rate * error * yhat * (1.0 -
yhat) * temp[j]
            gradients.append(coef)
        final_grads = dp_final_gradients(gradients)
        print('>epoch=%d, lrate=%.3f, error=%.3f, intercept=%.3f '%
(epoch, l_rate, sum_error,intercept))
    return final_grads
l_rate = 0.24
n_epoch = 4
print("Gradients using Normal SGD ")
coef = sgd(X_train[:10],y_train[:10],l_rate, n_epoch)
print("Gradients using Differentially Private SGD ")
coef = dp_sgd(X_train[:10],y_train[:10],l_rate, n_epoch)
print(coef)
```

```
Gradients using Normal SGD

>epoch=0, lrate=0.240, error=2.250, intercept=-0.030
>epoch=1, lrate=0.240, error=2.000, intercept=-0.030
>epoch=2, lrate=0.240, error=2.000, intercept=-0.030
>epoch=3, lrate=0.240, error=2.000, intercept=-0.030

[-136.44000000000003, -263.88000000000005, -0.5099999999999999]


Gradients using Differentially Private SGD
>epoch=0, lrate=0.240, error=2.146, intercept=-0.127
>epoch=1, lrate=0.240, error=1.654, intercept=-0.193
>epoch=2, lrate=0.240, error=1.478, intercept=-0.229
>epoch=3, lrate=0.240, error=1.396, intercept=-0.249

[-115.01700212848986, -222.44713076565455, -0.42992283117509383]
```

By combining SGD with differential privacy, as we have done here, we can develop machine learning models that not only provide accurate predictions but also offer privacy protection for sensitive data. It enables organizations to leverage large-scale datasets while adhering to privacy regulations and ethical considerations.

Clustering using differential privacy

Let's consider a scenario where we have a dataset of users' browsing behavior. The goal of k-means clustering in this context is to identify k points, referred to as cluster centers, that minimize the sum of squared distances of the data points from their nearest cluster center. This partitioning allows us to group the users based on their browsing patterns. Additionally, we can assign new users to a group based on the closest cluster center. However, the release of the cluster centers could potentially reveal sensitive information about specific users. For instance, if a particular user's browsing behavior is significantly different from the majority, the standard k-means clustering algorithm might assign a cluster center specifically to this user, thereby disclosing sensitive details about their browsing habits. To address this privacy concern, we will implement clustering with differential privacy. By doing so, we aim to protect the privacy of individual users while still providing meaningful clustering results based on their browsing behavior.

First, to illustrate the nature of the problem, let's generate our cluster centroids without differential privacy:

Source code: Clustering_Differential_Privacy_diffprivlib.ipynb

```
import numpy as np
from scipy import stats
from sklearn.cluster import KMeans

# Example dataset
data = np.array([[1, 2], [1.5, 1.8], [5, 8], [8, 8], [1, 0.6], [9,
11]])

#apply clustering on this dataset and cluster the data 2 clusters
kmeans = KMeans(n_clusters=2)
kmeans.fit(data)
clusters = kmeans.labels_
original_centroids = kmeans.cluster_centers_

# Print the original data points, clusters and centroids
print("Original Data Points:\n", data)
print("Clusters:\n", clusters)
print("Original Centroids:\n", original_centroids)
```

Original Data Points:

```
[[ 1.   2. ]
 [ 1.5  1.8]
 [ 5.   8. ]
 [ 8.   8. ]
 [ 1.   0.6]
```

```
[ 9.  11. ]]
```

Clusters:

```
[1 1 0 0 1 0]
```

Original Centroids:

```
[[7.33333333 9.         ]

 [1.16666667 1.46666667]]
```

In the preceding example, we have a synthetic dataset with two sensitive data attributes. We perform clustering using the KMeans algorithm from scikit-learn without incorporating differential privacy. We retrieve the cluster centroids using the `cluster_centers_` attribute.

The problem with this approach is that the cluster centroids, which represent the center of each cluster, can reveal sensitive information about the data. In this case, the cluster centroids could potentially expose the mean values of the sensitive attributes. To address this privacy concern, differential privacy techniques can be applied to add noise to the cluster centroids, making it more challenging to infer sensitive information. However, note that the application of differential privacy in clustering algorithms requires careful consideration of privacy-utility trade-offs and appropriate privacy parameter selection.

Let's generate the centroids by adding noise to the datasets.

Follow these steps:

1. Define the `add_noise` function, which takes the original data, privacy parameter epsilon, and sensitivity as inputs. It generates Laplace-distributed noise and adds it to the data points. The noise is scaled by the sensitivity and privacy parameter epsilon.

2. Calculate the sensitivity, which is the maximum change in data points due to the addition or removal of a single data point. In this case, we calculate the maximum absolute difference between any data point and the mean of the dataset.

3. Specify the privacy parameter epsilon, which determines the amount of noise to be added. Add noise to the data points using the `add_noise` function.

4. Perform clustering on the noisy data using k-means clustering with two clusters. Retrieve the cluster labels assigned by the algorithm. Print the original data points, noisy data points, and the clusters assigned to each point.

```python
def add_noise(data, epsilon, sensitivity):
    beta = sensitivity / epsilon
    laplace_noise = np.random.laplace(0, beta, data.shape)
    noisy_data = data + laplace_noise
```

```
        return noisy_data

# Sensitivity is the maximum change in data points due to the
addition or removal of a single data point
sensitivity = np.max(np.abs(data - np.mean(data, axis=0)))

# Privacy parameter epsilon determines the amount of noise to be added
epsilon = 0.1

# Add noise to the data points
noisy_data = add_noise(data, epsilon, sensitivity)

# Perform clustering on the noisy data
kmeans = KMeans(n_clusters=2)
kmeans.fit(noisy_data)
noisy_clusters = kmeans.labels_
noise_centroids = kmeans.cluster_centers_
print("Noisy Data Points:\n", noisy_data)
print("Noisy Clusters:\n", noisy_clusters)
print("Noisy Centroids :\n", noise_centroids)


Noisy Data Points:

 [[ -8.22894996 -25.09225801]
 [ 48.29852161 -93.63432789]
 [  2.61671234  86.87531981]
 [ 10.03114688   7.72529685]
 [-27.57009962  59.88763296]
 [ 16.99705384 -94.28428515]]

Noisy Clusters:
 [1 1 0 0 0 1]

Noisy Centroids :
 [[ -4.97408014  51.49608321]
 [ 19.0222085  -71.00362369]]
```

Generating differentially private centroids using IBM's diffprivlib framework as an alternative method to the same use case "browsing behaviour scenario".

The diffprivlib framework is a Python library that provides tools and algorithms for performing differentially private data analysis. Diffprivlib consists of four main components that contribute to its functionality:

- **Mechanisms**: These components serve as the fundamental building blocks of differential privacy and are utilized in all models implementing differential privacy. Mechanisms in diffprivlib are customizable and designed for use by experts who are implementing their own models.

- **Models**: This module encompasses machine learning models integrated with differential privacy. Diffprivlib provides a variety of models that implement differential privacy, including clustering, classification, regression, dimensionality reduction, and preprocessing. These models are designed to ensure privacy while performing their respective tasks.

- **Tools**: Diffprivlib offers a range of generic tools designed for differentially private data analysis. These tools provide functionalities such as differentially private histograms, which adhere to the same format as NumPy's histogram function. They enable users to perform various data analysis tasks while preserving privacy.

- **Accountant**: The accountant component includes the BudgetAccountant class, which facilitates the tracking of privacy budgets and calculation of total privacy loss using advanced composition techniques. This feature is crucial for managing and controlling privacy expenditure across multiple differentially private operations, ensuring that privacy guarantees are maintained. Together, these components in diffprivlib contribute to a comprehensive framework for implementing differential privacy in various data analysis scenarios. They provide the necessary tools, models, mechanisms, and privacy accounting capabilities to support privacy-preserving data analysis tasks.

Let's generate cluster centroids using the diffprivlib framework.

Install the diffprivlib framework using pip:

```
!pip3 install diffprivlib
```

Follow these steps:

1. Import the KMeans class from diffprivlib.models.

2. Specify the epsilon privacy parameter=, which determines the strength of privacy protection.

3. Create an instance of KMeans with the epsilon parameter and the desired number of clusters.

4. Fit the differentially private KMeans model to the data using the fit method.

5. Finally, retrieve the differentially private cluster centroids using the cluster_centers_ attribute of the KMeans model.

```
from diffprivlib.models import KMeans
epsilon = 1.0
```

```
# Perform clustering with differential privacy

dp_kmeans = KMeans(epsilon=epsilon, n_clusters=2)
dp_kmeans.fit(data)

# Get the differentially private cluster centroids

dp_centroids = dp_kmeans.cluster_centers_

# Print the differentially private cluster centroids

print("Differentially Private Cluster Centroids:\n", dp_centroids)
```

```
Differentially Private Cluster Centroids:
 [[8.71915573 9.51643083]
 [5.96366996 3.84980361]]
```

It's important to note that differential privacy in diffprivlib operates under the assumption of a trusted curator, where privacy guarantees are provided if the curator follows the privacy-preserving mechanisms correctly.

However, for a complete implementation of differential privacy, additional considerations, such as the choice of appropriate privacy parameters and the impact on data utility, should be carefully addressed.

Deep learning using differential privacy

In this section, we will focus on developing a fraud detection model using the PyTorch framework. Additionally, we will train deep learning models with differential privacy using open source frameworks such as PyTorch and Opacus. Using the PyTorch framework, we will develop a deep learning model specifically designed for fraud detection. PyTorch is a popular open source deep learning library that provides a flexible and efficient platform for building and training neural networks. Its rich set of tools and APIs make it well-suited for developing sophisticated machine learning models.

To incorporate differential privacy into the training process, we will utilize the Opacus library. Opacus is an open source PyTorch extension that provides tools for training deep learning models with differential privacy. It offers mechanisms such as gradient clipping, noise addition, and privacy analysis, which help ensure that the trained model preserves the privacy of individual data points.

By combining PyTorch and Opacus, we can train deep learning models with differential privacy for fraud detection. This approach allows us to benefit from the expressive power of deep learning while adhering to privacy regulations and protecting sensitive information. Throughout this section, we will explore techniques for data preprocessing, model architecture design, training, and evaluation. We will consider the unique challenges and considerations associated with fraud detection, such as imbalanced datasets, feature engineering, and performance evaluation metrics.

By the end of this section, you will have a comprehensive understanding of how to develop a fraud detection model using PyTorch and train it with differential privacy using frameworks such as Opacus. This knowledge will empower you to build robust and privacy-preserving machine learning models for fraud detection and similar applications.

Fraud detection model using PyTorch

In order to develop a deep learning model using PyTorch, we can follow the steps outlined next:

1. **Load the transaction data**: Start by loading the transaction data into a pandas DataFrame object.

2. **Split the data**: Split the loaded data into training and testing sets.

3. **Convert the data to PyTorch tensors**: To work with PyTorch, we need to convert the data into PyTorch tensors. PyTorch tensors are efficient data structures that allow us to perform computations using the GPU for accelerated training. We use the `torch.tensor` function to convert the data.

4. **Create a simple linear model**: Define a deep learning model architecture using PyTorch's `nn.Module` class. For a simple linear model, we use the `nn.Linear` module to create a linear layer. To classify transactions as fraud or not, we add a sigmoid layer at the end of the model using the `nn.Sigmoid` activation function.

5. **Train the model**: Set up the training loop to iterate over the training data and update the model's parameters based on the defined loss function and optimization algorithm. Use PyTorch's `nn.CrossEntropyLoss` as the loss function and select an appropriate optimizer such as `torch.optim.SGD` or `torch.optim.Adam` for updating the model's parameters.

6. **Monitor the loss**: During training, keep track of the loss at each step. The loss represents the discrepancy between the predicted outputs of the model and the true labels. By monitoring the loss, you can assess the progress of the model's training and make adjustments if necessary.

By following these steps, we develop a deep learning model using PyTorch for fraud detection. It's important to note that this is a simplified overview, and you may need to customize the model architecture, incorporate additional layers or techniques, and fine-tune hyperparameters based on the specific requirements of your fraud detection task.

Source Code: Fraud_Detection_Deep Learning.ipynb

 # unzip the fraud_transactions.csv.zip file (provided in the GitHub repo of this book as fraud_transactions.csv)

```
https://github.com/PacktPublishing/Privacy-Preserving-Machine-Learning/blob/main/Chapter%205/fraud_transactions.csv.zip
```

```
import pandas as pd
import torch
url ="fraud_transactions.csv"
```

```
df_actual = pd.read_csv(url, sep=",")
df_actual.head()
df_transactions = df_actual[['CUSTOMER_ID','TERMINAL_ID','TX_
AMOUNT','TX_FRAUD']]
df_transactions=df_transactions.head(50000)
df_transactions
from sklearn.model_selection import train_test_split
from sklearn.model_selection import StratifiedShuffleSplit


print("No of Fraud Transactions:", df_transactions['TX_FRAUD'].value_
counts()[0])
print("No of Non Fraud Transactions:", df_transactions['TX_FRAUD'].
value_counts()[1])
print('No Frauds', round(df_transactions['TX_FRAUD'].value_counts()
[0]/len(df_transactions) * 100,2), '% of the dataset')
print('Frauds', round(df_transactions['TX_FRAUD'].value_counts()[1]/
len(df_transactions) * 100,2), '% of the dataset')
```

No of Fraud Transactions: 37870
No of Non Fraud Transactions: 12130
No Frauds 75.74 % of the dataset
Frauds 24.26 % of the dataset

```
X = df_transactions.drop('TX_FRAUD', axis=1)
y = df_transactions['TX_FRAUD']
X_train, X_test, y_train, y_test = train_test_split(X, y, test_
size=0.3, random_state=42)


# Convert into Pytorch Tensors

x_train = torch.FloatTensor(X_train.values)
x_test = torch.FloatTensor(X_test.values)
y_train = torch.FloatTensor(y_train.values)
y_test = torch.FloatTensor(y_test.values)


if torch.cuda.is_available():
    DEVICE = "cuda"
else:
    DEVICE = "cpu"
print("Selected device is",DEVICE)
class FraudDataset(torch.utils.data.Dataset):
    def __init__(self, x, y):
```

```
            'Initialization'
            self.x = x
            self.y = y

    def __len__(self):
        'Returns the total number of samples'
        return len(self.x)

    def __getitem__(self, index):
        'Generates one sample of data'
        # Select sample index
        if self.y is not None:
            return self.x[index].to(DEVICE), self.y[index].to(DEVICE)
        else:
            return self.x[index].to(DEVICE)

train_loader_params = {'batch_size': 64,
        'shuffle': True,
        'num_workers': 0}
test_loader_params = {'batch_size': 64,
        'num_workers': 0}

# Loaders

training_set = FraudDataset(x_train, y_train)
testing_set = FraudDataset(x_test, y_test)
train_loader = torch.utils.data.DataLoader(training_set, **train_
loader_params)
test_loader = torch.utils.data.DataLoader(testing_set, **test_loader_
params)

class SimpleFraudMLP(torch.nn.Module):

    def __init__(self):
        super().__init__()

        self.first_sec = torch.nn.Sequential(
                    torch.nn.Linear(3, 450),
                    torch.nn.ReLU(),
        )
        self.second_sec = torch.nn.Sequential(
                    torch.nn.Linear(450, 450),
                    torch.nn.ReLU(),
                    torch.nn.Linear(450, 1),
```

```
                              torch.nn.Sigmoid(),
        )

    def forward(self, x):
        return self.second_sec(self.first_sec(x))

fraud_nn_model = SimpleFraudMLP().to(DEVICE)
from torch import nn, optim
loss_func = torch.nn.BCELoss().to(DEVICE)
optimizer = torch.optim.SGD(fraud_nn_model.parameters(), lr = 0.07)

def train(fraud_nn_mode,num_epochs):

    fraud_nn_model.train()

    for epoch in range(num_epochs):
            for x_batch, y_batch in train_loader:
                output = fraud_nn_model(x_train)
                print(output.squeeze())
                print(y_train)
                loss = loss_func(output.squeeze(), y_train)

# clear gradients for this training step
                optimizer.zero_grad()
# backpropagation, compute gradients
                loss.backward()
# apply gradients
                optimizer.step()
                print(epoch, loss.item())
        pass
train (fraud_nn_model, 10)


tensor([0., 0., 0.,  ..., 0., 0., 0.], grad_fn=<SqueezeBackward0>)
tensor([1., 1., 1.,  ..., 0., 0., 0.])
0 24.191429138183594
```

Fraud detection model with differential privacy using the Opacus framework

Opacus, an open source library, is the PyTorch implementation of the SGD-DP algorithm that supports differential privacy. Opacus preserves the privacy of each training sample while limiting the impact on the accuracy of the final model. In this way, the privacy of outliers is also preserved. Opacus adds noise to the gradients in every iteration to prevent the model from simply memorizing the training examples. Opacus adds noise at the right scale (too much noise will reduce the accuracy, and too little won't help to protect privacy) by looking at the norm of the gradients.

More details about Opacus can be found at `https://opacus.ai/`.

Installing the Opacus library is done with the following code (I used the following version in this example: `opacus==1.1.2`):

```
pip install opacus==1.1.2
```

We will develop the same deep learning model and train it with differential privacy using the Opacus framework with PyTorch.

Implement the following steps:

1. Load the transaction data.

2. Split the data into train and test data sets using pandas DataFrames.

3. Convert the data into PyTorch tensors.

4. Create a simple linear model and use the sigmoid layer at the end to classify transactions as fraud or not.

5. Make the model a private one (i.e., apply differential privacy to the model) using the instance of `PrivacyEngine` provided by Opacus to protect the training data.

6. Train the model and measure the epsilon (privacy budget).

Source : Fraud_Detection_DP.ipynb

```
import pandas as pd
url=" fraud_transactions.csv"
df_actual = pd.read_csv(url, sep=",")
df_actual.head()
df_transactions = df_actual[['CUSTOMER_ID','TERMINAL_ID','TX_
AMOUNT','TX_FRAUD']]
df_transactions=df_transactions.head(50000)
df_transactions
from sklearn.model_selection import train_test_split
from sklearn.model_selection import StratifiedShuffleSplit

print("No of Fraud Transactions:", df_transactions['TX_FRAUD'].value_
counts()[0])
print("No of Non Fraud Transactions:", df_transactions['TX_FRAUD'].
value_counts()[1])
print('No Frauds', round(df_transactions['TX_FRAUD'].value_counts()
[0]/len(df_transactions) * 100,2), '% of the dataset')
print('Frauds', round(df_transactions['TX_FRAUD'].value_counts()[1]/
```

```
len(df_transactions) * 100,2), '% of the dataset')

X = df_transactions.drop('TX_FRAUD', axis=1)
y = df_transactions['TX_FRAUD']
X_train, X_test, y_train, y_test = train_test_split(X, y, test_
size=0.3, random_state=42)
x_train = torch.FloatTensor(X_train.values)
x_test = torch.FloatTensor(X_test.values)
y_train = torch.FloatTensor(y_train.values)
y_test = torch.FloatTensor(y_test.values)

if torch.cuda.is_available():
    DEVICE = "cuda"
else:
    DEVICE = "cpu"
print("Selected device is",DEVICE)
class FraudDataset(torch.utils.data.Dataset):
    def __init__(self, x, y):
        'Initialization'
        self.x = x
        self.y = y

    def __len__(self):
        'Returns the total number of samples'
        return len(self.x)

    def __getitem__(self, index):
        'Generates one sample of data'
        # Select sample index
        if self.y is not None:
            return self.x[index].to(DEVICE), self.y[index].to(DEVICE)
        else:
            return self.x[index].to(DEVICE)
train_loader_params = {'batch_size': 64,
        'shuffle': True,
        'num_workers': 0}
test_loader_params = {'batch_size': 64,
        'num_workers': 0}

# Generators

training_set = FraudDataset(x_train, y_train)
testing_set = FraudDataset(x_test, y_test)
train_loader = torch.utils.data.DataLoader(training_set, **train_
```

```
loader_params)
test_loader = torch.utils.data.DataLoader(testing_set, **test_loader_
params)
fraud_nn_model = SimpleFraudMLP().to(DEVICE)

import warnings
warnings.simplefilter("ignore")
MAX_GRAD_NORM = 1.2
EPSILON = 90.0
DELTA = 1e-5
EPOCHS = 20
LR = 1e-3
from opacus.validators import ModuleValidator
errors = ModuleValidator.validate(fraud_nn_model, strict=False)
errors[-5:]
from torch import nn, optim

#loss_func = nn.CrossEntropyLoss()

loss_func = torch.nn.BCELoss().to(DEVICE)
optimizer = torch.optim.SGD(fraud_nn_model.parameters(), lr = 0.07)
from opacus import PrivacyEngine
fraud_nn_model.train()
privacy_engine = PrivacyEngine()
model, optimizer, train_loader = privacy_engine.make_private_with_
epsilon(
    module=fraud_nn_model,
    optimizer=optimizer,
    data_loader=train_loader,
    epochs=EPOCHS,
    target_epsilon=EPSILON,
    target_delta=DELTA,
    max_grad_norm=MAX_GRAD_NORM,
)
print(f"Using sigma={optimizer.noise_multiplier} and C={MAX_GRAD_
NORM}")

import numpy as np
import time

n_epochs = 10
#Setting the model in training mode
fraud_nn_model.train()
```

```
#Training loop
start_time=time.time()
epochs_train_losses = []
epochs_test_losses = []
for epoch in range(n_epochs):
    train_loss=[]
    train_loss1=0
    for x_batch, y_batch in train_loader:

        fraud_nn_model.train()

        y_pred = fraud_nn_model(x_batch)

        loss = loss_func(y_pred.squeeze(), y_batch)

        optimizer.zero_grad()

        loss.backward()

        optimizer.step()

        train_loss.append(loss.item())

        train_loss1 += loss.item()*x_batch.size(0)

    epsilon = privacy_engine.get_epsilon(DELTA)
    print('ε epsilon{}    : delta:{}'.format(epsilon, DELTA))
    epochs_train_losses.append(np.mean(train_loss))
    print('Epoch {}: train loss: {}'.format(epoch, np.mean(train_
loss)))
```

ε epsilon33.98911164791893   : delta:1e-05
Epoch 0: train loss: 22.66661006201338
ε epsilon38.786904746786384   : delta:1e-05
Epoch 1: train loss: 23.087044257350552
ε epsilon42.819749628256126   : delta:1e-05
Epoch 2: train loss: 23.234367423345226
ε epsilon46.852594509725854   : delta:1e-05
Epoch 3: train loss: 23.257508610022786
ε epsilon50.8854393911956   : delta:1e-05
Epoch 4: train loss: 23.949310037727983
ε epsilon54.91828427266533   : delta:1e-05
Epoch 5: train loss: 22.498504093839657

In each epoch, the training loss fluctuates, but at the same time, privacy loss (budget) ε increases.

In this way, deep learning models can be trained for differential privacy with minimal code changes using Opacus and protect the training data's privacy.

Differential privacy machine learning frameworks

The following are some of the popular differential privacy machine learning frameworks:

Framework	Implementation
Opacus	PyTorch.
Tensor Flow Privacy	TensorFlow.
Pyvacy	TensorFlow Privacy, but for PyTorch.
JAX(DP)	JAX is Autograd and XLA, brought together for high-performance machine learning research.
Pysyft	Pysyft is a Python library for private, secure machine learning using federated learning and differential privacy. It allows for secure and private training and inference of machine learning models across multiple devices.

Table 5.17 – DP ML frameworks

Limitations of differential privacy and strategies to overcome them

Differential privacy has gained significant attention and adoption in both academia and industry due to its ability to balance privacy and utility. However, like any other technique, differential privacy has its limitations and challenges that need to be addressed to ensure its effective implementation. Following are some of the major limitations of differential privacy and potential strategies to overcome them:

- **Noise and utility trade-off**: Differential privacy achieves privacy by adding noise to query responses, which introduces a trade-off between privacy and utility. The amount of noise added determines the level of privacy, but excessive noise can significantly reduce the utility of the released data. Striking the right balance between privacy and utility is a challenge.

  - **Strategy to overcome**: One approach to mitigating this limitation is to design better algorithms that minimize the impact of noise on utility. Researchers are constantly developing advanced mechanisms and techniques to optimize the noise injection process, such as adaptive noise calibration, privacy amplification through subsampling, or leveraging machine learning to generate more accurate and privacy-preserving responses.

- **Inference attacks**: Differential privacy focuses on protecting individual privacy by limiting the influence of a single record. However, adversaries may employ sophisticated inference attacks to glean information by combining multiple noisy queries or utilizing external side information. These attacks exploit patterns or correlations present in the data to infer sensitive details.

  - **Strategy to overcome**: To overcome inference attacks, additional privacy-preserving techniques can be combined with differential privacy. For instance, secure **multi-party computation (MPC)** protocols can be used to compute aggregate statistics without revealing individual data points, thereby enhancing privacy protection.

- **Privacy budget exhaustion**: Differential privacy employs a privacy budget, which represents the maximum allowable privacy loss over a sequence of queries. Each query consumes a portion of this budget, and once it is depleted, no further queries can be made while maintaining differential privacy guarantees. This limitation poses a challenge in scenarios where a large number of queries need to be answered.

  - **Strategy to overcome**: One approach to address privacy budget exhaustion is to allocate budgets dynamically based on the sensitivity of the data or the specific context of the queries. By adapting the budget allocation strategy, it is possible to optimize the utility of the released data and extend the number of allowable queries without compromising privacy. Additionally, advanced composition techniques, such as Rényi differential privacy, can be employed to manage privacy budgets more effectively and allow for finer-grained control.

- **External data and auxiliary information**: Differential privacy assumes that the released data is the only source of information available to an adversary. However, adversaries could potentially leverage external data sources or auxiliary information to improve their attacks. These external sources might reveal additional details about individuals or contain correlated data, making it challenging to maintain privacy guarantees.

  - **Strategy to overcome**: To overcome this limitation, it is crucial to carefully analyze the potential impact of external data sources and auxiliary information on privacy. Adapting data integration techniques, such as secure multiparty computation or cryptographic protocols, can help protect against attacks that exploit external information. Moreover, proactive measures such as data de-identification and minimizing data linkage can further enhance privacy protection against such threats.

- **Limited support for complex data types**: Differential privacy has predominantly focused on numerical or categorical data, which limits its applicability to more complex data types such as text, images, or graphs. Preserving privacy in these domains while maintaining meaningful utility poses a challenge.

  - **Strategy to overcome**: Researchers are actively exploring techniques to extend differential privacy to complex data types. For example, for text data, approaches such as differentially private text generation or privacy-preserving NLP models are being developed. For images, techniques such as differentially private deep learning or generative adversarial networks with privacy guarantees are being investigated. These advancements aim to provide privacy guarantees for a wider range of data types. LLMs trained with differential privacy also expose certain privacy leaks. For more information, we strongly suggest reading the research paper at `https://arxiv.org/pdf/2202.05520.pdf`.

- **Limited protection against insider attacks**: Differential privacy primarily focuses on protecting data against external adversaries. However, it may not be as effective in scenarios where insider attacks are a concern. Insiders with access to the raw data might intentionally modify the data or use their knowledge to breach privacy.

  - **Strategy to overcome**: Combining differential privacy with additional security measures can help mitigate insider attacks. Techniques such as secure enclaves or secure hardware can be employed to protect the privacy of sensitive data even from those with direct access. Employing access controls, audit logs, and strict data governance policies can also deter insider threats and ensure accountability. We will learn more about secure enclaves in *Chapter 9*.

- **Difficulty in preserving temporal privacy**: Differential privacy is primarily designed for static datasets, and it can be challenging to preserve privacy when dealing with temporal data or time-series analysis. Temporal correlations in the data can potentially lead to privacy breaches or inference attacks.

  - **Strategy to overcome**: To address temporal privacy concerns, researchers are exploring techniques such as personalized differential privacy, where privacy parameters are adjusted based on an individual's data history. Another approach involves introducing temporal consistency mechanisms that consider the correlation between consecutive queries or time intervals while preserving privacy. These techniques aim to protect privacy in dynamic and evolving datasets.

- **Limited support for machine learning models**: Differential privacy techniques often pose challenges when applied to machine learning models, especially deep learning architectures. The perturbation of model parameters or gradients may degrade the model's performance or introduce vulnerabilities.

  - **Strategy to overcome**: To overcome this limitation, researchers are developing privacy-preserving machine learning techniques tailored to differential privacy. Techniques such as federated learning, where models are trained on decentralized data without sharing sensitive information, can ensure privacy while maintaining utility. Additionally, advancements in privacy-preserving deep learning algorithms, such as differentially private stochastic gradient descent, aim to strike a balance between model performance and privacy guarantees. We will cover federated learning in more depth in the next chapter.

- **Lack of standardization and interoperability**: The absence of standardized frameworks and interoperability can hinder the widespread adoption of differential privacy. Different implementations and approaches make it challenging to compare results or integrate privacy-preserving techniques across different platforms or systems.

  - **Strategy to overcome**: Establishing standardized guidelines and frameworks can help address the interoperability challenge. Organizations and industry consortia can collaborate to develop common APIs, protocols, and evaluation metrics for differential privacy. Efforts in open source libraries and tools, along with community-driven initiatives, can facilitate knowledge sharing and enable seamless integration of differential privacy techniques into existing systems and workflows.

Overall, while differential algorithms have many advantages, they are not a one-size-fits-all solution, and careful consideration must be taken when using them in practical applications.

Summary

In summary, in this chapter, we went through open source frameworks including PyDP, PipelineDP, Tumult Analytics, and PySpark in order to implement differential privacy. We implemented fraud detection machine learning models with and without differential privacy by developing a private stochastic gradient descent algorithm. We also implemented deep learning models and trained the models with differential privacy using the Opacus framework, which is based on PyTorch. Finally, we covered the limitations of differential privacy and strategies to overcome these limitations.

In the next chapter, we'll learn about the need for federated learning as we deep dive into it, covering the algorithms used and the frameworks that support federated learning, and explore an end-to-end implementation of a fraud detection use case using federated learning.

Part 3: Hands-On Federated Learning

This part covers the need for federated learning (FL) and the implementation of FL using open source frameworks. It also touches upon FL benchmarks, start-ups, and future opportunities in the field.

We highlight the importance of FL as a privacy-preserving approach to machine learning and emphasize the need for FL in scenarios where data cannot be centrally aggregated due to privacy concerns or regulatory restrictions. Furthermore, we discuss the implementation of FL using open source frameworks, which provide accessible and customizable tools for deploying FL algorithms and models.

We explore the significance of FL benchmarks for evaluating and comparing FL algorithms and techniques and emphasize the need for standardized benchmarks to assess the performance and effectiveness of FL models across different scenarios. By leveraging FL benchmarks, researchers and practitioners can identify the strengths and limitations of various FL approaches, facilitating advancements in the field.

We explore the presence of start-ups working on FL and highlight their contributions and focus areas. We shed light on the innovative applications and solutions being developed by these start-ups, which further enhance the adoption and development of FL. Additionally, we touch upon future opportunities in FL, including advancements in privacy-preserving techniques, scalability of FL frameworks, and integration with emerging technologies.

Overall, this part of the book emphasizes the need for FL, its implementation using open source frameworks, the importance of FL benchmarks, and the contributions of start-ups in the FL space. It also acknowledges the potential future opportunities that lie ahead, driving further advancements and adoption of FL in various domains.

This part has the following chapters:

- *Chapter 6, Need for Federated Learning and Implementing Federated Learning Using Open Source Frameworks*

- *Chapter 7, Federated Learning Benchmarks, Start-Ups, and Next Opportunities*

6

Federated Learning and Implementing FL Using Open Source Frameworks

In this chapter, you will learn about **Federated Learning** (**FL**) and how to implement it using open source frameworks. We will cover why it is needed and how to preserve data privacy. We will also look at the definition of FL, as well as its characteristics and the steps involved in it.

We will cover the following main topics:

- FL

- FL algorithms

- The steps involved in implementing FL

- Open source frameworks for implementing FL

- An end-to-end use case of implementing fraud detection using FL

- FL with differential privacy

By exploring these topics, you will gain a comprehensive understanding of the need for FL and the open source frameworks for implementing FL.

Federated learning

FL has emerged as a solution to address the challenges of traditional centralized **Machine Learning** (**ML**) approaches in scenarios where data privacy and data locality are of paramount importance.

The key reasons that we need FL are as follows:

- **Preserving data privacy**: In many situations, data is sensitive and cannot be shared due to legal, ethical, or privacy concerns. FL enables you to train models directly on distributed data sources without sharing the raw data, ensuring privacy protection. By keeping data local and performing model updates locally, FL minimizes the risk of exposing sensitive information.

- **Data localization and regulatory compliance**: FL allows organizations to comply with data localization requirements and regulations. Instead of transferring data to a central server, data remains within the jurisdiction where it is generated or collected, addressing concerns related to cross-border data transfers.

- **Scalability and efficiency**: Centralized machine learning approaches often face challenges when dealing with large volumes of data, as aggregating and processing data from various sources can be time-consuming and resource-intensive. FL distributes the training process, allowing data to remain decentralized while benefiting from the collective intelligence of all participating devices or data sources. This decentralized approach improves scalability and computational efficiency.

- **Access to diverse data**: FL facilitates the pooling of data from multiple sources, enabling models to learn from diverse datasets without the need for direct data sharing. This is particularly beneficial in scenarios where data sources have distinct characteristics, such as different demographics, geographical regions, or user preferences. Access to a diverse range of data enhances the generalization and robustness of ML models.

- **Enhanced security and resilience**: With FL, the data remains distributed across devices or edge nodes, reducing the risk of a single point of failure or vulnerability. This distributed nature enhances the security and resilience of the overall system, making it less susceptible to attacks or breaches.

- **User empowerment and inclusion**: FL offers opportunities for user participation and control over their data. Instead of relinquishing data ownership and control to a centralized authority, users can actively contribute to the learning process while retaining control over their personal information. This empowers individuals and promotes a sense of inclusion and transparency.

The need for FL arises from the critical requirements of preserving data privacy, complying with regulatory frameworks, achieving scalability and efficiency, accessing diverse data sources, ensuring security, and empowering users.

By leveraging FL, organizations can overcome the limitations of centralized approaches and unlock the potential of distributed data for training robust and privacy-preserving ML models.

Preserving privacy

Let's consider the case of ARDHA Bank (a fictional bank for illustration purposes only). ARDHA Bank is a financial institution that has been operating in the United States for several years, adhering to country-specific regulations. The bank offers a range of services to its customers, including fraud prevention, loyalty programs, and digital payments. Initially, ARDHA Bank employed static rule-based systems to detect and prevent fraudulent activities. However, recognizing the need for more advanced approaches, they transitioned to utilizing ML algorithms for enhanced fraud detection and prevention.

With access to a comprehensive dataset comprising historical and current transaction data, ARDHA Bank developed ML and **Deep Learning** (**DL**) algorithms specifically tailored to their operations. These algorithms were trained on this extensive dataset, allowing the bank to effectively identify and prevent financial fraud with exceptional accuracy. By leveraging the power of ML and DL techniques, ARDHA Bank significantly improved its ability to detect and mitigate fraudulent digital transactions, thereby safeguarding its customers' financial interests.

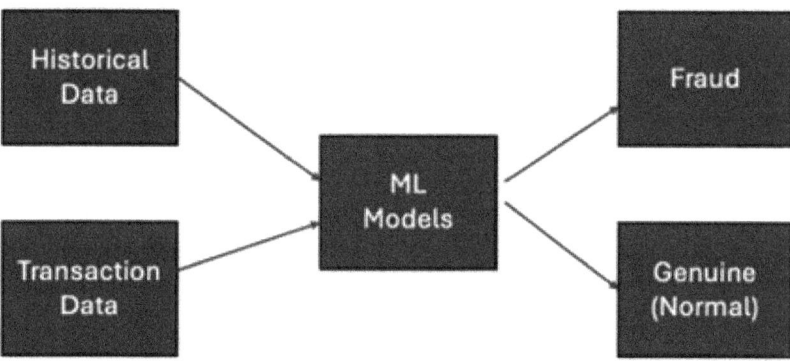

Figure 6.1 – A simple ML model in a financial bank

ARDHA Bank, having experienced success in the **United States** (**US**), made the strategic decision to expand its business and establish branches in two additional countries – France (for Europe) and India. With the expansion, ARDHA Bank aimed to offer the same suite of services to its customers in both regions. To provide digital payment services in France and India, one option considered by ARDHA Bank was to transmit periodic transaction data from both countries to their US servers. The US servers would then serve as the central location to run the ML models. After training the ML models on the combined data from all regions, the trained models would be deployed to the regional servers in France and India. By adopting this approach, ARDHA Bank sought to leverage the infrastructure of its US servers to process and analyze the transaction data efficiently. The centralized training of ML models allowed for a unified approach to fraud detection and prevention, ensuring consistency and accuracy across different regions. This strategy enabled ARDHA Bank to provide reliable and effective digital payment services in Europe and India while maintaining data security and privacy. By utilizing regional servers and deploying the trained ML models locally, the bank ensured swift and localized decision-making, catering to the specific needs and regulatory requirements of each region.

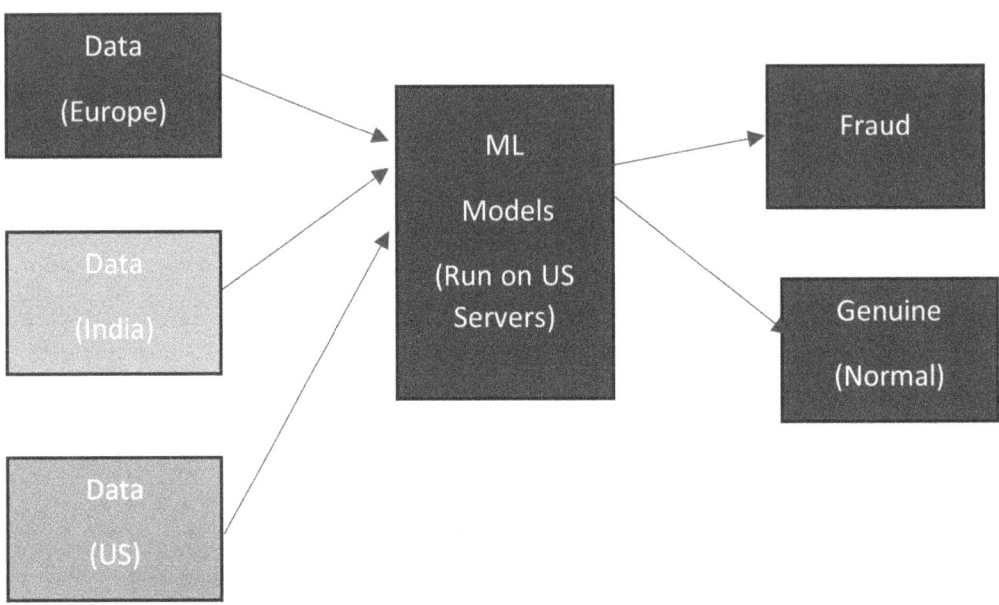

Figure 6.2 – A simple ML model in a financial bank in three locations

The proposed solution, which involves transferring data to a central server and running ML models on that data, faces challenges due to privacy regulations and data localization laws in Europe and India. These regulations, such as the **General Data Protection Regulation** (**GDPR**) in Europe and India's data localization requirements, stipulate that data generated within these countries must be stored within local data centers. Data must remain within the borders of the country where it was created.

Given these privacy and localization constraints, an alternative approach is necessary. One possible alternative is to run ML models locally at each branch or location of the bank. This approach entails deploying client models that utilize the local data available at each location. The local models would process the data within the boundaries of the respective country, ensuring compliance with privacy regulations. To implement this alternative, only the model weights and parameters, not the transaction data used by customers, would be shared with a central server. The central server, hosted in any country, would be responsible for running a global model using the aggregated model weights and parameters from each location. The resulting global model could then be regularly distributed back to the local clients in each country. This approach enables the bank to leverage the benefits of ML models while adhering to privacy regulations and data localization laws. By conducting ML computations locally and sharing only model-related information, the bank ensures compliance, data security, and privacy. Additionally, this distributed approach allows for local adaptation and customization while still benefiting from the insights gained through the global model.

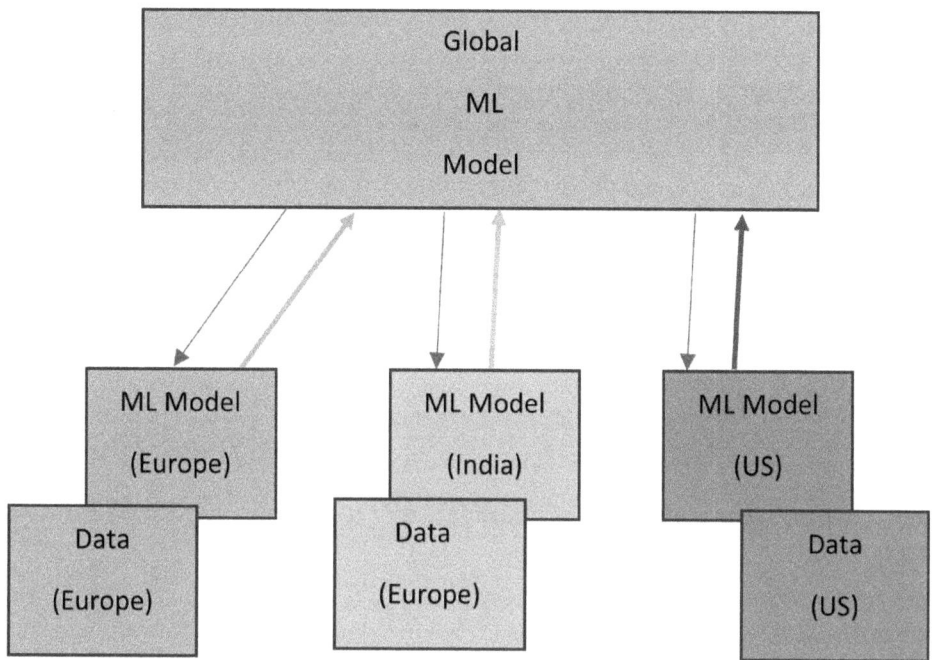

Figure 6.3 – Local model interactions with the global model

This approach is known as **Federated Machine Learning**, or FL. In FL, the traditional paradigm of moving data to a central location is reversed. Instead, the model and computation are brought to the data.

In FL, the ML model is deployed and executed directly on the local data sources or devices where the data resides. This eliminates the need to transfer raw data to a central server, addressing privacy concerns and regulatory requirements. The model is trained locally using the data on each device, and only the model updates, such as gradients or weights, are securely transmitted to a central aggregator.

By keeping the data decentralized and performing computations locally, FL ensures data privacy and reduces the risks associated with data transfer. It allows organizations to leverage the collective knowledge and insights from distributed data sources without compromising individual data privacy. This approach is particularly beneficial in scenarios where data cannot be easily shared due to legal, regulatory, or privacy constraints.

FL represents a paradigm shift in ML, enabling collaborative and privacy-preserving model training. It promotes a distributed approach where data remains under the control of the data owners while contributing to a shared model. This decentralized and privacy-conscious framework opens up possibilities to harness the power of large-scale data without sacrificing privacy and security.

FL definition

The following is the formal definition of FL proposed as per the *Advances and Open Problems in Federated Learning* paper published at `arxiv/1912.04977`:

> *"Federated learning is a machine learning setting where multiple entities (clients) collaborate in solving a machine learning problem, under the coordination of a central server or service provider. Each client's raw data is stored locally and not exchanged or transferred; instead, focused updates intended for immediate aggregation are used to achieve the learning objective"*

As per this definition, these are the characteristics of FL:

- Multiple clients (entities) collaborate to solve an ML problem.

- A service provider or central server coordinates with these entities.

- Raw data (data with samples) is stored locally at each client location and is not transferred to the servers.

- The learning objective (or loss function) is defined. To minimize the loss (predictions versus actual), focused updates (weights and biases) are sent to the server from clients, the aggregation of weights (either average or dynamic aggregation) is done at the server, and these updates are sent back to clients.

Let's delve further into each one of these in detail to understand them better.

Characteristics of FL

The following subsections will cover the characteristics of FL in depth.

Multiple clients (entities) collaborate to solve an ML problem

In FL, the participation requirement typically involves a minimum of two clients, while the maximum number of clients can vary based on the specific use cases and client types.

Clients participating in FL can be broadly classified into two categories – cross-device and cross-silo:

- **Cross-device** clients are individual devices, such as smartphones, laptops, or IoT devices, that contribute their local data for model training. These devices act as clients in the FL framework, allowing their data to be utilized while preserving privacy.

- **Cross-silo** clients, on the other hand, represent data sources that are distributed across different organizational silos or entities. These silos can be different departments within an organization, separate institutions, or even distinct geographical regions. Each silo acts as a client, contributing its local data for collaborative model training.

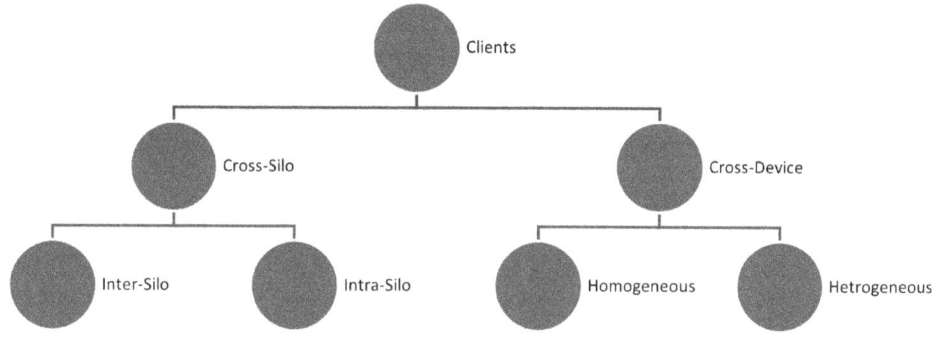

Figure 6.4 – A classification of FL clients

The maximum number of clients in an FL setup depends on the specific use cases and the scale of the distributed data sources. For instance, in scenarios where multiple organizations collaborate to build a global model while maintaining data privacy, the number of participating clients can be substantial. On the other hand, in more focused or localized use cases, the number of clients may be limited to a smaller group.

Cross-silo FL clients

Cross-silo clients are entities such as financial banks, institutions, hospitals, and pharmacy companies. These clients can be further categorized into two groups:

- **Different clients within the same institution**: This includes different branches of the same bank, different branches within a hospital network, and similar setups where multiple branches or divisions of a single institution participate in FL.

- **Different clients across different institutions**: This involves different organizations, such as different banks or hospitals, collaborating and contributing their data to the FL process. These clients represent inter-institutional collaborations.

The maximum number of clients in the cross-silo category can vary based on the specific use case, but typically, it ranges from tens to hundreds. The number of participating clients is usually limited due to the nature of collaborations and the scale of the institutions involved.

Cross-device FL clients

Cross-device clients, on the other hand, encompass various devices that participate as clients or nodes in FL. These devices can be either homogenous or heterogenous, and examples include devices such as Apple iPhones, Google phones, and the Brave browser.

In the case of cross-device FL clients, each device runs its own ML model based on the local data available on that specific device. Only the model weights and biases are transmitted to the server based on device conditions and other configuration settings.

In this scenario, the maximum number of clients can reach thousands or even millions, as it encompasses a wide range of devices participating in FL across different locations and user bases.

By accommodating both cross-silo and cross-device clients, FL enables collaboration and knowledge sharing while respecting data privacy and ensuring scalable participation across institutions and devices.

A service provider or central server coordinates with these entities

The server in FL makes decisions based on the network topology of the participating clients and the total number of clients involved in the process. The server determines when to distribute the initial model or updated models to the clients, considering factors such as the network structure and the specific number of participating clients. It decides whether to send the model updates to all clients or only a subset of them, based on the requirements of the learning task.

After the clients receive the model, they compute and update the weights and biases based on their local data. The clients then send these updated weights and biases back to the server. The server aggregates the received data and performs computations using an objective function to minimize the loss or optimize the learning objective.

Based on the aggregated information, the server generates an updated model. It decides which clients need to be updated with the new model and which clients can continue running the existing model without any changes. This decision is based on factors such as the learning progress, the need for updates, or the compatibility of clients with the updated model.

By carefully orchestrating these steps, the server manages the distribution of models, collects client updates, aggregates data, and ultimately, sends back the updated model to the appropriate clients. This iterative process in FL ensures collaborative model improvement while accounting for the individual requirements and capabilities of the participating clients.

Raw data (data with samples) is stored locally at each client location and is not transferred to the servers

In FL, raw data is stored locally at each client location instead of being centralized in a single server. This decentralized approach ensures that the data remains under the control and ownership of the respective clients, preserving privacy and complying with data regulations.

The data at each client location exhibits a specific distribution, which can vary across different clients. The distribution of the data refers to the statistical characteristics and patterns present within the dataset. The data samples within a client's dataset can be independent of each other, meaning that they are unrelated or do not rely on each other for their values or properties. Alternatively, the data samples can be dependent, indicating that there is some form of correlation or relationship between them.

Furthermore, the data distribution can be either identical or non-identical among the clients. Identical data distribution implies that the statistical properties of the datasets are the same across different clients. On the other hand, non-identical data distribution suggests that the datasets exhibit variations in their statistical characteristics, such as mean, variance, or other relevant parameters.

The presence of diverse data distributions, whether independent or dependent, identical or non-identical, introduces challenges and complexities in FL. Nevertheless, FL methods are designed to handle these variations and enable collaborative model training across decentralized data sources, leveraging the collective knowledge while respecting data privacy and distribution characteristics.

Datasets with IID and non-IID data

Independent and identically distributed (IID) data refers to a dataset in which the data samples are independent of each other, and the distribution of the data is identical across all samples. In this case, the outcomes of each data sample are not dependent on previous samples, and the statistical properties of the data remain consistent.

For example, consider a dataset where a coin is tossed five times and the number of times it turns up heads is recorded. In this scenario, each coin toss is independent of the previous tosses, and the probability of getting heads is identical for each toss. This results in an IID dataset where the distribution of outcomes is the same for every coin toss.

In FL, the data across different clients may exhibit **non-IID** characteristics. This means that the data samples are not identically distributed, and they may also be dependent on each other. Various factors can contribute to non-IID data, such as variations in the amount of labeled data, differences in the features present in the samples, data drift, concept drift, or imbalanced data.

For example, in the case of cross-silo entities within a company, each client may have the same kind of features and labels for classification. However, the number of data samples at each location may vary, resulting in imbalanced data. Additionally, each location may not have data for all classes or may exhibit different distributions of examples.

Raw data in cross-silo entities in FL

When dealing with cross-silo entities in FL, the raw data exhibits certain characteristics. Specifically, in the case of intra-company scenarios, the following can be observed:

- Each client within the cross-silo entities will possess the same kind of features. This means that the types of data attributes or variables available for analysis will be consistent across all clients.

- The labels or classes used for classification tasks will also be the same among the clients. This ensures that the target categories or outcomes for classification are consistent throughout the participating entities.

- The number of data samples at each client location may vary. This implies that the amount of available data may differ across different locations or branches within the same company. Some clients may have more extensive datasets, while others may have fewer samples.

- Not all classes or categories may be represented in each client's data. This results in imbalanced data, where certain classes may be overrepresented or underrepresented compared to others. Such imbalances can pose challenges for model training and evaluation.

- The distribution of examples may not be the same across all clients. This means that the statistical characteristics, such as the mean, variance, or other properties, may vary between different client locations. Each client's data may exhibit unique distributional patterns, which need to be accounted for during the FL process.

Considering these characteristics, FL techniques must address the variability in data samples, imbalanced class distributions, and divergent data distributions across the cross-silo entities.

In the context of the banking example we discussed, since it is the same bank operating in different countries, the features (such as customer ID, amount, transaction date, source account, destination account, and address) and labels (*fraud* or *non-fraud*) will be the same. However, the distribution of data samples and labels may vary at each location, based on factors such as the number of customers and the types of transactions. This introduces non-IID characteristics to the data, requiring careful handling in FL approaches.

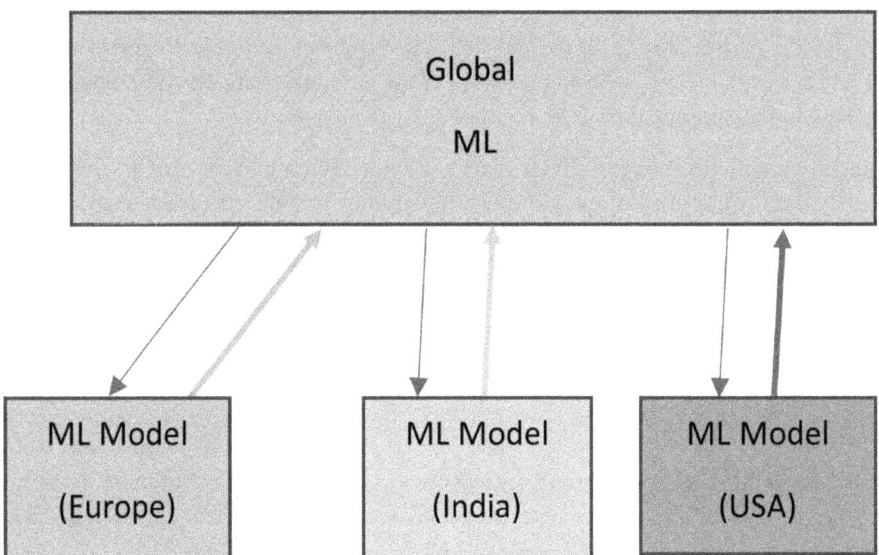

Figure 6.5 – The ML model in a financial bank in three locations

Data is distributed in the following way to each client. There is skewness in the label data but samples with all features exist in each location/client/entity.

Data at different clients	Features X={ X1, X2, X3, …Xn}				Label y = { y1, y2…, ym}
	X1	X2	X3	X4	Fraud data label counts
Europe (Client 1)	yes	yes	yes	yes	Fraud count = N, Non-fraud = 0
US (Client 2)	yes	yes	yes	yes	Fraud count = 0, Non-fraud = N
India (Client 3)	yes	yes	yes	yes	Fraud count = N/2, Non-fraud = N/4

Table 6.1 – Label data skewness

In the case of intra-institutions, where different institutions within the same industry participate in FL to offer similar ML services, the data may exhibit the following characteristics:

- Each client, representing a different institution, may or may not have the same kind of features. This means that the available data attributes or variables may differ between institutions, based on their specific contexts or data collection practices.

- The number of data samples at each client location may vary. This indicates that the amount of data available for analysis could differ between different institutions. Some institutions may have larger datasets, while others may have relatively smaller ones.

- Not all classes or categories may be present in each client's data. This can result in imbalanced data, where certain classes may be underrepresented or missing altogether in some institutions' datasets. Handling imbalanced data is an important consideration in the FL process.

- The distribution of examples may also differ among the participating institutions. Each institution's data may have its own unique distributional patterns, including variations in mean, variance, or other statistical properties. These differences need to be taken into account during the collaborative model training process.

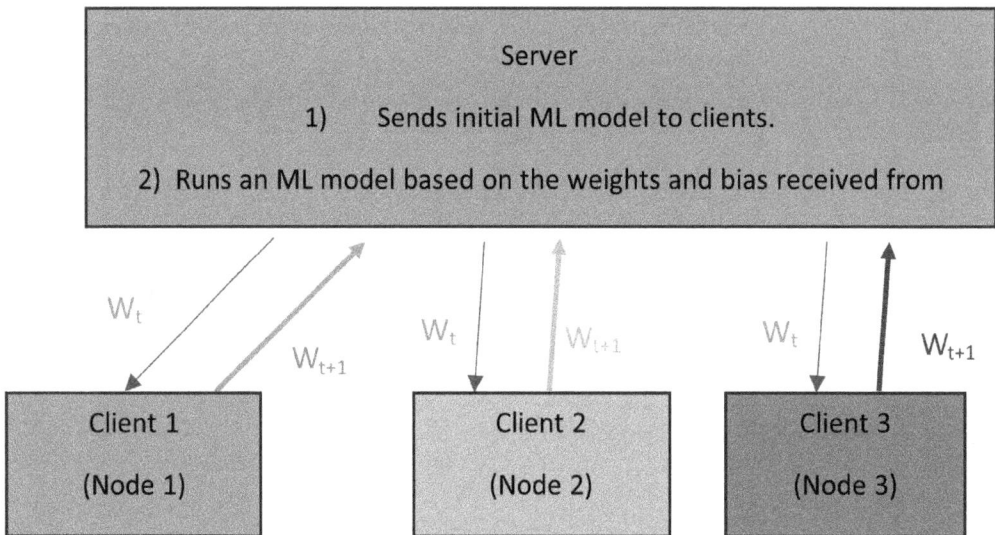

Figure 6.6 – FL client and server communication (send and receive) model parameters

Data is distributed in the following way to each client. In this scenario, there is skewness in features and label data.

Data at different clients	Features X={ X1, X2, X3, …Xn}				Label y = { y1, y2…, yn}
	X1	X2	X3	X4	Fraud data label counts
Client 1	yes		yes		No (Non-Fraud)= 70%
Client 2		yes		yes	Yes (Fraud)= 100%
Client 3	yes	yes	yes	yes	Yes (Fraud) = 50%, No (Non-Fraud)=50%

Table 6.2 – Feature and label data skewness at different clients

Learning objective

In FL, the central server takes on the responsibility of executing the learning objective and minimizing the loss function. It achieves this by leveraging the model weights (Wt) and biases received from the participating clients. The server determines the number of rounds of data it needs from the clients and the specific clients that need to participate.

Let's consider an example where there are three clients involved in the FL process. Each client sends its respective model weights and biases to the central server. The server then performs the following objective or learning function to minimize the loss:

minimize loss (Wt, biases)

The objective of the server is to optimize the model parameters, represented by the weights (*Wt*) and biases, to minimize the loss function. By utilizing the received weights and biases from the participating clients, the server performs iterative updates to refine the model and improve its performance.

The specific details of the learning objective and loss function depend on the specific ML algorithm and the task at hand. The central server orchestrates the aggregation of client updates, manages the training process, and sends back the updated model to the clients. This collaborative approach enables the clients to collectively contribute their local knowledge while benefiting from the improved global model provided by the server.

This is the objective function:

$$\text{Min } f(w) = \sum_{i=1}^{n} \text{£}i * Fi(w)$$

Here, *w* is the model parameters (weights, and so on), *f(w)* is the objective function, and *n* is the number of clients participating in FL.

A few more mathematical terms will be used in the next section including the following.

- *Wt*: Model weights in the communication round *t* (client to server)
- *Wt k*: Model weights in the communication round on client *k*
- *C*: The number of clients participating in each round to update the model and compute the weights
- *B*: The local clients' batch size of the data samples
- *Pk*: The set of data samples at client *k*
- *nk*: The number of data points at client *k*
- *fi (w): loss L (xi, yi, w)* – the loss function

On the server side, various objective functions can be implemented, depending on the specific requirements and goals of the FL process.

FL algorithms

FL algorithms, such as FedSGD, FedAvg, and Adaptive Federated Optimization, play a crucial role in the distributed training of ML models while ensuring privacy and security. In this section, we will explore these algorithms and their key characteristics.

FedSGD

Federated stochastic gradient descent (FedSGD) is a fundamental algorithm used in FL. It extends the traditional SGD optimization method to the federated setting. In FedSGD, each client (entity) computes the gradients on its local data and sends them to the central server. The server aggregates the gradients and updates the global model parameters accordingly. FedSGD is efficient for large-scale distributed training but may suffer from issues related to non-IID data and communication efficiency.

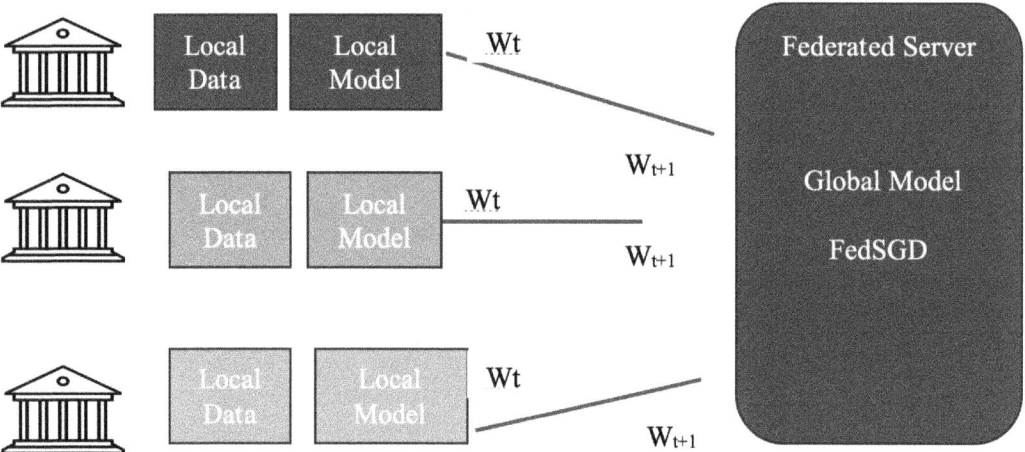

Figure 6.7 – The FedSGD model weights exchange with the server

Let's look at the FedSGD algorithm:

Server-side algorithm	Client-side algorithm
Initialize weights (w0)	Client-side function (k, w):
for each round t = 1,2, …	
m = max (C, K, 1)	• Split the data in k batches, with each batch based on the batch size B (complete local dataset)
st = random set of m clients	• For each batch:
for client k in st,	• $f_{i\,(w)\,=}$ loss L (xi, yi, w)
$w_{t+1\,=}$ client-side function (k, w_t)	• w = w – learning rate * loss
$w_{t+1\,=}$ average of weights	

Table 6.3 - FedSGD Algorithem

On the client side, each participating client performs the following steps:

1. **Data partitioning**: Clients have their own local datasets and partition them into smaller subsets to ensure privacy.

2. **Local model training**: Each client independently trains the shared model using its local data. This involves computing the gradients of the model parameters (weights and biases) on the local dataset using SGD or a variant.

3. **Model update**: After the local model training, the client sends the computed gradients to the server for aggregation.

On the server side, the central server performs the following steps:

1. **Aggregation**: The server receives the gradients from all participating clients and aggregates them using various aggregation techniques, such as averaging or weighted averaging.

2. **Model update**: The aggregated gradients are used to update the global model's parameters. The server applies the received gradients to the global model, adjusting its weights and biases to reflect the collective knowledge from all clients.

3. **Model distribution**: The updated global model is then sent back to the clients for the next round of training, ensuring that each client benefits from the collective knowledge while preserving data privacy.

FedSGD aims to minimize the communication overhead between the clients and the server by exchanging only the model gradients rather than the raw data. This allows for distributed model training while maintaining data privacy and security. However, it is important to address challenges such as data heterogeneity and non-IID data distribution, which can impact the convergence and performance of the FL process.

Overall, FedSGD enables collaborative model training in a decentralized manner, leveraging the computational resources of multiple clients while preserving data privacy. It serves as a foundational algorithm for FL and has paved the way for more advanced techniques to improve the efficiency and effectiveness of distributed ML.

FedAvg

Federated averaging (**FedAvg**) is a widely adopted FL algorithm designed to address the challenges of non-IID data and communication efficiency. In FedAvg, similar to FedSGD, each client computes the gradients on its local data. However, instead of directly updating the global model with the individual gradients, FedAvg employs weighted averaging to combine the client models' parameters. This approach allows for better handling of data heterogeneity and reduces the communication overhead between the clients and the server.

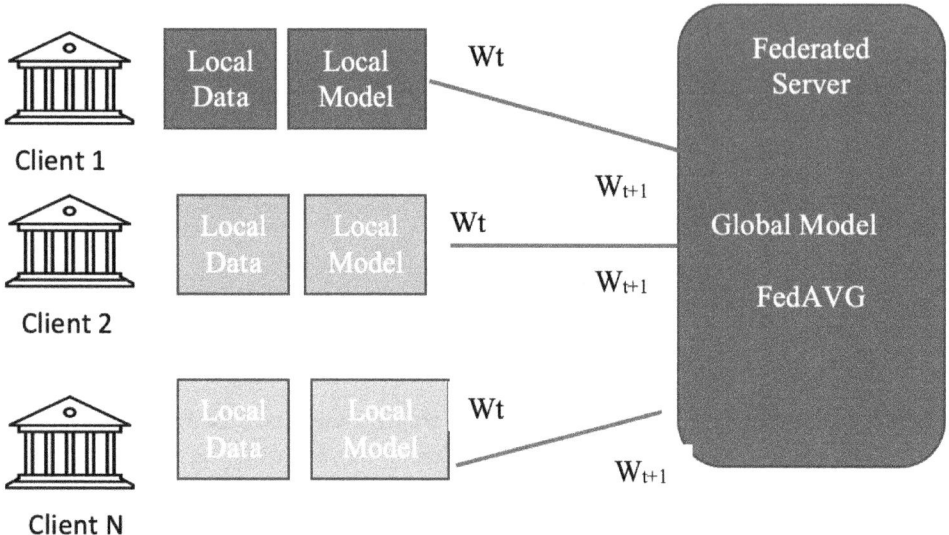

Figure 6.8 – The FedAvg model weights exchange with the server

Let's look at the FedAvg algorithm:

Server-side algorithm	Client-side algorithm
Initialize weights (w_0)	Client-side function (k, w):
for each round t = 1,2, …	• Split the data into *k* batches, with each batch based on the batch size *B*
m = max (C, K, 1)	
st – random set of m clients	• For each epoch in training E
for client k in st,	• For each batch:
$w_{t+1 =}$ client-side function (k, w_t)	▪ $f_{i(w) =}$ loss L (xi, yi, w)
$w_{t+1 =}$ average of gradients	▪ w = w – learning rate * loss

Table 6.4 – FedAVG Algorithm

FedAvg leverages the concept of averaging to combine the locally trained models of different clients, which helps mitigate the impact of data heterogeneity and non-IID data distribution. By averaging the model parameters, FedAvg effectively creates a global model that captures insights from all participating clients while preserving the privacy of individual data. The iterative nature of FedAvg allows the shared model to progressively improve with each round of training. As the process continues, the global model becomes more refined and represents the collective knowledge of all clients.

Overall, FedAvg enables collaborative training of a shared model in a privacy-preserving manner. It addresses challenges associated with data privacy and distribution, allowing multiple clients to contribute to the model's improvement without sharing their raw data. FedAvg has been instrumental in advancing the field of FL, enabling applications in various domains while maintaining data privacy and security.

Fed Adaptative Optimization

In cross-device FL, a multitude of clients communicate with a central server, and each client possesses a unique set of data.

For instance, in the context of next-word prediction on phones, different users' phones contain distinct word sets based on factors such as country, region, and language. However, traditional FL algorithms such as FedSGD and FedAvg may not perform optimally when confronted with heterogeneous data from diverse clients. The challenge arises from the inherent differences in data distribution and characteristics among the clients. Heterogeneous data introduces complexities that can impact the convergence and performance of FL algorithms. As a result, handling heterogeneous data poses a considerable obstacle compared to scenarios where the clients have homogeneous data.

Efforts are being made to address the challenges associated with heterogeneous data in cross-device FL.

In order to overcome this, researchers at Google (*Sashank J. Reddi et al., 2021*) proposed new adaptive optimizations in the research paper published at `arxiv.org/abs/2003.00295`.

Here is the detailed algorithm (the image is sourced from the preceding URL):

Algorithm 2 FedAdagrad, FedYogi, and FedAdam

1: Initialization: $x_0, v_{-1} \geq \tau^2$, decay parameters $\beta_1, \beta_2 \in [0, 1)$
2: **for** $t = 0, \cdots, T-1$ **do**
3:    Sample subset S of clients
4:    $x_{i,0}^t = x_t$
5:    **for** each client $i \in S$ **in parallel do**
6:       **for** $k = 0, \cdots, K-1$ **do**
7:          Compute an unbiased estimate $g_{i,k}^t$ of $\nabla F_i(x_{i,k}^t)$
8:          $x_{i,k+1}^t = x_{i,k}^t - \eta_l g_{i,k}^t$
9:       $\Delta_i^t = x_{i,K}^t - x_t$
10:    $\Delta_t = \frac{1}{|S|} \sum_{i \in S} \Delta_i^t$
11:    $m_t = \beta_1 m_{t-1} + (1 - \beta_1)\Delta_t$
12:    $v_t = v_{t-1} + \Delta_t^2$ **(FedAdagrad)**
13:    $v_t = v_{t-1} - (1 - \beta_2)\Delta_t^2 \, \text{sign}(v_{t-1} - \Delta_t^2)$ **(FedYogi)**
14:    $v_t = \beta_2 v_{t-1} + (1 - \beta_2)\Delta_t^2$ **(FedAdam)**
15:    $x_{t+1} = x_t + \eta \frac{m_t}{\sqrt{v_t} + \tau}$

Figure 6.9 – The Fed Adaptive Optimization algorithm proposed by Google researchers

Please refer to the article for a detailed explanation of the Adaptive Optimization algorithm. In a nutshell, the idea is to optimize the communication cost like FEDAVG and work in cross-device settings.

The steps involved in implementing FL

The following are the five steps that are typically followed to implement FL. There can be alternatives/changes to these steps, but initially, these are the steps that need to be followed:

The server side – the initialization of the global model: In this step, the server starts and accepts the client requests. Before actually starting the server, the model on the server side will be initiated with model parameters. Typically, model parameters will be initiated with zeros or from the previous checkpoint model.

The server sends model parameters to all or a subset of clients: In this step, the server sends the initial model parameters to all clients (for cross-silo FL clients, they will be within the same institutions and may only be numbered in the tens) or a subset of clients (in the case of cross-device FL where devices are in the millions, the server decides to select only a subset from the total devices). Each client will make use of these initial model parameters for the local training of the model.

The clients train the model and send the model weights/parameters back to the server: In this step, each client will train the model with their local data, making use of the entire local data in one shot, dividing the data into several batches, or splitting the data randomly and making use of the different splits for different rounds (a multiple rounds of exchanges of model parameters between the client and server). The clients will send the model parameters or weights only to the server.

The server executes one of the FL algorithms, updates the global model, and sends the updated weights to the client for the next round: In this step, the server will run one of the FL algorithms and make use of the weights received by the clients to update the global model. In the case of FedAvg, it will calculate the weighted average of the weights received from clients and send the updated weights back to the client for the next round.

Repeat steps 2 to 4 based on the number of rounds configured: Repeat *steps 2* to *4* for each round. If five rounds are configured, then repeat *steps 2* to *4* five times, and after the last round, clients will make use of the weights received by the server for the final ML model. Clients can make use of these model weights either at the end of the last round or in each round and evaluate the model's accuracy with the test data.

The following sequence diagram shows these steps in detail:

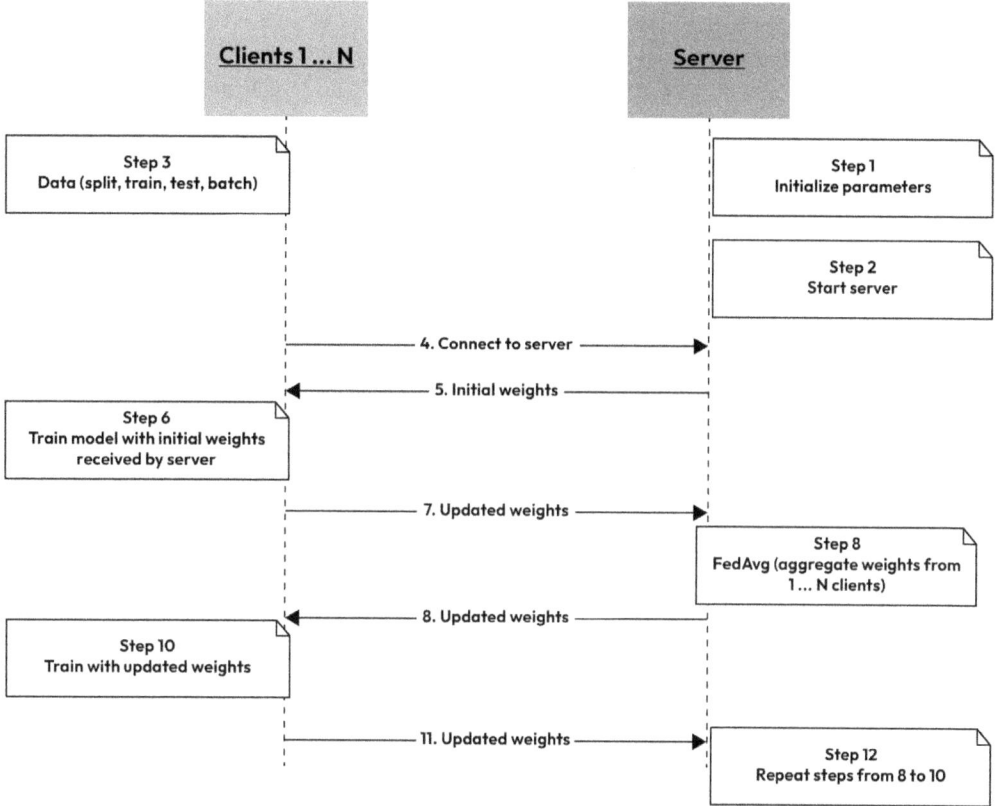

Figure 6.10 – The steps in the FL sequence diagram

The sequence diagram shows the detailed interactions between the server and the clients participating in the FL, performing four high-level steps as explained in this section.

Open source frameworks to implement FL

There are a few open source frameworks to implement FL at scale. The following are some of the most popular.

PySyft (`https://github.com/OpenMined/PySyft`), developed by OpenMined, is an open source stack that offers secure and private data science capabilities in Python. It introduces a separation between private data and model training, enabling functionalities such as FL, differential privacy, and encrypted computation. Initially, PySyft utilized the Opacus framework to support differential privacy, as discussed in the Differential privacy chapter. However, the latest version of PySyft incorporates its own differential privacy component to provide enhanced functionality and efficiency in preserving privacy while performing data analysis tasks.

TensorFlow Federated

TensorFlow Federated (**TFF**) is a library developed by Google that facilitates the training of shared ML models across multiple clients using their local data (`https://www.tensorflow.org/federated`). TFF consists of two layers – the Federated Core API and the Federated Learning API.

The Federated Core API offers low-level interfaces for tasks such as data serialization, distribution communication between the server and clients, and implementation of FL algorithms. It provides the foundational components necessary to build FL systems.

Conversely, the Federated Learning API provides a higher-level interface that allows users to easily construct FL models or wrap existing models as FL models. It offers a set of APIs for training and evaluating models using federated computations and datasets. This higher-level interface abstracts away some of the complexities involved in building and training FL models, making it more accessible and convenient for developers.

By providing these two layers, TFF empowers researchers and developers to leverage the power of FL in their projects. It simplifies the process of building and training models on decentralized data while ensuring privacy and data security.

Flower

Flower (`https://flower.dev/`) is an open source framework that aims to provide a user-friendly experience. It supports ML and DL models developed using various frameworks, such as scikit-learn, TensorFlow, PyTorch, PyTorch Lightning, MXNet, and JAX. Flower makes it easy to convert these models into FL models.

One of the key features of Flower is its communication implementation, which is built on top of bidirectional gRPC streams. This enables an efficient and seamless exchange of multiple messages between clients and the server without the need to establish a new connection for each message request.

Flower offers a range of strategies and implements several FL algorithms on the server side. These algorithms include FedAvg, FedSGD, Fault Tolerance FedAvg, FedProxy, and FedOptim (which consists of FedAdagrad, FedYogi, and FedAdam). These algorithms provide different approaches to model aggregation and training in FL scenarios.

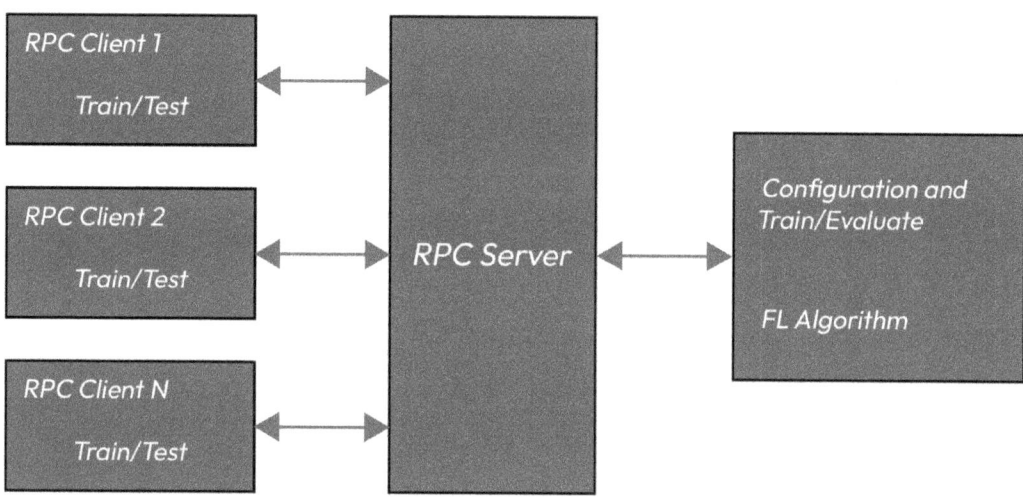

Figure 6.11 – The Flower framework architecture diagram (simplified)

To validate its performance, Flower has been extensively benchmarked. The framework has demonstrated the ability to scale up to 15 million clients using only two GPU servers. These experiments were compared with FedScale, another FL engine and benchmark suite, to evaluate Flower's performance and efficiency in large-scale FL settings.

An end-to-end use case of implementing fraud detection using FL

Fraud detection is a critical task for many industries, including finance, e-commerce, and healthcare. Traditional fraud detection methods often rely on centralized data collection, where sensitive customer information is gathered and analyzed in a single location. However, this approach raises concerns about data privacy and security, as well as compliance with regulations such as the GDPR.

FL offers a promising solution to address these challenges. By leveraging the power of distributed computing and collaborative learning, FL enables fraud detection models to be trained directly on the devices or local servers of individual institutions, without the need for data sharing. This decentralized approach ensures that sensitive customer data remains private and secure, as it never leaves the local environment.

Implementing fraud detection using FL involves several key steps. Firstly, a consortium of institutions or organizations, such as banks or e-commerce platforms, need to establish an FL framework that enables them to collaborate on model training while preserving data privacy. This may involve the adoption of FL libraries or platforms such as TFF or Flower.

Next, the participating institutions define a common fraud detection objective and develop a shared model architecture. Each institution then trains its local model using its own private data, which may include transaction records, user behavior patterns, and other relevant features. The models are trained locally, ensuring that sensitive data remains under the control of the respective institutions.

To facilitate collaborative learning, the institutions periodically share model updates with a central server. These updates, which typically include model weights and parameters, are aggregated using federated averaging or other aggregation techniques to create a global model that captures insights from all participants, while preserving the privacy of individual data.

The central server, which oversees the aggregation process, ensures that the global model is refined based on the collective knowledge of the participating institutions. This process allows the model to learn from a diverse range of fraud patterns and adapt to evolving fraudulent activities while maintaining data privacy and compliance with regulations.

Implementing fraud detection using FL offers several advantages. It allows institutions to leverage a larger and more diverse dataset, leading to improved fraud detection accuracy. It also reduces the risks associated with data breaches or unauthorized access, since sensitive data remains under the control of the respective institutions. Additionally, FL enables real-time updates and faster model deployment, allowing institutions to respond quickly to emerging fraud patterns.

Implementing fraud detection using FL offers a privacy-preserving and collaborative approach to combat fraud in various industries. By combining the power of distributed computing and shared learning, organizations can enhance fraud detection capabilities while safeguarding sensitive customer data.

Let's implement this use case using the Flower framework and the open source dataset.

Developing an FL model for fraud detection using the Flower framework

In this example, we will leverage the Flower framework to develop an FL model for fraud detection. The implementation will involve both server-side and client-side components. To illustrate the process, we will set up one server and two clients.

The communication between the server and clients will occur over several rounds, with the exchange of weights and parameters. The exact number of rounds may vary depending on the specific scenario, but typically, the communication continues until the weights converge or a predetermined convergence criterion is met.

On the server side, we will implement the FedAvg algorithm to aggregate the weights received from the clients. FedAvg is a widely used algorithm in FL that combines the knowledge from multiple clients to create a global model.

For the fraud detection task, we will develop an actual linear regression model using the scikit-learn library. This model will be trained using the data available at each client, which consists of transaction records and relevant features. The goal is to classify whether a transaction is fraudulent or not.

The client-side implementation will involve training the local linear regression models using the respective client's data. The clients will then communicate with the server, exchanging their model weights and parameters over the predefined rounds. This collaborative learning process allows the clients to contribute their local insights to the global model while preserving the privacy of their data.

The server will receive the model updates from the clients and perform the aggregation step using the FedAvg algorithm. This aggregation process ensures that the global model incorporates the knowledge learned from all the participating clients, resulting in an enhanced fraud detection capability.

Throughout the implementation, the Flower framework provides the necessary infrastructure for the communication between the server and clients. It abstracts the underlying complexities of distributed computing and handles the synchronization of model updates. By developing an FL model for fraud detection, we can leverage the distributed knowledge and data from multiple clients to improve the accuracy of fraud classification. The federated approach also addresses privacy concerns by keeping the sensitive transaction data local to each client.

In summary, this project demonstrates the implementation of an FL model using the Flower framework. The server and clients collaborate to train a global model for fraud detection, exchanging model weights and parameters over multiple communication rounds. By aggregating the client models using FedAvg, we can leverage the collective intelligence of multiple participants while ensuring data privacy.

The dataset used in the example

Lopez, Elmir, and Axelsson developed a mobile money dataset for fraud detection, and it is featured on Kaggle as well (E. A. Lopez-Rojas, A. Elmir, and S. Axelsson, *PaySim: A financial mobile money simulator for fraud detection*, 28th European Modeling and Simulation Symposium-EMSS, Larnaca, Cyprus. 2016).

We will make use of this dataset for the detection of fraud using FL, but the same can be extended to anti-money laundering use cases as well, with minor changes to the model. The dataset can be found at `https://github.com/EdgarLopezPhD/PaySim`.

Download this dataset and keep the file in the `Chapter 6` directory with the name `PS_20174392719_1491204439457_log.csv`.

This dataset consists of 6.3 million records of transactions and has the following features:

Field	Data type	Details
Step	Numerical	The unit of time in the real world. One step is 1 hour.
Type	Object	CASH-IN, CASH-OUT, DEBIT, PAYMENT, and TRANSFER.
Amount	Numerical	The amount of the transaction.

Field	Data type	Details
nameOrig	Object	The customer who started the transaction.
nameDest	Object	The recipient ID of the transaction.
oldbalanceOrg	Numerical	The initial balance before the transaction.
newbalanceOrig	Numerical	The customer's balance after the transaction.
oldbalanceDest	Numerical	The initial recipient's balance before the transaction.
newbalanceDest	Numerical	The recipient's balance after the transaction.
isFraud	Boolean	Identifies fraudulent (1) and non-fraudulent (0) transactions.
isFlaggedFraud	Boolean	Flags illegal attempts to transfer more than 200,000 amount in a single transaction.

Table 6.5 – The dataset features

The installation of Flower

Install Flower using the `python -m pip install flwr` command.

The implementation of a server

The server-side implementation of the FL model for fraud detection involves several high-level steps. We will utilize the sample code provided by the Flower framework and extend it to fit our specific use case.

The following steps outline the process:

1. **Initialize the model parameters**:

    A. Set the initial model weights to 0 and initialize the intercept as 0 (since we are working with a regression model).

    B. Determine the number of classes or labels for classification.

    C. Determine the number of features used in the model.

    D. Determine the number of participating clients.

2. **Define the supporting functions**: Develop additional functions to load the data from clients, define the loss function, and evaluate the model's performance. These functions will help facilitate data handling, calculate the loss during training, and assess the model's accuracy.

3. **Choose the server-side strategy**: Select the FedAvg algorithm as the strategy to aggregate the weights received from the clients. FedAvg is a popular choice to combine model updates from multiple clients and generate an updated global model.

4. **Start the server**:

   A. Initiate the server-side component, which will orchestrate the FL process.

   B. The server will communicate with the participating clients, receive their model updates, and aggregate the weights using the FedAvg algorithm.

   C. It will also handle the synchronization of the model updates between the clients and ensure the convergence of the global model.

By following these steps, we can implement the server-side functionality of our FL model. The initialization of model parameters, definition of supporting functions, selection of the server-side strategy (FedAvg), and starting the server itself are crucial in facilitating the collaborative training process among the clients. Through this implementation, the server will act as the central coordinator, receiving and aggregating the model updates from the clients. It plays a crucial role in ensuring the model's convergence and generating an updated global model that incorporates the knowledge from all participating clients.

Save the following code as `FL_AML_Server.py`:

```
import pandas as pd
url ="PS_20174392719_1491204439457_log.csv"
df_actual = pd.read_csv(url, sep=",")
df_actual
```

This results in the following output:

	step	type	amount	nameOrig	oldbalanceOrg	newbalanceOrig	nameDest	oldbalanceDest	newbalanceDest	isFraud	isFlaggedFraud
0	1	PAYMENT	9839.64	C1231006815	170136.00	160296.36	M1979787155	0.00	0.00	0	0
1	1	PAYMENT	1864.28	C1666544295	21249.00	19384.72	M2044282225	0.00	0.00	0	0
2	1	TRANSFER	181.00	C1305486145	181.00	0.00	C553264065	0.00	0.00	1	0
3	1	CASH_OUT	181.00	C840083671	181.00	0.00	C38997010	21182.00	0.00	1	0
4	1	PAYMENT	11668.14	C2048537720	41554.00	29885.86	M1230701703	0.00	0.00	0	0
...	...	...	...	...	...	...	...	...	...	...	...
6362615	743	CASH_OUT	339682.13	C786484425	339682.13	0.00	C776919290	0.00	339682.13	1	0
6362616	743	TRANSFER	6311409.28	C1529008245	6311409.28	0.00	C1881841831	0.00	0.00	1	0
6362617	743	CASH_OUT	6311409.28	C1162922333	6311409.28	0.00	C1365125890	68488.84	6379898.11	1	0
6362618	743	TRANSFER	850002.52	C1685995037	850002.52	0.00	C2080388513	0.00	0.00	1	0
6362619	743	CASH_OUT	850002.52	C1280323807	850002.52	0.00	C873221189	6510099.11	7360101.63	1	0

6362620 rows × 11 columns

Figure 6.12 – Dataset information with few rows and columns

Instead of using all 6 million records, we will use only the first 25,000 records in this example implementation:

```
df_transactions=df_actual.head(25000)
from sklearn.model_selection import train_test_split
from sklearn.model_selection import StratifiedShuffleSplit
print("No of Fraud Transactions:",
df_transactions["isFraud"].value_counts()[0])
print("No of Non Fraud Transactions:",
df_transactions["isFraud"].value_counts()[1])
print('No Frauds',
round(df_transactions['isFraud'].value_counts()[0]/len(df_t
ransactions) * 100,2), '% of the dataset')
print('Frauds',
round(df_transactions['isFraud'].value_counts()[1]/len(df_t
ransactions) * 100,2), '% of the dataset')
```

This results in the following output:

```
No of Fraud Transactions: 24917
No of Non Fraud Transactions: 83
No Frauds 99.67 % of the dataset
Frauds 0.33 % of the dataset
```

Let's find the data types of each field in the dataset:

```
df_transactions.dtypes
```

This results in the following output:

```
step              int64
type              object
amount            float64
nameOrig          object
oldbalanceOrg     float64
newbalanceOrig    float64
nameDest          object
oldbalanceDest    float64
newbalanceDest    float64
isFraud           int64
isFlaggedFraud    int64
dtype: object
```

Encode the object data type fields as labels using `LabelEncoder`:

```python
from sklearn.preprocessing import LabelEncoder
encoder = {}
for i in df_transactions.select_dtypes('object').columns:
    encoder[i] = LabelEncoder()
    df_transactions[i] = encoder[i].fit_transform(df_transactions[i])
X = df_transactions.drop('isFraud', axis=1)
y = df_transactions['isFraud'
from typing import Tuple, Union, List
import numpy as np
from sklearn.linear_model import LogisticRegression
XY = Tuple[np.ndarray, np.ndarray]
Dataset = Tuple[XY, XY]
LogRegParams = Union[XY, Tuple[np.ndarray]]
XYList = List[XY]


def get_model_parameters(model: LogisticRegression) -> LogRegParams:
    if model.fit_intercept:
        params = [
            model.coef_,
            model.intercept_,
        ]
    else:
        params = [
            model.coef_,
        ]
    return params



def set_model_params(
    model: LogisticRegression, params: LogRegParams
) -> LogisticRegression:

    model.coef_ = params[0]
    if model.fit_intercept:
        model.intercept_ = params[1]
    return model

def shuffle(X: np.ndarray, y: np.ndarray) -> XY:

    rng = np.random.default_rng()
    idx = rng.permutation(len(X))
    return X[idx], y[idx]
```

```python
def partition(X: np.ndarray, y: np.ndarray, num_partitions: int) ->
XYList:

    return list(
        zip(np.array_split(X, num_partitions), np.array_split(y, num_
partitions))
    )

def set_initial_params(model: LogisticRegression):

    n_classes = 2  # Fraud Detection has only  classes
    n_features = 9  # Number of features in dataset
    model.classes_ = np.array([i for i in range(n_classes)])

    model.coef_ = np.zeros((n_classes, n_features))
    if model.fit_intercept:
        model.intercept_ = np.zeros((n_classes,))


import flwr as fl
from sklearn.metrics import log_loss
from sklearn.linear_model import LogisticRegression
from typing import Dict


def fit_round(server_round: int) -> Dict:
    """Send round number to client."""
    return {"server_round": server_round}


def get_evaluate_fn(model: LogisticRegression,X_test,y_test):

# The `evaluate` function will be called after every round
    def evaluate(server_round, parameters: fl.common.NDArrays,
config):
        # Update model with the latest parameters
        set_model_params(model, parameters)
        loss = log_loss(y_test, model.predict_proba(X_test))
        accuracy = model.score(X_test, y_test)
        return loss, {"accuracy": accuracy}

    return evaluate
```

```
# Start Flower server for five rounds of federated learning
def Server():
    model = LogisticRegression(max_iter=10000)
    set_initial_params(model)
    strategy = fl.server.strategy.FedAvg(
        min_available_clients=2,
        evaluate_fn=get_evaluate_fn(model,X_test,y_test),
        on_fit_config_fn=fit_round,
    )
    fl.server.start_server(
        server_address="0.0.0.0:8080",
        strategy=strategy,
        config=fl.server.ServerConfig(num_rounds=5),
    )

Server()
```

Open a terminal and run this program (python3 FL_AML_Server.py)

This results in the following output:

```
bash-3.2$ python3 FL_AML_Server.py
No of Fraud Transactions: 24917
No of Non Fraud Transactions: 83
No Frauds 99.67 % of the dataset
Frauds 0.33 % of the dataset
FL_AML_Server.py:51: SettingWithCopyWarning:
A value is trying to be set on a copy of a slice from a DataFrame.
Try using .loc[row_indexer,col_indexer] = value instead

See the caveats in the documentation: https://pandas.pydata.org/pandas-docs/stable/user_guide/indexing.html#returning-a-view-versus-a-copy
  df_transactions[i] = encoder[i].fit_transform(df_transactions[i])
FL_AML_Server.py:51: SettingWithCopyWarning:
A value is trying to be set on a copy of a slice from a DataFrame.
Try using .loc[row_indexer,col_indexer] = value instead

See the caveats in the documentation: https://pandas.pydata.org/pandas-docs/stable/user_guide/indexing.html#returning-a-view-versus-a-copy
  df_transactions[i] = encoder[i].fit_transform(df_transactions[i])
FL_AML_Server.py:51: SettingWithCopyWarning:
A value is trying to be set on a copy of a slice from a DataFrame.
Try using .loc[row_indexer,col_indexer] = value instead

See the caveats in the documentation: https://pandas.pydata.org/pandas-docs/stable/user_guide/indexing.html#returning-a-view-versus-a-copy
  df_transactions[i] = encoder[i].fit_transform(df_transactions[i])
INFO flwr 2023-03-14 18:48:55,252 | app.py:139 | Starting Flower server, config: ServerConfig(num_rounds=5, round_timeout=None)
INFO flwr 2023-03-14 18:48:55,273 | app.py:152 | Flower ECE: gRPC server running (5 rounds), SSL is disabled
INFO flwr 2023-03-14 18:48:55,273 | server.py:86 | Initializing global parameters
INFO flwr 2023-03-14 18:48:55,273 | server.py:270 | Requesting initial parameters from one random client
INFO flwr 2023-03-14 18:50:32,770 | server.py:274 | Received initial parameters from one random client
INFO flwr 2023-03-14 18:50:32,771 | server.py:88 | Evaluating initial parameters
INFO flwr 2023-03-14 18:50:32,777 | server.py:91 | initial parameters (loss, other metrics): 0.14332985986961255, {'accuracy': 0.996266666666666
6}
INFO flwr 2023-03-14 18:50:32,777 | server.py:101 | FL starting
```

Figure 6.13 – Server startup logs

Server will run and wait for data from clients to process.

The implementation of clients

The client-side implementation of the FL model for fraud detection involves the following steps. We will utilize the provided `NumPyClient` from the Flower samples. The steps are as follows:

1. **Load the data**:

    A. Load the relevant data to train and test the fraud detection model.

    B. Ensure the data is properly formatted and available for processing.

2. **Split the data**:

    Split the loaded data into training and testing sets. This division allows you to evaluate the model's performance on unseen data.

3. **Shuffle/partition the data**:

    A. Shuffle or partition the training data into batches.

    B. Randomly select a partition for each round of communication with the server. This ensures that different subsets of the training data are used in each round.

4. **Create the linear regression model**:

    A. Develop a simple linear regression model using the chosen framework (for example, scikit-learn).

    B. Configure the model with appropriate settings for the fraud detection task.

5. **Establish a connection with the server**:

    A. Establish a connection with the server to send and receive model weights.

    B. Utilize the provided communication protocol (for example, gRPC) to exchange information.

6. **Train the model**:

    A. Initialize the model with the initial weights received from the server.

    B. Train the model using the client's local data and the weights updated by the server for each round.

    C. Apply appropriate optimization techniques (for example, gradient descent) to update the model parameters.

7. **Test the model**:

    A. Evaluate the trained model using the testing data to assess its performance.

    B. Calculate relevant metrics such as accuracy, precision, recall, or F1 score.

    C. Determine the model's effectiveness in detecting fraudulent transactions.

By following these steps, we can implement the client-side functionality of the FL model. The client will load and partition the data, create the linear regression model, establish a connection with the server, train the model using local data and updated weights, and evaluate its performance. The client's role is crucial in contributing local knowledge while preserving data privacy. By training on their respective local data and participating in the FL process, clients collectively improve the global fraud detection model without sharing sensitive information.

Creating a non-IID dataset

To transform the dataset into a non-IID setting, we can apply the following approach:

- **First client (client 1):**

  - Apply the **Synthetic Minority Oversampling Technique** (**SMOTE**) to oversample the fraud transactions

  - This technique generates synthetic examples of the minority class (fraudulent transactions) to balance the dataset

  - As a result, client 1 will have a training dataset with 50,000 samples, consisting of 25,000 original transactions and 25,000 synthetic fraud examples created using SMOTE

  - The distribution of fraud versus non-fraud transactions will be balanced at 50% for each class

- **Second client (client 2):**

  - Leave the transactions as they are without any oversampling or modification

  - Client 2 will have a training dataset with the last 25,000 transactions

  - The class distribution will reflect the original distribution, with only 2% of the transactions classified as fraud

By employing this approach, we introduce non-identical and imbalanced datasets across the two clients. Client 1 will have a balanced dataset with equal representation of fraud and non-fraud transactions, while client 2 will have a dataset that mirrors the original distribution.

This non-IID setup allows us to simulate real-world scenarios where different clients may have varying distributions of data. Through FL, both clients can contribute their local knowledge while training their models on distinct datasets, ultimately improving the overall fraud detection model.

	Client 1	Client 2
Original transactions	25,000	25,000
Transactions generated using SMOTE	25,000	0
Total transactions	50,000	25,000

	Client 1	Client 2
Fraud versus non-fraud	50% and 50%	Fraud: 2.43% (608) Non-fraud: 97.57% (24,392)
Train and test split	70:30	70:30
Number of partitions after shuffling the train data with equal size in each partition	10	10

Table 6.6 – Training data distribution on the client side as non-IID data

Here is the code for client 1. Save this code as `FL_AML_Client1.py`:

```
import pandas as pd
import torch
url ="PS_20174392719_1491204439457_log.csv"
df_actual = pd.read_csv(url, sep=",")
df_actual
df_transactions=df_actual.head(25000)
from sklearn.model_selection import train_test_split
from sklearn.model_selection import StratifiedShuffleSplit
print("No of Fraud Transactions:", df_transactions["isFraud"].value_
counts()[0])
print("No of Non Fraud Transactions:", df_transactions["isFraud"].
value_counts()[1])
print('No Frauds', round(df_transactions['isFraud'].value_counts()[0]/
len(df_transactions) * 100,2), '% of the dataset')
print('Frauds', round(df_transactions['isFraud'].value_counts()[1]/
len(df_transactions) * 100,2), '% of the dataset')
```

This results in the following output:

```
No of Fraud Transactions: 24917
No of Non Fraud Transactions: 83
No Frauds 99.67 % of the dataset
Frauds 0.33 % of the dataset
```

In this dataset, fraudulent transactions account for 0.33% of the total data, indicating a highly imbalanced dataset. This imbalance is typical in real-world scenarios, where fraud transactions are much less frequent compared to genuine (non-fraud) transactions.

```
df_transactions.dtypes
```

This results in the following output:

```
step                int64
type               object
amount            float64
nameOrig           object
oldbalanceOrg     float64
newbalanceOrig    float64
nameDest           object
oldbalanceDest    float64
newbalanceDest    float64
isFraud             int64
isFlaggedFraud      int64
dtype: object
```

Encode the object types as labels using sci-kit learn's `LabelEncoder`:

```
from sklearn.preprocessing import LabelEncoder
encoder = {}
for i in df_transactions.select_dtypes('object').columns:
    encoder[i] = LabelEncoder()
    df_transactions[i] = encoder[i].fit_transform(df_transactions[i])

X = df_transactions.drop('isFraud', axis=1)
y = df_transactions['isFraud']
```

Apply SMOTE to generate synthetic data:

```
from imblearn.over_sampling import SMOTE

over_sample = SMOTE(random_state=0)
X,y = over_sample.fit_resample(X,y)
y.value_counts()
```

This results in the following output:

```
0    24917
1    24917

Name: isFraud, dtype: int64

X = df_transactions[['step', 'type', 'amount','nameOrig',
'oldbalanceOrg', 'newbalanceOrig','nameDest', 'oldbalanceDest',
'isFlaggedFraud']]

y= df_transactions['isFraud']
```

```python
X_train, X_test, y_train, y_test = train_test_split(X, y, test_
size=0.3, random_state=42)
X_train = X_train.values
X_test = X_test.values
y_train = y_train.values
y_test = y_test.values

from typing import Tuple, Union, List
import numpy as np
from sklearn.linear_model import LogisticRegression
XY = Tuple[np.ndarray, np.ndarray]
Dataset = Tuple[XY, XY]
LogRegParams = Union[XY, Tuple[np.ndarray]]
XYList = List[XY]
def get_model_parameters(model: LogisticRegression) -> LogRegParams:

    if model.fit_intercept:
        params = [
            model.coef_,
            model.intercept_,
        ]
    else:
        params = [
            model.coef_,
        ]
    return params

def set_model_params(
    model: LogisticRegression, params: LogRegParams
) -> LogisticRegression:

    model.coef_ = params[0]
    if model.fit_intercept:
        model.intercept_ = params[1]
    return model

def shuffle(X: np.ndarray, y: np.ndarray) -> XY:

    rng = np.random.default_rng()
    idx = rng.permutation(len(X))
    return X[idx], y[idx]

def partition(X: np.ndarray, y: np.ndarray, num_partitions: int) ->
XYList:
```

```python
    return list(
        zip(np.array_split(X, num_partitions), np.array_split(y, num_
partitions))
    )

def set_initial_params(model: LogisticRegression):

    n_classes = 2  # only 2 classes  Fraud or Genuine
    n_features = 9  # Number of features in dataset
    model.classes_ = np.array([i for i in range(n_classes)])

    model.coef_ = np.zeros((n_classes, n_features))
    if model.fit_intercept:
        model.intercept_ = np.zeros((n_classes,))

partition_id = np.random.choice(10)
(X_train, y_train) = partition(X_train, y_train, 10)[partition_id]

from sklearn.linear_model import LogisticRegression
model = LogisticRegression(
        penalty="l2",
        max_iter=1,  # local epoch
        warm_start=True,  # prevent refreshing weights when fitting
    )
model.fit(X_train, y_train)

class AML_Detection_Client(fl.client.NumPyClient):
        def get_parameters(self, config):
            return get_model_parameters(model)

        def fit(self, parameters, config):
            set_model_params(model, parameters)
            with warnings.catch_warnings():
                warnings.simplefilter("ignore")
                model.fit(X_train, y_train)
            print(f"Training finished for round {config['server_
round']}")
            return get_model_parameters(model), len(X_train), {}

        def evaluate(self, parameters, config):
            set_model_params(model, parameters)
            loss = log_loss(y_test, model.predict_proba(X_test))
            accuracy = model.score(X_test, y_test)
```

```
              print(loss,accuracy)
              return loss, len(X_test), {"accuracy": accuracy}


fl.client.start_numpy_client(server_address="0.0.0.0:8080",
client=AML_Detection_Client())
```

Open a second terminal and run the client 1 code (python3 FL_AML_Client1.py)

This results in the following output:

```
bash-3.2$ python3 FL-AML-Client1.py
[/Library/Frameworks/Python.framework/Versions/3.8/bin/python3: can't open file 'FL-AML-Client1.py': [Errno 2] No such file or directory ]
bash-3.2$ python3 FL_AML_Client1.py
No of Fraud Transactions: 24917
[No of Non Fraud Transactions: 83
No Frauds 99.67 % of the dataset
Frauds 0.33 % of the dataset
FL-AML_Client1.py:49: SettingWithCopyWarning:
A value is trying to be set on a copy of a slice from a DataFrame.
Try using .loc[row_indexer,col_indexer] = value instead

See the caveats in the documentation: https://pandas.pydata.org/pandas-docs/stable/user_guide/indexing.html#returning-a-view-versus-a-co
py
    df_transactions[i] = encoder[i].fit_transform(df_transactions[i])
FL-AML_Client1.py:49: SettingWithCopyWarning:
A value is trying to be set on a copy of a slice from a DataFrame.
Try using .loc[row_indexer,col_indexer] = value instead

See the caveats in the documentation: https://pandas.pydata.org/pandas-docs/stable/user_guide/indexing.html#returning-a-view-versus-a-co
py
    df_transactions[i] = encoder[i].fit_transform(df_transactions[i])
FL-AML_Client1.py:49: SettingWithCopyWarning:
A value is trying to be set on a copy of a slice from a DataFrame.
Try using .loc[row_indexer,col_indexer] = value instead

See the caveats in the documentation: https://pandas.pydata.org/pandas-docs/stable/user_guide/indexing.html#returning-a-view-versus-a-co
py
    df_transactions[i] = encoder[i].fit_transform(df_transactions[i])
[[6.00000000e+00 3.00000000e+00 6.94584000e+03 ... 1.53530000e+04
  0.00000000e+00 0.00000000e+00]
 [8.00000000e+00 1.00000000e+00 2.03546940e+05 ... 7.60000000e+02
  3.78437160e+05 0.00000000e+00]
 [7.00000000e+00 3.00000000e+00 7.28363000e+03 ... 7.87200000e+03
  0.00000000e+00 0.00000000e+00]
 ...
 [8.00000000e+00 1.00000000e+00 5.02507120e+05 ... 2.05400000e+03
  1.53204355e+06 0.00000000e+00]
 [8.00000000e+00 1.00000000e+00 8.13244000e+03 ... 1.38600000e+03
  2.59444167e+06 0.00000000e+00]
 [4.00000000e+00 3.00000000e+00 1.96760000e+02 ... 7.88600000e+03
  0.00000000e+00 0.00000000e+00]]
/Library/Frameworks/Python.framework/Versions/3.8/lib/python3.8/site-packages/sklearn/linear_model/_logistic.py:458: ConvergenceWarning:
lbfgs failed to converge (status=1):
STOP: TOTAL NO. of ITERATIONS REACHED LIMIT.

Increase the number of iterations (max_iter) or scale the data as shown in:
    https://scikit-learn.org/stable/modules/preprocessing.html
Please also refer to the documentation for alternative solver options:
    https://scikit-learn.org/stable/modules/linear_model.html#logistic-regression
  n_iter_i = _check_optimize_result(
INFO flwr 2023-03-14 18:50:32,730 | grpc.py:50 | Opened insecure gRPC connection (no certificates were passed)
DEBUG flwr 2023-03-14 18:50:32,756 | connection.py:38 | ChannelConnectivity.IDLE
DEBUG flwr 2023-03-14 18:50:32,759 | connection.py:38 | ChannelConnectivity.CONNECTING
DEBUG flwr 2023-03-14 18:50:32,769 | connection.py:38 | ChannelConnectivity.READY
```

Figure 6.14 – The execution of client 1 and the logs

Now, let's look at the code for client 2. Save this code as FL_AML_Client2.py. The client 2 code will be the same as client 1, but fraud transactions are not increased using the SMOTE method. For thoroughness, here is the complete code for the second client:

```
import pandas as pd
import torch
url ="PS_20174392719_1491204439457_log.csv"
```

```
df_actual = pd.read_csv(url, sep=",")
df_actual
df_transactions=df_actual.head(25000)
from sklearn.model_selection import train_test_split
ffrom sklearn.model_selection import StratifiedShuffleSplit

print("No of Fraud Transactions:", df_transactions["isFraud"].value_
counts()[0])
print("No of Non Fraud Transactions:", df_transactions["isFraud"].
value_counts()[1])
print('No Frauds', round(df_transactions['isFraud'].value_counts()[0]/
len(df_transactions) * 100,2), '% of the dataset')
print('Frauds', round(df_transactions['isFraud'].value_counts()[1]/
len(df_transactions) * 100,2), '% of the dataset')
```

This results in the following output:

```
No of Fraud Transactions: 24917
No of Non Fraud Transactions: 83
No Frauds 99.67 % of the dataset
Frauds 0.33 % of the dataset
df_transactions.dtypes
step                  int64
type                 object
amount              float64
nameOrig             object
oldbalanceOrg       float64
newbalanceOrig      float64
nameDest             object
oldbalanceDest      float64
newbalanceDest      float64
isFraud               int64
isFlaggedFraud        int64
dtype: object
```

Encode the object types as labels using sci-kit learn's LabelEncoder:

```
from sklearn.preprocessing import LabelEncoder
encoder = {}
for i in df_transactions.select_dtypes('object').columns:
    encoder[i] = LabelEncoder()
    df_transactions[i] = encoder[i].fit_transform(df_transactions[i])
X = df_transactions.drop('isFraud', axis=1)
y = df_transactions['isFraud']
y.value_counts()
```

This results in the following output:

```
0      24392
1        608
Name: isFraud, dtype: int64
```

```python
X = df_transactions[['step', 'type', 'amount','nameOrig',
'oldbalanceOrg', 'newbalanceOrig','nameDest', 'oldbalanceDest',
'isFlaggedFraud']]
y= df_transactions['isFraud']
X_train, X_test, y_train, y_test = train_test_split(X, y, test_
size=0.3, random_state=42)
X_train = X_train.values
X_test = X_test.values
y_train = y_train.values
y_test = y_test.values
from typing import Tuple, Union, List

import numpy as np
from sklearn.linear_model import LogisticRegression
XY = Tuple[np.ndarray, np.ndarray]
Dataset = Tuple[XY, XY]
LogRegParams = Union[XY, Tuple[np.ndarray]]
XYList = List[XY]
def get_model_parameters(model: LogisticRegression) -> LogRegParams:

    if model.fit_intercept:
        params = [
            model.coef_,
            model.intercept_,
        ]
    else:
        params = [
            model.coef_,
        ]
    return params

def set_model_params(
    model: LogisticRegression, params: LogRegParams
) -> LogisticRegression:

    model.coef_ = params[0]
    if model.fit_intercept:
        model.intercept_ = params[1]
    return model
```

```python
def shuffle(X: np.ndarray, y: np.ndarray) -> XY:

    rng = np.random.default_rng()
    idx = rng.permutation(len(X))
    return X[idx], y[idx]

def partition(X: np.ndarray, y: np.ndarray, num_partitions: int) ->
XYList:

    return list(
        zip(np.array_split(X, num_partitions), np.array_split(y, num_
partitions))
    )

def set_initial_params(model: LogisticRegression):

    n_classes = 2  # only 2 classes  Fraud or Geninue
    n_features = 9  # Number of features in dataset
    model.classes_ = np.array([i for i in range(n_classes)])
    model.coef_ = np.zeros((n_classes, n_features))
    if model.fit_intercept:
        model.intercept_ = np.zeros((n_classes,))

partition_id = np.random.choice(10)
(X_train, y_train) = partition(X_train, y_train, 10)[partition_id]
from sklearn.linear_model import LogisticRegression
model = LogisticRegression(
        penalty="l2",
        max_iter=1,  # local epoch
        warm_start=True,  # prevent refreshing weights when fitting
    )
model.fit(X_train, y_train)


class AML_Detection_Client(fl.client.NumPyClient):
        def get_parameters(self, config):
            return get_model_parameters(model)

        def fit(self, parameters, config):
            set_model_params(model, parameters)
            with warnings.catch_warnings():
                warnings.simplefilter("ignore")
                model.fit(X_train, y_train)
```

```python
            print(f"Training finished for round {config['server_
    round']}")
            return get_model_parameters(model), len(X_train), {}

        def evaluate(self, parameters, config):
            set_model_params(model, parameters)
            loss = log_loss(y_test, model.predict_proba(X_test))
            accuracy = model.score(X_test, y_test)
            print(loss,accuracy)
            return loss, len(X_test), {"accuracy": accuracy}


    fl.client.start_numpy_client(server_address="0.0.0.0:8080",
    client=AML_Detection_Client())
```

Open another terminal and run the client 2 code (python3 FL_AML_Client2.py):

```
bash-3.2$ python3 FL-AML-Client2.py
No of Fraud Transactions: 24392
No of Non Fraud Transactions: 608
No Frauds 97.57 % of the dataset
Frauds 2.43 % of the dataset
FL-AML-Client2.py:49: SettingWithCopyWarning:
A value is trying to be set on a copy of a slice from a DataFrame.
Try using .loc[row_indexer,col_indexer] = value instead

See the caveats in the documentation: https://pandas.pydata.org/pandas-docs/stable/user_guide/indexing.html#returning-a-view-versus-a-copy
  df_transactions[i] = encoder[i].fit_transform(df_transactions[i])
FL-AML-Client2.py:49: SettingWithCopyWarning:
A value is trying to be set on a copy of a slice from a DataFrame.
Try using .loc[row_indexer,col_indexer] = value instead

See the caveats in the documentation: https://pandas.pydata.org/pandas-docs/stable/user_guide/indexing.html#returning-a-view-versus-a-copy
  df_transactions[i] = encoder[i].fit_transform(df_transactions[i])
FL-AML-Client2.py:49: SettingWithCopyWarning:
A value is trying to be set on a copy of a slice from a DataFrame.
Try using .loc[row_indexer,col_indexer] = value instead

See the caveats in the documentation: https://pandas.pydata.org/pandas-docs/stable/user_guide/indexing.html#returning-a-view-versus-a-copy
  df_transactions[i] = encoder[i].fit_transform(df_transactions[i])
[[6.9200000e+02 0.0000000e+00 8.5202500e+03 ... 6.7770000e+03
  1.2096596e+06 0.0000000e+00]
 [7.1800000e+02 3.0000000e+00 6.3866400e+03 ... 1.7309000e+04
  0.0000000e+00 0.0000000e+00]
 [6.9400000e+02 3.0000000e+00 1.1976710e+04 ... 2.1263000e+04
  0.0000000e+00 0.0000000e+00]
 ...
 [7.0500000e+02 0.0000000e+00 1.9403522e+05 ... 7.5570000e+03
  0.0000000e+00 0.0000000e+00]
 [7.1000000e+02 1.0000000e+00 7.7252730e+04 ... 9.9600000e+02
  0.0000000e+00 0.0000000e+00]
 [6.9200000e+02 3.0000000e+00 2.9527900e+03 ... 1.9761000e+04
  0.0000000e+00 0.0000000e+00]]
/Library/Frameworks/Python.framework/Versions/3.8/lib/python3.8/site-packages/sklearn/linear_model/_logistic.py:458: ConvergenceWarning: lbfgs failed to converge (s
tatus=1):
STOP: TOTAL NO. of ITERATIONS REACHED LIMIT.

Increase the number of iterations (max_iter) or scale the data as shown in:
    https://scikit-learn.org/stable/modules/preprocessing.html
Please also refer to the documentation for alternative solver options:
    https://scikit-learn.org/stable/modules/linear_model.html#logistic-regression
  n_iter_i = _check_optimize_result(
INFO flwr 2023-03-14 18:52:11,716 | grpc.py:50 | Opened insecure gRPC connection (no certificates were passed)
DEBUG flwr 2023-03-14 18:52:11,717 | connection.py:38 | ChannelConnectivity.IDLE
DEBUG flwr 2023-03-14 18:52:11,734 | connection.py:38 | ChannelConnectivity.CONNECTING
DEBUG flwr 2023-03-14 18:52:11,748 | connection.py:38 | ChannelConnectivity.READY
Training finished for round 1
0.7942011907124793 0.9745333333333334
Training finished for round 2
0.5924659189564626 0.9470666666666666
Training finished for round 3
0.8851423211477334 0.9745333333333334
Training finished for round 4
0.6293475886678264 0.9745333333333334
Training finished for round 5
0.31233391627359897 0.9745333333333334
DEBUG flwr 2023-03-14 18:52:11,967 | connection.py:109 | gRPC channel closed
INFO flwr 2023-03-14 18:52:11,967 | app.py:153 | Disconnect and shut down
bash-3.2$ ▊
```

Figure 6.15 – Running client 2 and the logs

Once you run the client 2 code, pay close attention to the log statements on the server side. The server will initiate communication with both clients, enabling the exchange of the initial parameters and, subsequently, the updated weights for each round. Monitoring the server logs will provide insights into the progress of the FL process and the information shared between the clients and the server:

```
●  ●  ●                              arrao — bash — 138×69
No of Non Fraud Transactions: 83
No Frauds 99.67 % of the dataset
Frauds 0.33 % of the dataset
FL_AML_Server.py:51: SettingWithCopyWarning:
A value is trying to be set on a copy of a slice from a DataFrame.
Try using .loc[row_indexer,col_indexer] = value instead

See the caveats in the documentation: https://pandas.pydata.org/pandas-docs/stable/user_guide/indexing.html#returning-a-view-versus-a-copy
  df_transactions[i] = encoder[i].fit_transform(df_transactions[i])
FL_AML_Server.py:51: SettingWithCopyWarning:
A value is trying to be set on a copy of a slice from a DataFrame.
Try using .loc[row_indexer,col_indexer] = value instead

See the caveats in the documentation: https://pandas.pydata.org/pandas-docs/stable/user_guide/indexing.html#returning-a-view-versus-a-copy
  df_transactions[i] = encoder[i].fit_transform(df_transactions[i])
FL_AML_Server.py:51: SettingWithCopyWarning:
A value is trying to be set on a copy of a slice from a DataFrame.
Try using .loc[row_indexer,col_indexer] = value instead

See the caveats in the documentation: https://pandas.pydata.org/pandas-docs/stable/user_guide/indexing.html#returning-a-view-versus-a-copy
  df_transactions[i] = encoder[i].fit_transform(df_transactions[i])
INFO flwr 2023-03-14 18:48:55,252 | app.py:139 | Starting Flower server, config: ServerConfig(num_rounds=5, round_timeout=None)
INFO flwr 2023-03-14 18:48:55,273 | app.py:152 | Flower ECE: gRPC server running (5 rounds), SSL is disabled
INFO flwr 2023-03-14 18:48:55,273 | server.py:86 | Initializing global parameters
INFO flwr 2023-03-14 18:48:55,273 | server.py:270 | Requesting initial parameters from one random client
INFO flwr 2023-03-14 18:50:32,770 | server.py:274 | Received initial parameters from one random client
INFO flwr 2023-03-14 18:50:32,771 | server.py:88 | Evaluating initial parameters
INFO flwr 2023-03-14 18:50:32,777 | server.py:91 | initial parameters (loss, other metrics): 0.14332985986961255, {'accuracy': 0.996266666
6666666}
INFO flwr 2023-03-14 18:50:32,777 | server.py:101 | FL starting
DEBUG flwr 2023-03-14 18:52:11,734 | server.py:215 | fit_round 1: strategy sampled 2 clients (out of 2)
DEBUG flwr 2023-03-14 18:52:11,778 | server.py:229 | fit_round 1 received 2 results and 0 failures
WARNING flwr 2023-03-14 18:52:11,782 | fedavg.py:242 | No fit_metrics_aggregation_fn provided
INFO flwr 2023-03-14 18:52:11,798 | server.py:116 | fit progress: (1, 0.06257641314657952, {'accuracy': 0.9962666666666666}, 99.0195622230
0001)
DEBUG flwr 2023-03-14 18:52:11,799 | server.py:165 | evaluate_round 1: strategy sampled 2 clients (out of 2)
DEBUG flwr 2023-03-14 18:52:11,816 | server.py:179 | evaluate_round 1 received 2 results and 0 failures
WARNING flwr 2023-03-14 18:52:11,816 | fedavg.py:273 | No evaluate_metrics_aggregation_fn provided
DEBUG flwr 2023-03-14 18:52:11,816 | server.py:215 | fit_round 2: strategy sampled 2 clients (out of 2)
DEBUG flwr 2023-03-14 18:52:11,835 | server.py:229 | fit_round 2 received 2 results and 0 failures
INFO flwr 2023-03-14 18:52:11,848 | server.py:116 | fit progress: (2, 0.5546253194605769, {'accuracy': 0.9577333333333333}, 99.069346225)
DEBUG flwr 2023-03-14 18:52:11,848 | server.py:165 | evaluate_round 2: strategy sampled 2 clients (out of 2)
DEBUG flwr 2023-03-14 18:52:11,856 | server.py:179 | evaluate_round 2 received 2 results and 0 failures
DEBUG flwr 2023-03-14 18:52:11,856 | server.py:215 | fit_round 3: strategy sampled 2 clients (out of 2)
DEBUG flwr 2023-03-14 18:52:11,864 | server.py:229 | fit_round 3 received 2 results and 0 failures
INFO flwr 2023-03-14 18:52:11,878 | server.py:116 | fit progress: (3, 0.0891901385044784, {'accuracy': 0.9962666666666666}, 99.09927336199
999)
DEBUG flwr 2023-03-14 18:52:11,878 | server.py:165 | evaluate_round 3: strategy sampled 2 clients (out of 2)
DEBUG flwr 2023-03-14 18:52:11,895 | server.py:179 | evaluate_round 3 received 2 results and 0 failures
DEBUG flwr 2023-03-14 18:52:11,896 | server.py:215 | fit_round 4: strategy sampled 2 clients (out of 2)
DEBUG flwr 2023-03-14 18:52:11,917 | server.py:229 | fit_round 4 received 2 results and 0 failures
INFO flwr 2023-03-14 18:52:11,924 | server.py:116 | fit progress: (4, 0.04778146274879876, {'accuracy': 0.9962666666666666}, 99.145138024)
DEBUG flwr 2023-03-14 18:52:11,924 | server.py:165 | evaluate_round 4: strategy sampled 2 clients (out of 2)
DEBUG flwr 2023-03-14 18:52:11,932 | server.py:179 | evaluate_round 4 received 2 results and 0 failures
DEBUG flwr 2023-03-14 18:52:11,932 | server.py:215 | fit_round 5: strategy sampled 2 clients (out of 2)
DEBUG flwr 2023-03-14 18:52:11,943 | server.py:229 | fit_round 5 received 2 results and 0 failures
INFO flwr 2023-03-14 18:52:11,950 | server.py:116 | fit progress: (5, 0.029942986518022303, {'accuracy': 0.9962666666666666}, 99.170831520
00001)
DEBUG flwr 2023-03-14 18:52:11,950 | server.py:165 | evaluate_round 5: strategy sampled 2 clients (out of 2)
DEBUG flwr 2023-03-14 18:52:11,958 | server.py:179 | evaluate_round 5 received 2 results and 0 failures
INFO flwr 2023-03-14 18:52:11,959 | server.py:144 | FL finished in 99.179840135
INFO flwr 2023-03-14 18:52:11,960 | app.py:202 | app_fit: losses_distributed [(1, 0.4283888041973114), (2, 0.5735456347465515), (3, 0.4871
6623336076736), (4, 0.33856448344886303), (5, 0.17113844864070415)]
INFO flwr 2023-03-14 18:52:11,960 | app.py:203 | app_fit: metrics_distributed {}
INFO flwr 2023-03-14 18:52:11,960 | app.py:204 | app_fit: losses_centralized [(0, 0.14332985986961255), (1, 0.06257641314657952), (2, 0.55
46253194605769), (3, 0.0891901385044784), (4, 0.04778146274879876), (5, 0.029942986518022303)]
INFO flwr 2023-03-14 18:52:11,960 | app.py:205 | app_fit: metrics_centralized {'accuracy': [(0, 0.9962666666666666), (1, 0.996266666666666
6), (2, 0.9577333333333333), (3, 0.9962666666666666), (4, 0.9962666666666666), (5, 0.9962666666666666)]}
bash-3.2$
```

Figure 6.16 – The server-side logs

Observe the logs on both clients as well as the server side. These metrics provide an overview of the loss (indicating the model's performance) and accuracy (representing the model's correctness) for each client across multiple rounds of the FL process:

Server	Client 1	Client 2
INFO flwr 2023-03-14 17:46:49,202 | app.py:139 | Starting Flower server, config: ServerConfig(num_rounds=5, round_timeout=None) INFO flwr 2023-03-14 17:51:21,810 | server.py:101 | FL starting		
	DEBUG flwr 2023-03-14 17:51:21,778 | connection.py:38 | ChannelConnectivity. READY	
		ChannelConnectivity. DEBUG flwr 2023-03-14 17:53:46,338 | connection.py:38 | ChannelConnectivity. READY
DEBUG flwr 2023-03-14 17:53:46,338 | server.py:215 | fit_round 1: strategy sampled 2 clients (out of 2) DEBUG flwr 2023-03-14 17:53:46,351 | server.py:229 | fit_round 1 received 2 results and 0 failures WARNING flwr 2023-03-14 17:53:46,354 | fedavg.py:242 | No fit_metrics_aggregation_fn provided INFO flwr 2023-03-14 17:53:46,362 | server.py:116 | fit progress: (1, 0.06756539217831908, {'accuracy': 0.9962666666666666}, 144.549737353) DEBUG flwr 2023-03-14 17:53:46,363 | server.py:165 | evaluate_round 1: strategy sampled 2 clients (out of 2) DEBUG flwr 2023-03-14 17:53:46,377 | server.py:179 | evaluate_round 1 received 2 results and 0 failures		

Server	Client 1	Client 2
INFO flwr 2023-03-14 17:53:46,400 | server.py:116 | fit progress: (2, 0.40485776608772656, {'accuracy': 0.9633333333333334}, 144.58791799899996)		
INFO flwr 2023-03-14 17:53:46,432 | server.py:116 | fit progress: (3, 0.11833075507570899, {'accuracy': 0.9962666666666666}, 144.61946266499996)		
INFO flwr 2023-03-14 17:53:46,465 | server.py:116 | fit progress: (4, 0.1145626928425223, {'accuracy': 0.9962666666666666}, 144.65267561899998)		
INFO flwr 2023-03-14 17:53:46,497 | server.py:116 | fit progress: (5, 0.27867744042157033, {'accuracy': 0.9861333333333333}, 144.68508043599996)		
INFO flwr 2023-03-14 17:53:46,511 | app.py:202 | app_fit: losses_distrib- uted [(1, 0.4398987330496311), (2, 0.4606742262840271), (3, 0.5105149038136005), (4, 0.5070083439350128), (5, 0.5951354652643204)]		

Server	Client 1	Client 2
	Training finished for round 1:	Training finished for round 1:
	0.06756539217831908 0.9962666666666666	0.8122320748323023 0.9745333333333334
	Training finished for round 2:	Training finished for round 2:
	0.40485776608772656 0.9633333333333334	0.5164906830160562 0.9541333333333334
	Training finished for round 3:	Training finished for round 3:
	0.11833075507570899 0.9962666666666666	0.9026990471833415 0.9745333333333334
	Training finished for round 4:	Training finished for round 4:
	0.1145626928425223 0.9962666666666666	0.8994540131249842 0.9745333333333334
	Training finished for round 5:	Training finished for round 5:
	0.27867744042157033 0.9861333333333333	0.9115935132282235 0.9736
	DEBUG flwr 2023-03-14 17:53:46,521 | connection.py:109 | gRPC channel closed	DEBUG flwr 2023-03-14 17:53:46,521 | connection.py:109 | gRPC channel closed
	INFO flwr 2023-03-14 17:53:46,522 | app.py:153 | Disconnect and shut down	INFO flwr 2023-03-14 17:53:46,522 | app.py:153 | Disconnect and shut down

Table 6.8 – Log data at Server and Clients

As per the log, there are 5 rounds of communication between clients and the server, and in each round, accuracy results and loss change based on the weights.

Client 1	Loss	Accuracy
Round 1	0.06756539217831908	0.9962666666666666
Round 2	0.40485776608772656	0.9633333333333334

Client 1	Loss	Accuracy
Round 3	0.11833075507570899	0.9962666666666666
Round 4	0.1145626928425223	0.9962666666666666
Round 5	0.27867744042157033	0.9861333333333333

Table 6.9 – Accuracy and Loss metrics at Client 1

As per the debug logs, loss and accuracy vary on client 1. Let's observe the loss and accuracy results on Client 2 as well.

Client 2	Loss	Accuracy
Round 1	0.8122320748323023	0.9745333333333334
Round 2	0.5164906830160562	0.9541333333333334
Round 3	0.9026990471833415	0.9745333333333334
Round 4	0.8994540131249842	0.9745333333333334
Round 5	0.9115935132282235	0.9736

Table 6.10 – Accuracy and Loss metrics at Client 2

We have implemented a sample fraud detection application using Federated Learning and made use of open-source frameworks like Flower. In the next section, let's try to learn and implement federated learning using differential privacy.

FL with differential privacy

Federated Learning with Differential Privacy (FL-DP) is an approach that combines the principles of FL and **Differential Privacy (DP)** to ensure privacy and security in distributed ML systems. FL-DP aims to protect sensitive data while enabling collaborative model training across multiple devices or entities.

The goal of FL-DP is to achieve accurate model training without compromising the privacy of individual data contributors. It addresses the challenge of preventing data leakage during the aggregation of model updates from different participants. By incorporating DP techniques, FL-DP provides strong privacy guarantees by adding noise or perturbation to the model updates or gradients before aggregating them.

There are different approaches to implementing FL-DP. One common approach involves each client training a local ML model using their own data. The client applies techniques such as clipping and noise addition to the gradients or weights of the model. The client then sends the updated data to the server. On the server side, the updates are aggregated while preserving privacy using techniques such as secure aggregation or privacy-preserving FL algorithms. This ensures that individual client data remains private while enabling collaborative model training.

FL-DP algorithms may vary depending on the specific differential privacy mechanisms used, such as Gaussian noise addition, subsampling, or advanced techniques such as **Private Aggregation of Teacher Ensembles (PATE)**. The choice of techniques depends on the level of privacy required and the characteristics of the distributed dataset.

Implementing FL-DP requires careful consideration of privacy, accuracy, and computational overhead. It involves striking a balance between preserving privacy and maintaining model utility. Various frameworks and libraries, such as Flower and TensorFlow Privacy, provide tools and techniques to facilitate the implementation of FL-DP.

FL-DP has the potential to unlock the benefits of collaborative ML in scenarios where data privacy and security are paramount. By preserving privacy, FL-DP enables organizations and individuals to collaborate on model training while safeguarding sensitive information.

FL-DP provides a way to implement privacy-preserving techniques in the FL process, ensuring that client-side data remains protected.

In this section, we will explore two general approaches to implementing FL-DP, although specific frameworks and implementations may have slight variations.

Approach one

This approach shares similarities with **Differentially Private Federated Averaging (DP-FedAvg)**, which was introduced by the Google research team. By following these approaches, FL-DP allows you to train ML models on client data while preserving privacy through techniques such as clipping and noise addition.

Each client does the following:

1. Trains an ML/DL model using its local data.
2. Computes gradients/weights using a standard SGD algorithm.
3. Applies clipping to the weights to limit their sensitivity.
4. Adds noise to the weights to introduce randomness and privacy.
5. Sends the modified weights to the server.

The server does the following:

1. Computes the average of the weights received from each client.
2. Broadcasts back the updated weights to the clients.
3. Alternatively to step 1, applies clipping and adds noise to the final weights before broadcasting.

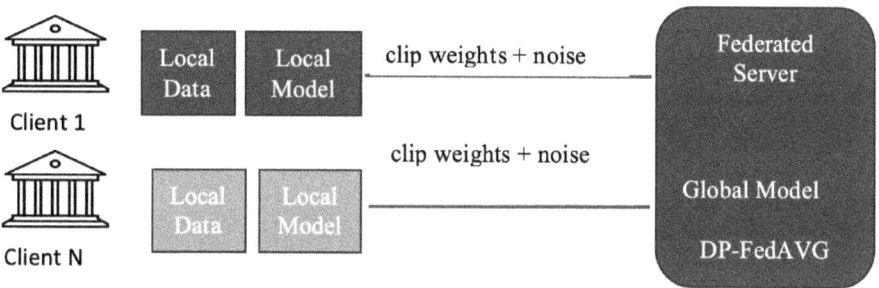

Figure 6.17 – The DP-FedAvg model weights exchanged with the server

In this approach, each client trains its model on local data and work on the updated average weights sent by the server.

Approach two

In this approach, each client trains its model on local data and applies privacy-preserving techniques to compute the gradients/weights. The server then incorporates these noisy weights and performs aggregation using the FD-SGD algorithm, ensuring privacy is maintained throughout the FL process.

Each client does the following:

1. Trains an ML model using its local data.

2. Computes the gradients/weights using either noisy SGD or DP-SGD (DP stochastic gradient) algorithms, which incorporate noise during gradient computation to preserve privacy.

3. Sends the weights to the server.

The server does the following:

1. Utilizes the noisy weights received from the clients.

2. Follows the **Federated Differential SGD (FD-SGD)** algorithm, which incorporates privacy-preserving techniques during the aggregation process on the server.

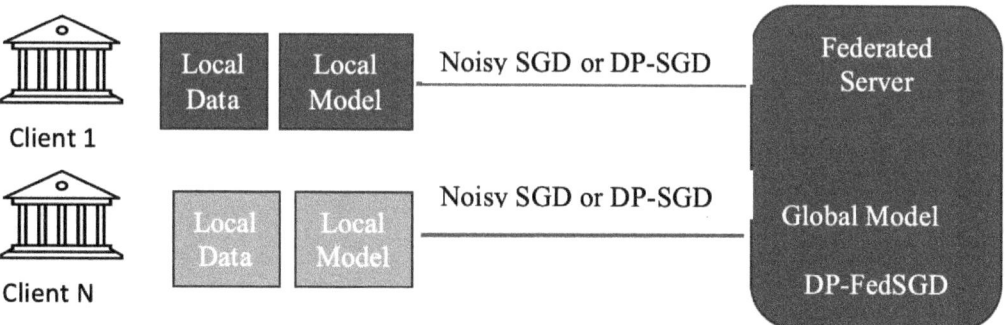

Figure 6.18 – The DP-FedSGD model weights exchanged with the server

There are various variants of Differential Privacy Federated Learning (FL-DP) algorithms designed to address different scenarios, such as cross-device and cross-silo FL, with both homogeneous and heterogeneous data. In our implementation, we will apply FL-DP to the same example as before, ensuring privacy preservation throughout the FL process.

A sample application using FL-DP

At the time of writing, the Flower framework (version 1.3) currently offers experimental support for FL-DP. It provides a strategy class (similar to FedAvg, FedYogi, and so on) specifically designed to support FL-DP. The class name designed to support this is DPFedAvg.

The DPFedAvg class in the Flower framework is a component specifically designed to support FL-DP. It extends the functionality of the FedAvg algorithm by incorporating differential privacy techniques to protect the privacy of individual client data during model aggregation.

DPFedAvg implements a privacy-preserving mechanism that ensures the privacy of client updates while enabling collaborative model training. It achieves this by adding noise or perturbation to the model updates or gradients received from each client, before aggregating them on the server side.

The key features and functionalities of the DPFedAvg class include the following:

- **DP**: DPFedAvg integrates DP techniques into the FL process, ensuring that the privacy of individual client data is preserved during model training.

- **Noise addition**: DPFedAvg applies noise to the gradients or model updates received from each client before aggregating them. The amount of noise added is determined based on privacy parameters and privacy budget allocation.

- **Privacy budget management**: DPFedAvg incorporates mechanisms to manage and allocate the privacy budget effectively, ensuring that the desired privacy guarantees are maintained throughout the training process.

- **Privacy parameters**: DPFedAvg allows users to customize the privacy parameters such as privacy budget, noise distribution, and sensitivity of the model updates. These parameters enable fine-grained control over the level of privacy and utility trade-off.

- **Model aggregation**: DPFedAvg performs the aggregation of client updates using the DP averaging algorithm. This ensures that the privacy of individual updates is preserved while generating an updated global model.

- **Compatibility with the Flower framework**: DPFedAvg is designed to seamlessly integrate with the Flower framework, allowing users to incorporate DP into their FL pipelines using the existing Flower infrastructure.

By using the `DPFedAvg` class in the Flower framework, developers and ML engineers can implement FL-DP straightforwardly and efficiently. It provides a powerful tool to ensure privacy in distributed ML scenarios while maintaining the collaborative benefits of FL.

Let's walk through the Flower-provided class in detail.

The DPFedAvgFixed class

This class is a wrapper class and adds clipping and Gaussian noise to the weights. The constructor of this class supports parameters to set server-side noise, a clip norm value, and the noise multiplier.

Let's use this class on the server side. The server code is as follows:

```
def DP_Fed_Server():
    model = LogisticRegression(max_iter=10000)
    set_initial_params(model)
    strategy = fl.server.strategy.FedAvg(
        min_available_clients=2,
        evaluate_fn=get_evaluate_fn(model,X_test,y_test),
        on_fit_config_fn=fit_round,
    )


    dps = DPFedAvgFixed(strategy,
    num_sampled_clients=2,
    clip_norm=0.03,
    noise_multiplier=0.5)


    fl.server.start_server(
            server_address="0.0.0.0:8080",
              strategy=dps,
             config=fl.server.ServerConfig(num_rounds=5),
        )
DP_Fed_Server()
```

The source code of the Jupyter notebooks for the server and clients is located in the `Chapter 6` folder:

- Server code: `Fed-DP-AML-Server.ipynb`
- Client 1 code: `DP-FL-AML_Client1.ipynb`
- Client 2 code: `DP-FL-AML-Client2.ipynb`

Let's look at the accuracy of the model for the clients and server:

Client 1	Client 2	Server
0.9962666666666666	0.9745333333333334	0.9962666666666666
0.7162666666666667	0.5978666666666667	0.7162666666666667
0.9876	0.9693333333333334	0.9876
0.9372	0.9448	0.9372
0.7714666666666666	0.7708	0.7714666666666666

Table 6.11 – Accracy results at Sever and Clients

Applying DP to FL introduces some overhead in terms of computational cost, communication overhead, and potentially reduced model performance. In our example case, by the fourth round, the accuracy was 93%, but in the fifth round, the accuracy suddenly dropped. This tells us that we need to monitor the accuracy during training to help us decide on the number of rounds each client needs to participate in and stop further rounds when the accuracy drops.

Summary

In this chapter, we explained why FL is needed and looked at its definition and characteristics in detail. We covered the steps involved in implementing FL and discussed IID and non-IID datasets and FL algorithms. We implemented a sample application using an open source FL framework. Finally, we converted the same application using DP.

In the next chapter, we will learn about FL benchmarks and look at key start-ups that are working on or already have FL products.

7

Federated Learning Benchmarks, Start-Ups, and the Next Opportunity

This chapter focuses on the importance of **Federated Learning** (**FL**) benchmarks and highlights products offered by start-up companies in the field.

We will cover the following main topics:

- FL benchmarks:

  - An introduction to FL benchmarks, including their significance

  - Considerations when designing FL benchmarks

  - An overview of FL datasets

  - A high-level overview of various FL benchmark suites

  - Selecting the appropriate FL framework for a project

- State-of-the-art research in FL

- Netxt Opportunity and Key start-up company products in FL

By exploring these topics, you will gain a comprehensive understanding of the need for FL benchmarks and the latest advancements in the field. Additionally, we will showcase notable products developed by start-up companies that are closely related to FL.

FL benchmarks

FL is a machine learning technique that allows multiple devices/clients or servers to collaboratively train a model and keep their data private. There has been an increasing need for standardized benchmarks to evaluate the performance of different FL algorithms and frameworks/platforms.

The IEEE 3652.1-2020 standard, officially titled *IEEE Guide for Architectural Framework and Application of Federated Machine Learning*, is a comprehensive guide that provides an architectural framework for **Federated Machine Learning (FML)**. More details about the IEEE 3652.1-2020 standard can be found at `https://ieeexplore.ieee.org/document/9382202`.

FL benchmarks are datasets and evaluation metrics that are used to compare and evaluate the performance of different FL algorithms and frameworks/platforms. These benchmarks can help engineers and researchers to identify the strengths and weaknesses of different algorithms and the pros and cons of different FL frameworks.

Next, we will discuss the importance of FL benchmarks and the key considerations that should be taken into account when designing them.

The importance of FL benchmarks

FL benchmarks are essential for various reasons.

Firstly, they help researchers evaluate the performance of different FL algorithms on standard public datasets, which can help to identify which algorithms are most effective for particular use cases. Comparing the performance of different algorithms can help drive progress in the field by identifying areas where new algorithms are needed, or where existing algorithms need to be improved.

Secondly, these benchmarks also help to compare the performance of different FL frameworks and identify which framework is best suited to solving business use cases using FL.

Finally, benchmarks can help to improve the reproducibility of research results by providing a standardized set of metrics and evaluation procedures.

Key considerations when designing FL benchmarks: When designing FL benchmarks, several key considerations should be taken into account. Some of them are as follows:

- **Data distribution**: FL algorithms are designed to work with data that is distributed across multiple devices or servers. Therefore, benchmarks should include datasets that are representative of the types of data that might be encountered in real-world use cases.

- **Heterogeneity**: FL involves training models on data that is distributed across different devices or servers, each with potentially different hardware capabilities and network conditions. Therefore, benchmarks should include datasets that are diverse and reflect different types of devices and network conditions.

- **Privacy**: One of the main benefits of FL is its ability to protect the privacy of data by keeping it on users' devices. Therefore, benchmarks should ensure that the data used for training is representative of real-world use cases while still protecting users' privacy.

- **Evaluation metrics**: Benchmarks should include standard metrics to evaluate the performance of FL algorithms. These metrics should be carefully selected to ensure that they accurately reflect the performance of the algorithm and apply to the specific use case.

- **Reproducibility**: FL benchmarks should be designed to be reproducible, meaning that the datasets and evaluation procedures should be clearly documented and made available to others.

We have discussed the key considerations in designing the FL benchmarks. Let's explore further the datasets that need to be considered in FL benchmarks.

FL datasets

One of the key aspects of FL is the selection of datasets that will be used to train the model. Next, we will explore the various considerations that need to be taken into account when selecting datasets for FL:

- **Data privacy**: The most critical consideration when selecting datasets for FL is data privacy. Since the data is stored locally on each device, it is essential to ensure that the data is not exposed to any unauthorized parties. Data privacy can be ensured by using encryption techniques or by anonymizing the data. Additionally, it is important to have proper consent and permission from the device owners to use their data for FL. Support for differential privacy also needs to be considered because data anonymization alone may not be sufficient.

- **Data diversity**: Another important consideration when selecting datasets for FL is data diversity. The datasets should be diverse enough to capture a wide range of data patterns and features. This is essential to ensure that the model is not biased toward a particular set of data. For example, if the datasets only include data from a particular region or demographic, the model may not be able to generalize well to other regions or demographics.

- **Data quality**: Data quality is also a critical consideration when selecting data sets for FL. The datasets should be clean and free from errors or inconsistencies. Data cleaning and preprocessing techniques may be required to ensure that the data is suitable for training the model. Additionally, the data should be representative of the real-world scenario to ensure that the model can perform well in practical situations.

- **Data size**: The size of the datasets is also an important consideration for FL. The datasets should be large enough to train the model effectively but not too large that they become unwieldy to work with. A good balance needs to be struck between the size of the datasets and the computational resources available to train the model.

- **Data distribution**: Finally, the distribution of data across devices is also an important consideration for FL. The data should be distributed in such a way that each device has access to a representative sample of the data. This ensures that each device contributes to the training of the model in a meaningful way. Additionally, the data distribution should be balanced to avoid any devices being overloaded with data.

Datasets are critical in benchmarking FL systems. Next, we will explore the following popular standard datasets that are used to benchmark FL systems:

- FLAIR
- Federated EMNIST
- Shakespeare
- CIFAR-10
- OpenStreetMap
- Medical image analysis datasets

FLAIR

Federated Learning Annotated Image Repository (**FLAIR**) is an open source project that aims to provide a centralized repository of annotated images for FL. The repository contains a diverse set of images, including medical images, satellite images, and natural images, with corresponding annotations. FLAIR provides a platform for researchers and developers to evaluate FL algorithms on image-based tasks, such as image classification and object detection. The repository allows you to share and collaborate on annotated images, reducing the need for individual organizations to collect and label their own data. The annotations in FLAIR are designed to be privacy-preserving, with sensitive information removed or obfuscated. The privacy-preserving annotations and standardized format make FLAIR a valuable resource for evaluating FL algorithms on image-based tasks, enabling easy integration with popular machine learning frameworks and tools.

FLAIR has become a popular resource in the FL community, with several FL frameworks integrating it into their evaluation pipeline. The repository is continually updated, with new datasets and annotations added regularly. It consists of 430,000 images from 51,000 Flickr users, which are mapped to 17 coarse-grained labels such as art, food, plant, and outdoor.

Visit the following GitHub URL for more information: `https://github.com/apple/ml-flair`.

Federated EMNIST

Federated Extended MNIST (**Federated EMNIST**) is a benchmark dataset for FL developed by Google. It consists of a set of images of handwritten digits and letters, similar to the MNIST dataset. The dataset is distributed across multiple devices, and the goal is to train a model that can classify images correctly while keeping data private.

Shakespeare

This dataset consists of a collection of Shakespearean texts, which can be used to train a language model for text generation.

CIFAR-10

This is a widely used benchmark dataset for computer vision tasks, consisting of a set of 60,000 color images of 10 different classes. It is used to train a model to classify images correctly while keeping data private.

OpenStreetMap

This is a dataset of geographic data, including maps, satellite imagery, and GPS coordinates. The dataset can be distributed across multiple devices, and the goal is to train a model that can predict traffic patterns or other features of the environment while preserving the privacy of data.

Medical image analysis datasets

There are several medical imaging datasets that can be used for FL, such as the **Brain Tumor Segmentation (BraTS)** challenge dataset and the **Prostate MR Image Segmentation (PROMISE12)** challenge dataset. These datasets consist of medical images that can be used to train models for diagnosis and treatment planning while preserving the privacy of patient data.

Frameworks for FL benchmarks

There are several FL benchmark frameworks available in the open source community, as well as commercially. FL frameworks provide libraries/platforms to benchmark FL systems and applications.

Some of the most popular FL systems and benchmarks are the following:

- LEAF
- FedML
- FATE
- FedScale
- PySyft
- MLCommons – MLPerf
- MedPerf
- TensorFlow Federated (TFF)
- Flower (we covered this in the previous chapter so won't go into detail again in this chapter)

We'll look at these in more detail in the following subsections.

The LEAF FL benchmarks suite

The LEAF FL benchmarks suite is a project initiated by Carnegie Mellon University, which aims to provide a standardized benchmarking suite for FL algorithms (https://leaf.cmu.edu/).

The LEAF benchmarks enable researchers and developers to evaluate the performance of FL algorithms across different domains and configurations. LEAF provides a standardized set of tasks and datasets to evaluate the performance of FL algorithms on various metrics, including convergence speed, communication efficiency, and accuracy. The broad range of tasks includes image classification, text classification, and regression tasks.

The datasets used in the benchmarks are designed to be representative of real-world scenarios, with varying levels of data distribution, data size, and data complexity. The benchmarks are conducted under various settings, such as varying the number of clients, the communication rounds, and the amount of data available to each client. The benchmarks also evaluate the algorithms' robustness to adversarial attacks and noisy data.

LEAF's core components are as follows:

- Datasets
- Reference implementation
- Metrics

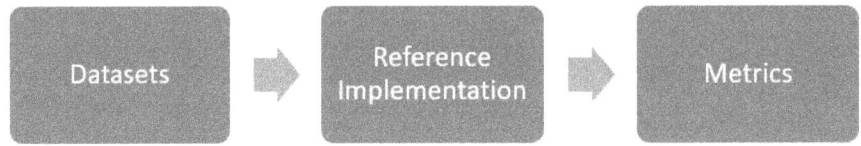

Figure 7.1 – LEAF's core components and flow

More details about the benchmarks of LEAF can be obtained at https://arxiv.org/abs/1812.01097.

FedML

FedML (https://github.com/FedML-AI/FedML) is a benchmark suite for FL that includes several datasets and tasks, such as image classification, language modeling, and speech recognition. FedML evaluates FL algorithms on their robustness in **non-independent and identically distributed (non-iid)** data distributions, scalability to large datasets, and communication efficiency.

The framework supports various types of FL setups, such as horizontal, vertical, and federated transfer learning, as well as different types of optimization algorithms, including federated averaging, FedProx, and FedAdapt. The FedML framework is designed to be modular, flexible, and scalable, allowing users to customize and extend the existing algorithms or develop new ones easily. It also provides a set of benchmark datasets and evaluation metrics to facilitate the comparison of different algorithms.

The FedML community actively develops and maintains the framework and organizes workshops and challenges to promote research and innovation in FL. The community also collaborates with industry partners to apply FL to real-world use cases, such as healthcare, finance, and telecommunications. FedML is an important initiative that addresses the challenges and opportunities of FL and enables the development of secure, privacy-preserving, and efficient machine learning models in a distributed setting. FedML supports three types of computing platforms – IoT/mobile, distributed computing, and standalone simulation.

More details can be found at the following reference research URL: `https://arxiv.org/abs/2007.13518`.

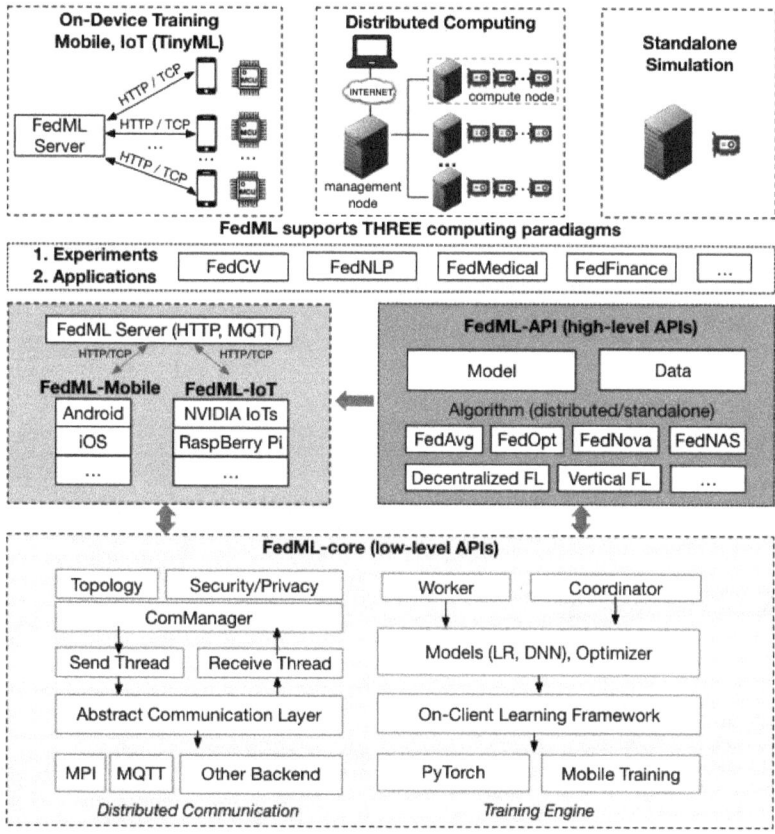

Figure 7.2 – FedML architecture

FATE

Federated AI Technology Enabler (**FATE**) is an open source project that provides a secure and privacy-preserving way to collaborate on **artificial intelligence** (**AI**) models. FATE is designed to enable FL, which is a distributed machine learning approach that allows multiple parties to train a shared model without sharing their data. FATE is developed by the WeBank AI department, a subsidiary of Tencent. `https://github.com/FederatedAI/FATE`

FATE provides a platform for developers to implement FL techniques in their AI applications. It supports different types of FL, such as horizontal and vertical, and provides tools to manage the communication, synchronization, and aggregation of models across different devices. FATE uses a variety of cryptographic techniques to ensure the privacy and security of the FL process. For example, FATE employs differential privacy to add noise to training data, preventing the leakage of sensitive information about individual users. FATE also uses homomorphic encryption to enable computation on encrypted data, which allows parties to collaborate without revealing their data to each other. We will learn about homomorphic encryption in the next chapter.

FATE is built on top of Kubernetes, a popular open source container orchestration system. Kubernetes provides FATE with the ability to manage containers and automate the deployment, scaling, and monitoring of the FL infrastructure. FATE leverages cryptographic techniques and Kubernetes to ensure the privacy and security of the FL process. FATE provides a web-based **graphical user interface (GUI)** for developers to interact with the platform, monitor the training process, and visualize the results. It has been adopted by various companies and organizations to build privacy-preserving AI applications. For example, Tencent has used FATE to develop a privacy-preserving recommendation system for its e-commerce platform. Huawei has also used FATE to build an FL platform for the healthcare industry. FATE's popularity is expected to grow as FL becomes increasingly important in the AI industry.

The architecture of FATE

The following figure shows the FATE architecture:

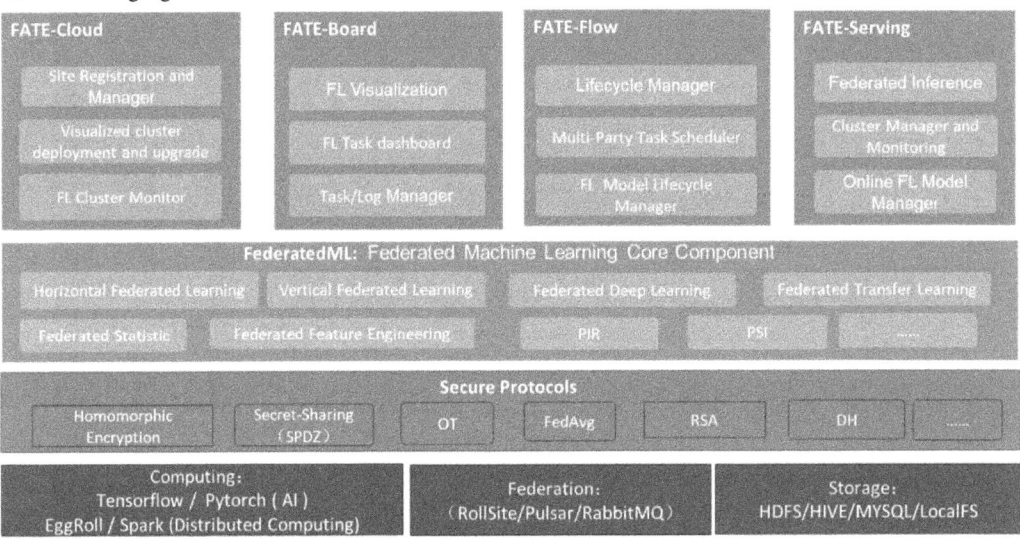

Figure 7.3 – FATE architecture

The preceding figure was sourced from `https://fate.readthedocs.io/en/latest/architecture/`.

The FedScale benchmarking platform

The FedScale benchmark suite includes several datasets and workloads that are designed to evaluate the performance of FL algorithms and systems. The datasets are selected to reflect the heterogeneity and non-IIDness of real-world data. The workloads are designed to simulate the communication overhead and computation complexity of real-world FL scenarios, as well as evaluate the performance of FL algorithms and systems in these scenarios.

One of the datasets included in the FedScale benchmark suite is the Synthetic IID dataset, which consists of synthetic data that is randomly generated and distributed evenly among the clients. This dataset is designed to evaluate the scalability and efficiency of FL algorithms and systems. Another dataset included in the FedScale benchmark suite is the Heterogeneous Image dataset, which consists of images from different sources and domains. This dataset is designed to evaluate the robustness of FL algorithms and systems to heterogeneous data.

The FedScale benchmark suite also includes more complex workloads, such as the Federated Meta-Learning workload, which involves clients learning from each other's data.

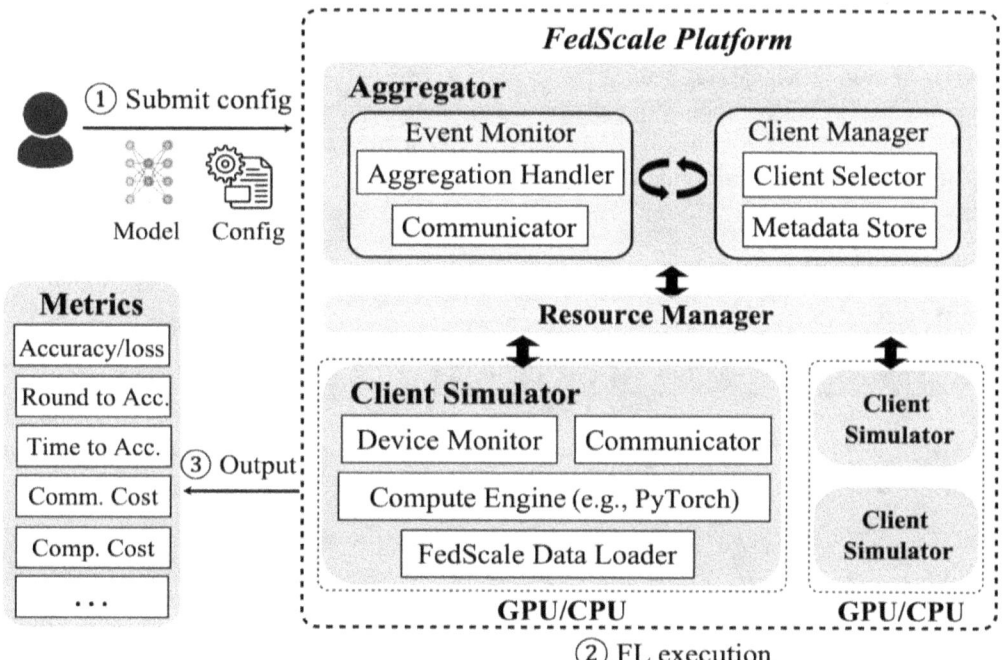

Figure 7.4 – The FedScale runtime to run FL benchmarks

The preceding figure was sourced from https://github.com/SymbioticLab/FedScale/blob/master/docs/fedscale-sim-mode.png.

MLCommons

MLCommons is a non-profit organization that aims to accelerate machine learning innovation and development by providing a platform for collaboration and the sharing of resources among its members. It was founded in June 2020 by a group of leading researchers, engineers, and entrepreneurs in the field of machine learning.

The organization's primary focus is on developing and promoting best practices for designing and deploying machine learning systems, including hardware and software frameworks, datasets, benchmarks, and evaluation metrics. One of the key initiatives of MLCommons is the MLPerf benchmark suite, which is designed to measure the performance of machine learning systems across a range of applications, including computer vision, natural language processing, and recommendation systems. The benchmarks are developed and maintained by a global community of researchers and engineers, and they are used to evaluate the performance of hardware and software platforms for machine learning. MLPerf has quickly become the industry standard for benchmarking machine learning performance and is used by leading technology companies, researchers, and government agencies around the world.

MedPerf

MedPerf is an open benchmarking platform for medical AI using FL. The MLCommons team piloted multiple use cases in collaboration with multiple institutions and universities to run FL models using MDPerf benchmarks. The use cases are brain tumor segmentation, pancreas segmentation, surgical workflow phase recognition, and cloud experiments.

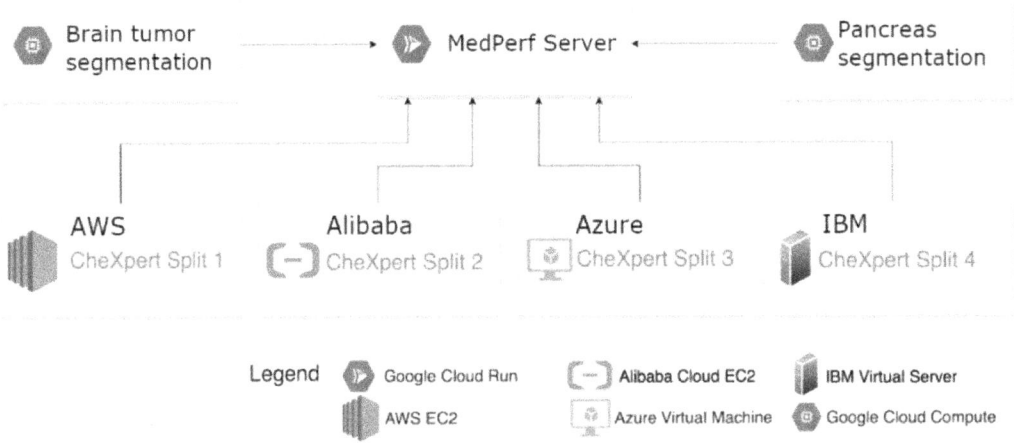

Figure 7.5 – MedPerf architecture for cloud experiments

The preceding figure was sourced from `https://github.com/mlcommons/medperf`.

Selecting an FL framework for a project

Selecting the best FL framework can be a challenging task, since there are several factors to consider, such as the features, scalability, security, ease of use, documentation, the machine learning algorithms to support (out of the box), neural network support, and the learning curve.

The following are some of the key considerations when selecting an FL framework:

- **Features**: It is important to consider the features that the framework offers. Some of the critical features to look for include support for different machine learning algorithms, distributed training, data privacy, and data validation. You should choose a framework that has features that match your use case needs.

- **Scalability**: Another critical consideration is scalability. The framework should be able to handle large datasets and a high number of participants. You should also consider whether the framework can scale horizontally or vertically. In the case of cross-device (mobile, web, etc.), the framework should be able to handle a large number of clients.

- **Security and privacy**: FL involves multiple parties sharing data, so it's essential to select a framework that offers robust security features. The framework should have encryption and authentication capabilities to protect the data from unauthorized access. You should also consider whether the framework has a transparent security model, allowing you to audit the security measures in place. You need to consider whether or not the framework supports privacy protection by default, such as differential privacy.

- **Ease of use**: The ease of use of the framework is also a critical consideration. The framework should have an intuitive user interface that makes it easy to set up, configure, and use. You should also consider whether the framework has good documentation and community support, as this can help you troubleshoot any issues that arise.

- **Learning curve**: You should consider the learning curve of the framework as well in terms of user documentation, deployment setup, programming language support, the examples and samples provided, and so on.

- **Compatibility**: Finally, you should consider whether the framework is compatible with your existing technology stack and programming language. For example, if you're using TensorFlow for machine learning, you may want to select an FL framework that is compatible with TensorFlow.

In conclusion, selecting the best framework requires careful consideration of several factors, including features, scalability, security, privacy, ease of use, the learning curve, and compatibility with your existing technology stack. By considering these factors, you can select a framework that meets your requirements to solve the business use case using FL.

A comparison of FedScale, FATE, Flower, and TensorFlow Federated

FedScale, FATE, Flower, FedML, and **TensorFlow Federated** (**TFF**) are some of the popular FL frameworks that offer unique features and capabilities. In the following subsections, we will compare these features and capabilities.

Features

FedScale is a framework that supports FL for a wide range of machine learning models, including deep neural networks, decision trees, and **Support Vector Machines** (**SVMs**). It also supports various optimization algorithms such as federated **stochastic gradient descent** (**SGD**) and federated proximal algorithms. FATE, conversely, offers a comprehensive suite of FL tools and supports various machine learning algorithms, including logistic regression, decision trees, and deep learning models. Flower is a lightweight framework that supports FL for various machine learning models, including neural networks, decision trees, and logistic regression. TFF is a widely used FL framework that supports TensorFlow-based machine learning models.

Scalability

FedScale and FATE are designed for large-scale deployments and can handle a high number of participants. Flower is a lightweight framework that is ideal for small to large-scale FL projects. TFF is a scalable framework that can handle large-scale FL projects and can be used to deploy FL algorithms in production environments.

Security

All four frameworks offer robust security features such as encryption and authentication to protect data during the FL process. FATE has additional features such as differential privacy and homomorphic encryption to enhance data privacy. TFF also offers secure **multi-party computation** (**MPC**) protocols that enable multiple parties to securely collaborate on FL projects.

Ease of use

FedScale and FATE have a steeper learning curve than Flower and TFF, as they offer more complex features and tools. Flower and TFF are both easy to use, with intuitive APIs and documentation that make it easy to get started. TFF also offers a high-level API that enables users to build FL algorithms without requiring deep knowledge of the underlying details.

Community support

All four frameworks have active open source contributors that provide support and documentation. TFF, being developed by Google, has the largest community and support, followed by Flower, FedScale, and FATE.

Each of these frameworks has its unique features and capabilities, making them suitable for different use cases. FedScale and FATE are ideal for large-scale FL deployments, while Flower and TFF are more suited for small to medium-scale projects. TFF has the largest community and support, making it an excellent choice for developers looking for an FL framework with extensive support and resources. Ultimately, the choice of framework will depend on the specific use case, requirements, and familiarity with the framework.

Research papers

Each framework published its own benchmark and results in the research journals, with a list of supported features and accuracy metrics. Currently, there are no industry standard benchmarks similar to `TPC.Org` or `SPEC.org` that are widely accepted.

State-of-the-art research in FL

FL is a rapidly growing technology, and there are several state-of-the-art research directions in this space. The following are some of the recent advances in FL research.

Here is a high-level comparison of the different FL frameworks:

High-level capabilities	FATE	FedML	Flower	FedScale	TFF
Regression models	Y	Y	Y	Y	Y
Neural networks	Y	Y	Y	Y	Y
Tree-based models	Y	N	N	N	N
Communication protocol	Customized	MPI	gRPC	gRPC	gRPC
Support for differential privacy	N	N	Y	Y	Y
Single host deployment	Y	Y	Y	Y	Y
Cross-device deployment	N	Y	Y	Y	N
Research papers	Y	Y	Y	Y	Y

Table 7.1 – Comparison of the different FL frameworks

Communication-efficient FL

One of the main challenges in FL is the high communication cost of transmitting the model updates between the devices and the central server. Recent research has focused on reducing the communication cost in FL by developing new compression techniques, such as quantization, sparsification, and differential compression. These techniques can significantly reduce the communication cost without sacrificing the model's accuracy. Several researchers and research groups are actively working on **communication-efficient FL (CE-FL)** research:

- **Google Research**: Google Research has been actively working on CE-FL research and has developed several techniques to reduce communication costs in FL. One of its recent works is *Communication-Efficient Learning of Deep Networks from Decentralized Data*, which proposes a new compression technique called **Quantized SGD (QSGD)** that can significantly reduce the communication cost in FL. This is a family of compression schemes that allow the compression of gradient updates at each node, while guaranteeing convergence under standard assumptions. See the Arxiv paper at `https://arxiv.org/abs/1602.05629`.

- **The Federated Learning Community**: The Federated Learning Community is an open community of researchers and practitioners who are working on FL research. The community has several working groups, including a CE-FL working group, which is focused on developing new techniques to reduce communication costs in FL.

- **Carnegie Mellon University**: The Machine Learning department at Carnegie Mellon University has several researchers who are working on CE-FL research. One of its recent works is *Communication-Efficient Distributed Learning with Feature-Selective Sampling*, which proposes a new sampling technique that can reduce the communication cost in FL.

- **IBM Research**: IBM researchers and the University of Michigan are working on CE-FL research, and their most recent work is *Federated Learning with Matched Averaging*, which proposes a new aggregation technique that can reduce the communication cost in FL. See the Arxiv paper at `https://arxiv.org/abs/2002.06440`.

These are just a few examples of the many researchers and research groups who are actively working on CE-FL research. As the field of FL continues to evolve, it is expected that more researchers and research groups will focus on developing new techniques to reduce the communication cost in FL.

Privacy-preserving FL

Privacy is a critical concern in FL, since data is distributed across multiple devices, and each device's owner wants to protect their data privacy. Recent research has focused on developing new privacy-preserving techniques, such as differential privacy and secure multi-party computation, to protect the data privacy of the devices' owners while enabling collaborative model training. These techniques can provide strong privacy guarantees while preserving the model's accuracy. Several researchers and research groups are actively working on **privacy-preserving FL (PP-FL)** research.

Here are some notable examples:

- **OpenMined**: OpenMined is an open source community dedicated to advancing PP-FL research. It has developed several PP-FL frameworks, including PySyft, which is a Python library for PP-FL.

- **Google Research**: Google Research has been actively working on PP-FL research and has developed several techniques to protect the privacy of user data in FL. One of its recent works is *Privacy-Preserving Federated Learning with Byzantine Robust Aggregation*, which proposes a new aggregation technique that can protect user privacy in the presence of malicious actors.

- **Carnegie Mellon University**: The CyLab Security and Privacy Institute at Carnegie Mellon University has several researchers who are working on PP-FL research. One of its recent works is *Privacy-Preserving Federated Learning via Randomized Smoothing*, which proposes a new technique that can protect user privacy by adding noise to the user data.

- **The University of Oxford**: The University of Oxford has several researchers who are working on PP-FL research. One of their recent works is *Secure Federated Learning on Curves*, which proposes a new technique that can protect user privacy by using elliptic curves to encrypt user data.

- **Microsoft Research**: Microsoft Research has been actively working on PP-FL research and has developed several techniques to protect user privacy in FL. One of its recent works is *Private Federated Learning with Secure Aggregation*, which proposes a new aggregation technique that can protect user privacy by using homomorphic encryption.

These are some examples of the many researchers and research groups who are actively working on PP-FL research. As the field of FL continues to evolve, it is expected that more researchers and research groups will focus on developing new techniques to protect the privacy of user data in FL.

Federated Meta-Learning

Federated Meta-Learning (FML) is a new research direction that combines FL and meta-learning to learn from multiple data sources and tasks while preserving privacy. FML enables collaborative model training across multiple devices and organizations while preserving the data privacy of the devices' owners. Recent research has explored new FML techniques and applications, such as personalized medicine and personalized recommendations. FML is a relatively new research direction in FL, and only a few research groups are working actively on it.

Here are some examples:

- **Google Research**: Google Research has been actively working on FML research and has developed several techniques to learn from multiple clients' meta-knowledge while preserving their privacy. One of its research works is *Federated Meta-Learning for Recommendation*, which proposes a new framework for FML for recommendation tasks.

- **Carnegie Mellon University**: The Machine Learning department at Carnegie Mellon University has several researchers who are working on FML research. One of its recent works is *Federated Meta-Learning for Fast Model Adaptation in Healthcare*, which proposes a new approach for learning from multiple healthcare institutions while preserving their privacy.

- **The University of Oxford**: The University of Oxford has several researchers who are working on FML research. One of its recent works is *Federated Meta-Learning for Recommendation with Private and Communication-Efficient Model Aggregation*, which proposes a new approach for FML for recommendation tasks that preserves privacy and reduces communication costs.

- **The University of California, Berkeley**: The RISELab at the University of California, Berkeley, has several researchers who are working on FML research. One of its recent works is *Federated Meta-Learning for Few-Shot Learning Across Heterogeneous Devices*, which proposes a new approach for FML that can learn from data across heterogeneous devices.

These are a few examples of the research groups that are actively working on FML research. As the field of FML continues to grow, it is expected that more researchers and research groups will focus on developing new techniques for FML in various applications.

Adaptive FL

In FL, devices' data distribution can change over time, leading to a concept drift problem that can degrade a model's performance. Recent research has focused on developing new adaptive learning techniques that can efficiently update the model while preserving privacy. These techniques can adapt the model to the changing data distribution and improve the model's performance over time. **Adaptive federated learning** (AFL) is an emerging research direction in FL that focuses on dynamically adapting the FL system to changing conditions.

Here are some notable examples of researchers and research groups who are working actively on AFL:

- **Carnegie Mellon University**: The Machine Learning department at Carnegie Mellon University has several researchers who are working on AFL research. One of its recent works is *Adaptive Federated Optimization*, which proposes a new framework for AFL that adapts the learning rate and aggregation strategy to network conditions.

- **Google Research**: Google Research has been actively working on AFL research and has developed several techniques to adapt the FL system to changing conditions. One of its recent works is *Adaptive Federated Optimization with Local Descent*, which proposes a new approach for AFL that adapts the learning rate and aggregation strategy to the client's local data distribution.

- **The University of Southern California**: The Department of Computer Science at the University of Southern California has several researchers who are working on AFL research. One of its recent works is *Decentralized Adaptive Federated Learning with Gradient Compression*, which proposes a new approach for AFL that adapts the learning rate and aggregation strategy to network conditions while compressing the gradients to reduce communication costs.

- **Tsinghua University**: The Department of Computer Science and Technology at Tsinghua University has several researchers who are working on AFL research. One of its recent works is *Adaptive Federated Learning via Second-Order Information Exchange*, which proposes a new approach for AFL that adapts the learning rate and aggregation strategy based on the second-order information exchange among clients.

These are a few examples of the research groups that are actively working on AFL research. As the field of AFL continues to grow, it is expected that more researchers and research groups will focus on developing new techniques to adapt the FL system to changing conditions.

Federated reinforcement learning

Reinforcement learning (**RL**) is a machine learning paradigm that enables agents to learn through trial and error. **Federated reinforcement learning** (**FRL**) is a new research direction that combines FL and RL to enable privacy-preserving collaborative learning for decision-making tasks. Recent research has explored new FRL algorithms and applications, such as autonomous driving and robotics.

There are many researchers and organizations actively working on FRL to improve machine learning techniques that protect user privacy and data security.

Here are a few examples:

- Google's **Federated Learning of Cohorts** (**FLoC**) team is working on developing FRL to create privacy-preserving user models for ad targeting.

- Researchers at Stanford University have been exploring the use of FRL to train autonomous robots. A team of researchers from Carnegie Mellon University is investigating FRL as a way to improve the efficiency and privacy of healthcare machine learning models.

- The OpenMined project is a community-driven initiative that aims to create an ecosystem for privacy-preserving machine learning, including FRL.

- Microsoft Research is working on developing FRL methods for distributed robotics applications.

These are just a few examples of the many researchers and organizations working on FRL, which is a rapidly growing field with many exciting developments and applications.

Federated Tumour Segmentation (FeTs) is a real-world medical FL platform developed by Intel and the **University of Pennsylvania (UPenn)**. They make use of OpenFL as a backend for the platform.

Key company products related to FL

As we have seen FL is a rapidly growing field that has gained significant attention from start-up companies. Here is a summary of some of the companies that are working or providing FL products:

- **DynamoFL**: DynamoFL is built by privacy and machine learning experts from **MIT (Massachusetts Institute of Technology)** and Harvard, who built leading FL solutions at Google AI and privacy-enhanced technologies at Microsoft. As per its website, the current **large language models (LLMs)** are not private, but the LLMs from DynamoFL are. They provide personalized FL, which is another key area of research. LLMs are covered in *Chapter 10* in the *Privacy-preserved generative AI* section.

- **NVIDIA FLARE**: NVIDIA **Federated Learning Application Runtime Environment (FLARE)** is an SDK for FL that is open sourced by NVIDIA. FLARE supports various FL algorithms (FedAvg, FedProx, FedOpt, etc.), neural networks, and ML algorithms. It supports differential privacy and homomorphic encryption features as part of the security and privacy preservation stack. The SDK also provides a simulator that can be used to start servers and clients and execute FL notebooks and FL applications.

- **OpenMined**: OpenMined is an open source community that provides a platform for privacy-preserving ML, including FL tools and libraries for secure aggregation and differential privacy.

- **DataFleets**: DataFleets provides a platform for privacy-preserving data access and analytics, including FL tools for secure data processing and analysis.

- **Scaleout**: Scaleout provides a platform for distributed ML, including FL tools for privacy-preserving and collaborative model training.

- **Edge Impulse**: Edge Impulse provides a platform for developing and deploying ML models on edge devices, including FL tools for privacy-preserving model training and inference.

- **Decentralized Machine Learning**: Decentralized Machine Learning provides a platform for privacy-preserving and decentralized ML, including FL tools for secure aggregation and differential privacy.

- **PySyft**: PySyft is an open source library for FL and secure multi-party computation, enabling privacy-preserving and collaborative ML on distributed data.

- **Intellegens**: Intellegens' product originated from the University of Cambridge and Ichnite and is an FL platform product that can be deployed in the cloud or on-premises (`https://intellegens.com/`).

These companies are actively contributing to the development and advancement of FL, providing tools and platforms for privacy-preserving and collaborative ML on distributed data. As FL continues to evolve, it is expected that more start-up companies will emerge, contributing to the growth and innovation of this field.

Summary

In this chapter, we explored FL benchmarks, their importance, and how to design benchmarks, among other things. We also discussed various FL frameworks and looked at what to consider when choosing a framework to implement FL applications. Finally, we covered the state-of-the-art research undertaken by various enterprises in collaboration with key universities, highlighting some of the key companies that are actively working and offering products/platforms to support FL.

In the next chapter, we will learn about homomorphic encryption and secure multi-party computation and how they help in achieving privacy in ML models.

Part 4: Homomorphic Encryption, SMC, Confidential Computing, and LLMs

This part covers homomorphic encryption and SMC. It covers, in detail, privacy preservation in large language models (LLMs).

We cover the concepts of homomorphic encryption (HE) and SMC as privacy-enhancing techniques. We highlight the significance of these cryptographic approaches in enabling secure computations on encrypted data without compromising privacy.

SMC enables multiple parties to collaboratively compute results while keeping their individual input private. Our summary emphasizes the importance of HE and SMC in scenarios where sensitive data needs to be analyzed or processed in a privacy-preserving manner. By utilizing these techniques, organizations and individuals can protect their data while still gaining valuable insights and outcomes from computations.

The second chapter in this part explores the concept of confidential computing, which aims to provide a trusted and secure environment for processing sensitive data. This chapter also provides an overview of the current state of confidential computing, highlighting the advancements and ongoing research in this field. It acknowledges the emergence of trusted execution environments (TEEs) and hardware-based solutions that provide secure enclaves for confidential computing.

The final chapter introduces generative AI and the fundamentals of LLMs, the privacy vulnerabilities associated with them, and the technologies and approaches to preserve privacy while using these models. This chapter covers developing LLM applications using open source LLMs and protecting them from privacy attacks (prompt injection attacks, membership inference attacks, etc.) and ends with state-of-the-art privacy research on LLMs.

Overall, this part of the book emphasizes the importance of HE, SMC, and confidential computing in safeguarding sensitive data. It underlines the current state of these technologies, highlighting their potential to address privacy concerns and enable secure data processing in various contexts.

This part has the following chapters:

- *Chapter 8, Homomorphic Encryption and Secure Multiparty Computation*
- *Chapter 9, Confidential Computing – What, Why, and Current State*
- *Chapter 10, Privacy Preserving in Large Language Models*

8

Homomorphic Encryption and Secure Multiparty Computation

Homomorphic encryption is a cryptographic technique that allows computation on encrypted data without decrypting it. It has the potential to revolutionize data privacy and security, enabling the secure computation of sensitive data without revealing the data itself. In this chapter, you will learn about homomorphic encryption and secure multiparty computation.

We will cover the following main topics in this chapter:

- Encryption, anonymization, and de-identification
- Homomorphic encryption and the mathematics behind

  - Open source Python frameworks for homomorphic encryption and Paillier schemes
  - Machine learning using homomorphic encryption (HE)
  - Federated learning with **Partially homomorphic encryption PHE**
  - Limitations of homomorphic encryption

- **Secure Multiparty Computation (SMC)** and its use cases
- A use case implementation using the **Private Set Interaction (PSI)** SMC technique
- A high-level overview of zero-knowledge proofs

Encryption, anonymization, and de-identification

Encryption, anonymization, and de-identification are all techniques used to protect sensitive data, but they differ in their approach and limitations.

Encryption

Encryption is the process of transforming data into a form that can only be read by authorized parties with access to a decryption key. The purpose of encryption is to ensure the confidentiality and integrity of data. Encrypted data remains readable by those who have the appropriate decryption key, but it is unintelligible to anyone who intercepts it without the key. Encryption is widely used to protect sensitive data in transit and data at rest, such as credit card numbers, passwords, and personally identifiable information.

Here's some simple Python code to implement basic encryption

Source code: Encryption_Example.ipynb

Develop a function to encrypt the given text using a basic encryption algorithm.

```
def simple_encryption(text, shift):
    """
    Encrypts the given text using the Caesar Cipher algorithm.
    """
    result = ""
# Loop through each character in the text
    for i in range(len(text)):
        char = text[i]
# If the character is a letter, shift its position in the alphabet
        if char.isalpha():
            if char.isupper():
                result += chr((ord(char) + shift - 65) % 26 + 65)
            else:
                result += chr((ord(char) + shift - 97) % 26 + 97)
        else:
            result += char
    return result


simple_encryption("Privacy Preserved Machine Learning",5)
```

The output of the preceding code is as follows:

```
'Uwnafhd Uwjxjwaji Rfhmnsj Qjfwsnsl'
```

This function takes two arguments: `text`, which is the plaintext to be encrypted, and `shift`, which is the number of positions by which each letter in the plaintext should be shifted in the alphabet. The function returns the encrypted text.

For example, if you call simple_encryption ("Privacy Preserved Machine Learning ", 5), then the function will return the string "Uwnafhd Uwjxjwaji Rfhmnsj Qjfwsnsl", which is the encrypted version of "Privacy Preserved Machine Learning" with a shift of 5.

Encryption algorithms

There are many popular encryption algorithms used today for securing data and communications. The most common are the following:

- **Advanced Encryption Standard** (**AES**): A symmetric-key encryption algorithm widely used for securing data in transit and at rest

- **Rivest-Shamir-Adleman** (**RSA**): An asymmetric-key encryption algorithm used for secure communication and digital signatures

- **Data Encryption Standard** (**DES**): A symmetric-key encryption algorithm widely used in the past, but now considered less secure

- **Blowfish**: A symmetric-key encryption algorithm designed for high-speed and efficient encryption of large amounts of data

- **Twofish**: A symmetric-key encryption algorithm that is a successor to Blowfish, designed for higher security and flexibility

- **ChaCha20**: A symmetric-key encryption algorithm that is becoming increasingly popular due to its high security and performance

- **Elliptic Curve Cryptography** (**ECC**): An asymmetric-key encryption algorithm that uses elliptic curves instead of prime numbers, resulting in smaller key sizes and faster performance

These are just a few of the most popular encryption algorithms. There are many others, and new ones are constantly being developed as computing power and security needs evolve.

Encryption example using AES

Source code: Encryption_Example.ipynb

Develop a function to encrypt the given text using the AES encryption algorithm:

```
from Crypto.Cipher import AES
from Crypto.Random import get_random_bytes
def aes_encrypt(key, plaintext):

# initialization vector to random bytes
    ini_vec = get_random_bytes(AES.block_size)
# Create the AES cipher object with the given key
    aes_ciper = AES.new(key, AES.MODE_CBC, ini_vec)
# Pad the plaintext to a multiple of the block size
    plaintext = pad_plain_text(plaintext, AES.block_size)
# Encrypt the plaintext using the AES cipher object
    ciphertext = aes_ciper.encrypt(plaintext)
    return ini_vec + ciphertext
```

```
def aes_decrypt(key, ciphertext):

    ini_vec = ciphertext[:AES.block_size]
    cipher = AES.new(key, AES.MODE_CBC, ini_vec)
    plaintext = cipher.decrypt(ciphertext[AES.block_size:])
    plaintext = unpad_decrypted_text(plaintext, AES.block_size)
    return plaintext.decode('utf-8')

def pad_plain_text(data, block_size):
    padding_length = block_size - (len(data) % block_size)
    padding = bytes([padding_length] * padding_length)
    return data + padding

def unpad_decrypted_text(data, block_size):
    padding_length = data[-1]
    if padding_length < 1 or padding_length > block_size:
        raise ValueError("Padding is not Valid ")
    padding = data[-padding_length:]
    if not all(padding[i] == padding_length for i in range(padding_
length)):
        raise ValueError("Padding is not Valid ")
    return data[:-padding_length]

key = b'ThisIsASecretKey'
plaintext = b'Privacy Preserved Machine Learning'
ciphertext = aes_encrypt(key, plaintext)
print('Encrypted message:', ciphertext)
decrypted_plaintext = aes_decrypt(key, ciphertext)
print('Decrypted message:', decrypted_plaintext)
```

This results in the following output:

```
Encrypted message: b'\xe3\x00\xe8\x10\xc6E\x0cn\x1bO,\x89-\x8d\xee\
xb3\xc6\x1a\xbf\x95\\l\x0e\x8d\xb0\xaa\x93\xf4_$?h\x1a`O\xf4N\x89!4\
xf7(\xd3\x8e\xde\xc7\xf7\xb8\x87\xea\n5W\x8e\xa5D\xec]\x80\xa8+\x92?\
xa9'
Decrypted message: Privacy Preserved Machine Learning
```

In this example program, we used the PyCrypto library to implement AES encryption with **cipher-block chaining (CBC)** mode.

The aes_encrypt function takes a 128-bit key and plaintext message as the input and returns the encrypted ciphertext. The aes_decrypt function takes a key and ciphertext as input and returns the decrypted plaintext message. We used the PKCS7 padding scheme to pad the plaintext to a multiple of the block size and to remove the padding after decryption. It generates a random initialization vector for each encryption operation to add an additional layer of security.

Limitations of encryption

Encryption doesn't provide anonymity or de-identification of data. Encrypted data may contain sensitive information that could be used to identify individuals. Encryption only protects data while it is being transmitted (data in motion) or stored (data at rest, i.e., in a persistent store), but it doesn't control who has access to it or how it is used once decrypted.

Data anonymization

Anonymization is the process of removing identifiable information from data so that individuals cannot be identified. The main purpose of anonymization is to protect the privacy of individuals while still allowing the data to be used for analysis or research use cases.

Here's some example Python code for data anonymization:

Source code: Example_Data_annomization.ipynb

```
import hashlib
import random

# Example dataset
dataset = [
    {"name": "Alice", "age": 23, "email": "alice@example.com"},
    {"name": "Bob", "age": 32, "email": "bob@example.com"},
    {"name": "Gandhi", "age": 55, "email": "gandhi@example.com"},
]

# Create a random salt value for anonymization
salt = str(random.getrandbits(128))

# Loop through the dataset and anonymize each record
for record in dataset:
# Hash the name and email using the salt value name_hash = hashlib.
sha256((record['name'] +salt).encode()).hexdigest() email_hash =
hashlib.sha256((record['email'] + salt).encode()).hexdigest()

# Replace the original name and email with the hashed values
record['name'] = name_hash
record['email'] = email_hash
print(dataset)
```

This results in the following output:

```
[{'name':
'e807ef3ca985de8ef481ddf8af4d3ac4c34089519cf225a7863306ced0a691fa',
'age': 23, 'email':
'474f2b3dce2701b08651d64364ab5e83575e9bd8ff7e0e14d654afbdf19f6683'},
{'name':
'36cbc209f7afaba2a3a4d8830c2b85b3813f467f4bf442bb439b3f112be00bd0',
'age': 32, 'email':
'6a7b3de0488fc3a81436b5c70e943ca7139c93d1832430db2e6bac3f2c25cce5'},
{'name':
'096aca9f2b7872c89b9c71ff1a649e7fe53fc7eb1a04354792ea58aaee3bbecf',
'age': 55, 'email':
'e15e4e48f0161e7a64b8ef604e6d9bd5d9d6f2da3c8d848268aeec0ff5da2ef8'}]
```

In the preceding code, we used Python's `hashlib` library to hash the name and email fields of each record in the dataset using a random salt value. The hashed values are then stored in place of the original values. This technique can be used to protect the privacy of sensitive information in a dataset.

Is anonymization alone sufficient to protect sensitive data? In short, no, it is not sufficient, as we'll see in the following case study from the real world.

Real-world case study (Netflix user data anonymization)

Our Netflix data anonymization case study dates back to 2006 when Netflix launched the Netflix Prize, a competition aimed at improving the accuracy of the company's recommendation algorithm.

As part of the competition, Netflix released a dataset containing the viewing history of 500,000 subscribers, with personally identifiable information removed. The dataset was made available to researchers to develop better algorithms. In 2009, researchers published a paper that showed that it was possible to re-identify individuals in the Netflix dataset by using information from the **Internet Movie Database (IMDb)**. The researchers were able to match the anonymized viewing history with the reviews that users had posted on IMDb, allowing them to re-identify individuals with high accuracy.

Following this discovery, Netflix took steps to improve its anonymization techniques, but in 2010, researchers once again showed that it was possible to re-identify individuals in the Netflix dataset. This time, researchers used external data sources such as social networks and movie rating websites to re-identify individuals.

In response to these privacy concerns, Netflix stopped releasing anonymized datasets to researchers in 2010. Instead, the company created an internal research program where researchers could analyze Netflix data without any personally identifiable information being shared. However, in 2020, a group of researchers from the University of Texas at Austin and the University of California, Irvine showed that they could still re-identify Netflix users by analyzing their viewing history and correlating it with publicly available datasets. The researchers were able to accurately re-identify users even when no personally identifiable information was present in the Netflix dataset.

These findings showed that the anonymization techniques alone were not enough to protect users' privacy as they are prone to data linkage attacks, which we learned about in *Chapter 1*.

Limitations of anonymization

Anonymization may not completely eliminate the risk of re-identification. If enough data is available, it may be possible to identify individuals by linking the anonymized data with other available information such as public datasets.

Anonymization may result in loss of data quality or accuracy, as certain data elements may be removed or masked in the process.

De-identification

De-identification is the process of removing or masking identifiable information in data so that it cannot be used to identify individuals. The purpose of de-identification is to protect the privacy of individuals while still allowing the data to be used for research, analysis, or other applications. De-identified data may be used for healthcare research, marketing analysis, financial use cases, IoT use cases, and other applications where sensitive data is required.

De-identification algorithms

One-way hashing algorithms can be used to run de-identification processes on data. One-way hashing algorithms, also known as cryptographic hash functions, are mathematical functions that take an input message of any length and generate a fixed-size output, which is referred to as a hash or message digest. The key property of one-way hash algorithms is that they are designed such that it should be computationally infeasible to reverse-engineer the original input message from the hash value.

There are many different hashing algorithms, and new ones are continually being developed.

The following are some of the most widely used hashing algorithms:

- **Message Digest 5 (MD5):** This is a widely used hashing algorithm that produces a 128-bit hash value. It is now considered to be insecure due to vulnerabilities in its design.

- **Secure Hash Algorithm 1 (SHA-1):** This is another widely used hashing algorithm that produces a 160-bit hash value. It is also now considered to be insecure.

- **Secure Hash Algorithm 2 (SHA-2):** This is a family of hashing algorithms that includes SHA-224, SHA-256, SHA-384, and SHA-512. These algorithms produce hash values of 224, 256, 384, and 512 bits, respectively.

- **Secure Hash Algorithm 3 (SHA-3):** This is a family of hashing algorithms that includes SHA3-224, SHA3-256, SHA3-384, and SHA3-512. These algorithms are designed as a replacement for SHA-2 and produce hash values of 224, 256, 384, and 512 bits, respectively.

- **BLAKE2**: This is a family of hashing algorithms that includes BLAKE2b and BLAKE2s. These algorithms were designed to be faster than SHA-3 while still providing strong security.

- **RACE Integrity Primitives Evaluation Message Digest (RIPEMD)**: This is a family of hashing algorithms that includes RIPEMD-128, RIPEMD-160, RIPEMD-256, and RIPEMD-320. These algorithms were designed as a replacement for the MD4 and MD5 algorithms and produce hash values of 128, 160, 256, and 320 bits, respectively.

- **Whirlpool**: This is a hashing algorithm that produces a 512-bit hash value. It was designed to be an alternative to the SHA-2 family of algorithms.

These are some of the examples of hashing algorithms available. The choice of which algorithm to use will depend on the specific application and the desired balance between security, speed, and other factors.

Example Python code to de-identify the data

As discussed, De-identification is the process of removing or masking identifiable information in data so that it cannot be used to identify the original data. In the code below, let's de-identify a given data (in this case a string) using the SHA256 hashing algorithm.

Source code: De-Identify_Exmple.ipynb

```
import hashlib

# Define a function to hash a given string using SHA256 Algorithm
def hash_string(s):
    return hashlib.sha256(s.encode()).hexdigest()

# Define a function to de-identify a given input data
def deidentify_data(data):

# Remove leading/trailing whitespace and convert to lowercase
    data = data.strip().lower()

# Hash the name using SHA-256
    hashed_data = hash_string(name)

# Return the hashed data
    return hashed_data

# input
person_name = "Sundar P"
hashed_name = deidentify_data(person_name)
print(hashed_name)
```

This results in the following output:

```
"6cea57c2fb6cbc2a40411135005760f241fffc3e5e67ab99882726431037f908""
```

In this example, the `hashlib` library is used to compute a SHA-256 hash of a given input string. We define a `deidentify_data` function that takes data as input, removes any leading/trailing whitespace, converts the data to lowercase, and then hashes it using the `hash_string` function.

This ensures that the same input name will always produce the same output hash. Note that because one-way hashing is an irreversible process, there is no way to retrieve the original name from the hashed value.

The `hashlib` library in Python provides a collection of hashing algorithms. The following methods are available in the `hashlib` library:

- `hashlib.md5()`: This method returns an MD5 hash object
- `hashlib.sha1()`: This method returns a SHA-1 hash object
- `hashlib.sha224()`: This method returns a SHA-224 hash object
- `hashlib.sha256()`: This method returns a SHA-256 hash object
- `hashlib.sha384()`: This method returns a SHA-384 hash object
- `hashlib.sha512()`: This method returns a SHA-512 hash object
- `hashlib.blake2s()`: This method returns a BLAKE2s hash object
- `hashlib.blake2b()`: This method returns a BLAKE2b hash object
- `hashlib.sha3_224()`: This method returns a SHA3-224 hash object
- `hashlib.sha3_256()`: This method returns a SHA3-256 hash object
- `hashlib.sha3_384()`: This method returns a SHA3-384 hash object
- `hashlib.sha3_512()`: This method returns a SHA3-512 hash object

The purpose of de-identification is to make it difficult or impossible to identify individuals based on their personal information.

Limitations of de-identification

De-identification may not completely eliminate the risk of re-identification. If enough data is available, it may be possible to identify individuals based on other available information. De-identification may result in the loss of data quality or accuracy, as certain data elements may be removed or masked in the process.

In summary, each technique has its own strengths and limitations, and the appropriate technique should be selected based on the specific use case and requirements. Encryption is the most secure way to protect data, but it does not provide anonymity or de-identification. Anonymization and de-identification can both provide privacy protection, but they may not be sufficient in all cases and may result in data quality or accuracy issues.

Exploring Homomorphic encryption

Homomorphic Encryption (**HE**) is a cryptographic technique that allows computation on encrypted data without decrypting it. In other words, it is possible to perform operations on ciphertexts, generating new ciphertexts that are decrypted to the result of the operation on the plaintexts. HE has the potential to revolutionize data privacy and security, enabling secure computation of sensitive data without revealing the data itself. HE is based on mathematical concepts such as algebraic structures, number theory, and polynomial theory. The most common types of HE are based on the following algebraic structures.

Ring-based

In the context of cryptography, an algebraic structure refers to a set of mathematical operations that can be performed on elements of the set in a specific way. In the case of a ring, the set of elements is closed under addition and multiplication, and the operations satisfy certain properties, such as associativity, commutativity, and the existence of an identity element and inverse elements. In the context of HE, an algebraic structure is used to perform computations on encrypted data without first decrypting it. The underlying algebraic structure is a ring, which is a set of elements with two binary operations (addition and multiplication) that satisfy certain properties. The most widely used ring-based HE is the **Brakerski-Gentry-Vaikuntanathan** (**BGV**) scheme.

The BGV scheme is a **fully homomorphic encryption** (**FHE**) scheme based on the **learning with errors** (**LWE**) problem.

Let's now examine a mathematical description of the scheme.

Key generation

To generate a public key, the following steps are performed:

- Choose two integers n and q, where q is a prime number.
- Generate a random matrix A with integer entries in the range $[-q/2, q/2]$.
- Generate a random vector with integer entries in the range $[-q/2, q/2]$.
- Compute the vector $b = As + e$, where e is a random vector with entries in the range $[-B/2, B/2]$ for some integer B.
- Set the public key to be (A, b).

To generate a secret key, a random vector s is chosen with integer entries in the range $[-q/2, q/2]$.

Encryption

To encrypt a message m, the following steps are performed:

- Represent the message m as a polynomial m(x) with integer coefficients modulo q.

- Choose a random polynomial r(x) with integer coefficients modulo q.

- Compute c = (A*r + b + m(x)*t)/q, where t is a scaling factor that controls the noise in the ciphertext.

The resulting ciphertext c consists of a matrix and a polynomial.

Decryption

To decrypt a ciphertext c, the following steps are performed:

- Compute c' = c*s/q.

- Round each entry of c' to the nearest integer.

- Recover the polynomial m(x) by subtracting A*r from c'.

Homomorphic operations

The BGV scheme allows homomorphic addition and multiplication on ciphertexts. These operations are performed as follows:

Homomorphic addition: To add two ciphertexts c1 and c2, add their corresponding polynomials modulo q, and add their matrices element-wise modulo q.

Homomorphic multiplication: To multiply two ciphertexts c1 and c2, compute their product modulo q using the polynomial multiplication algorithm, and multiply their matrices using matrix multiplication modulo q.

Bootstrapping

To perform more than one homomorphic multiplication, the BGV scheme uses a technique called bootstrapping to "refresh" the ciphertext. The bootstrapping process involves decoding the ciphertext, performing a homomorphic operation, and then re-encrypting the result with a new set of keys. The bootstrapping process allows an arbitrary number of homomorphic operations to be performed on the ciphertext while still maintaining the security of the scheme.

Lattice-based

HE based on lattices uses the mathematical concept of lattices, which are geometric structures in higher dimensions that allow the efficient computation of certain types of problems. The most widely used lattice-based HE is the FHE scheme.

Elliptic curve-based

HE based on elliptic curves uses elliptic curves over finite fields to create the cryptographic scheme. This type of HE is relatively new, and not as widely used as the other two types.

Exploring the mathematics behind HE

The mathematics behind HE is based on two main concepts: encryption and homomorphism.

Encryption

Encryption is the process of transforming plaintext into ciphertext using an encryption algorithm and a secret key. The ciphertext can then be transmitted over a network or stored in a database without fear of unauthorized access. To decrypt the ciphertext and obtain the plaintext, the recipient must possess the secret key that was used to encrypt the data.

Homomorphism

Homomorphism is a mathematical property that allows an operation to be performed on ciphertexts, generating a new ciphertext that is the result of the operation on the plaintexts. This means that if we have two plaintexts x and y, and their respective ciphertexts $C(x)$ and $C(y)$, we can perform an operation on $C(x)$ and $C(y)$ to obtain a new ciphertext $C(x+y)$, which can be decrypted to obtain the result of the operation on x and y.

The most commonly used homomorphic operations are addition and multiplication, but other operations such as subtraction and division can also be performed. The level of homomorphism determines how many operations can be performed on the ciphertexts before the noise introduced during the encryption process becomes too high and the ciphertext becomes unusable.

HE is based on the concept of adding noise to the ciphertext to make it impossible to recover the plaintext without the secret key. The noise is added in such a way that homomorphic operations can be performed on the ciphertext without revealing the plaintext. The amount of noise added determines the level of security of the encryption. The noise is also the reason why the level of homomorphism is limited in HE, as too many homomorphic operations can cause the noise to become too high and the ciphertext to become unusable.

Types of HE

There are three types of implementation for HE. Let's examine them now.

Fully Homomorphic Encryption (FHE)

FHE is a type of encryption that allows computation to be performed on encrypted data without decrypting it. It means that the ciphertext can be used as input for a computation and the result will be a ciphertext as well, which can be decrypted to get the result of the computation. FHE is a powerful tool that can be used in various applications such as cloud computing, machine learning, and secure outsourcing. One of the main challenges of FHE is its computational complexity, which makes it impractical for many applications. However, recent advancements in FHE have led to the development of more efficient algorithms that reduce the computational overhead of FHE.

The Gentry scheme is one of the earliest and most well-known FHE schemes, but it has a high computational cost. More recent schemes, such as the CKKS scheme and the BFV scheme, offer more efficient FHE algorithms that are suitable for practical applications.

Somewhat Homomorphic Encryption (SHE)

SHE is a type of encryption that allows limited computations on encrypted data. Unlike FHE, SHE cannot perform arbitrary computations on encrypted data. Instead, it can only perform a limited set of operations, such as addition and multiplication, on the encrypted data. While SHE is less powerful than FHE, it is still useful in many applications, such as secure voting, secure messaging, and secure database queries. SHE is less computationally intensive than FHE, which makes it more practical for certain applications.

Partially Homomorphic Encryption (PHE)

PHE is a type of encryption that allows for computations on encrypted data, but only for one type of operation, either addition or multiplication. PHE is less powerful than both FHE and SHE but is still useful in some applications, such as secure key generation, secure function evaluation, and secure scalar product calculation. PHE is less computationally intensive than both FHE and SHE, which makes it more practical for some applications. However, its limited functionality means that it is less flexible than FHE and SHE and cannot be used in as many applications.

FHE, SHE, and PHE are three related encryption schemes that provide different levels of functionality and computational complexity. FHE is the most powerful but also the most computationally intensive, while PHE is the least powerful but also the least computationally intensive. SHE provides a middle ground between FHE and PHE in terms of functionality and computational complexity.

Paillier scheme

The Paillier scheme is a public key cryptosystem that is used for the encryption and decryption of data. It is based on the mathematical assumption i.e **decisional composite residuality assumption (DCRA)**, which is a hard computational problem. The scheme is named after its creator, Pascal Paillier, who introduced it in 1999.

The Paillier scheme is an asymmetric encryption algorithm, which means that it uses two different keys: a public key for encryption and a private key for decryption. The scheme is designed to be probabilistic, which means that each encryption of a given plaintext results in a different ciphertext. The scheme is also homomorphic, which means that it supports certain types of operations on the encrypted data.

Key generation

To use the Paillier scheme, a user first generates a public key and a corresponding private key. The public key consists of two large prime numbers, p and q, which are kept secret. The user then calculates n = p * q and lambda = LCM(p - 1, q - 1), where LCM is the least common multiple function. The value of lambda is used to generate the private key.

Encryption

To encrypt a message, the sender uses the recipient's public key. The plaintext message is represented as an integer m, where 0 <= m < n. The sender then chooses a random number r, where 0 <= r < n, and computes c = g^m * r^n mod n^2, where g is a random generator modulo n^2. The ciphertext c is then sent to the recipient.

Decryption

To decrypt the ciphertext, the recipient uses their private key. The recipient first computes the value of

mu = (L(g^lambda mod n^2)^-1 mod n)^(lambda^-1 mod n)

where

L(x) = (x - 1) / n

The value of mu is used to calculate the plaintext message m as follows:

m = L(c^lambda mod n^2) * mu mod n

The recipient can then recover the original message from m.

Homomorphic properties

The Paillier scheme supports two homomorphic properties: additive and multiplicative.

The additive property allows the recipient to perform addition on the encrypted data. Given two ciphertexts c1 and c2 that correspond to plaintext messages m1 and m2, the recipient can compute a new ciphertext c3 that corresponds to the sum of m1 and m2 by multiplying c1 * c2 mod n^2.

The multiplicative property allows the recipient to perform multiplication on the encrypted data. Given a ciphertext c that corresponds to a plaintext message m, the recipient can compute a new ciphertext c' that corresponds to the product of m and a constant k by raising c to the power of k mod n^2.

Security

The Paillier scheme is also resistant to known plaintext attacks and chosen plaintext attacks. The scheme is used in various applications, including electronic voting, privacy-preserving data mining, and secure multiparty computation.

Python frameworks for HE the following section, we will cover the open source Python frameworks that are used to implement HE. We will implement a few examples in detail to understand homomorphic operations in a better way.

Pyfhel

Pyfhel is a Python library for FHE. It provides an easy-to-use interface for performing FHE operations on encrypted data.

URL: `https://pyfhel.readthedocs.io/en/latest/`

SEAL Python

SEAL Python is a Python wrapper for the **Simple Encrypted Arithmetic Library** (**SEAL**) C++ library. SEAL Python provides a high-level interface for performing HE and decryption operations on data using the Brakerski/Fan-Vercauteren (BFV) and Cheon-Kim-Kim-Song (CKKS) schemes.

URL: `https://github.com/Huelse/SEAL-Python`

TenSEAL

TenSEAL is a Python library for homomorphic encryption on n tensors, built on top of Microsoft SEAL and using the CKKS scheme. It provides an easy-to-use interface for performing FHE operations on encrypted data with support for batching and batching rotations.

The main features offered by TenSEAL are the following:

- Encryption/decryption of the vectors of integers using BFV
- Encryption/decryption of the vectors of real numbers using CKKS
- Element-wise addition, subtraction, and multiplication of encrypted-encrypted vectors and encrypted-plain vectors
- Dot product and vector-matrix multiplication

URL: `https://github.com/OpenMined/TenSEAL`

phe

phe is a Python library for **partially homomorphic encryption** (PHE) using the Paillier scheme. It provides an API for performing PHE operations on encrypted data.

It is a simple and easy-to-use library. It only supports three operations (addition, subtraction, and scalar multiplication) on HE out of four operations.

URL: `https://pypi.org/project/phe/`

The homomorphic properties of the Paillier cryptographic system work as follows.

Encrypted numbers can be added together:

$$\text{Enc}(m1) + \text{Enc}(m2) \equiv \text{Enc}(m1+m2)$$

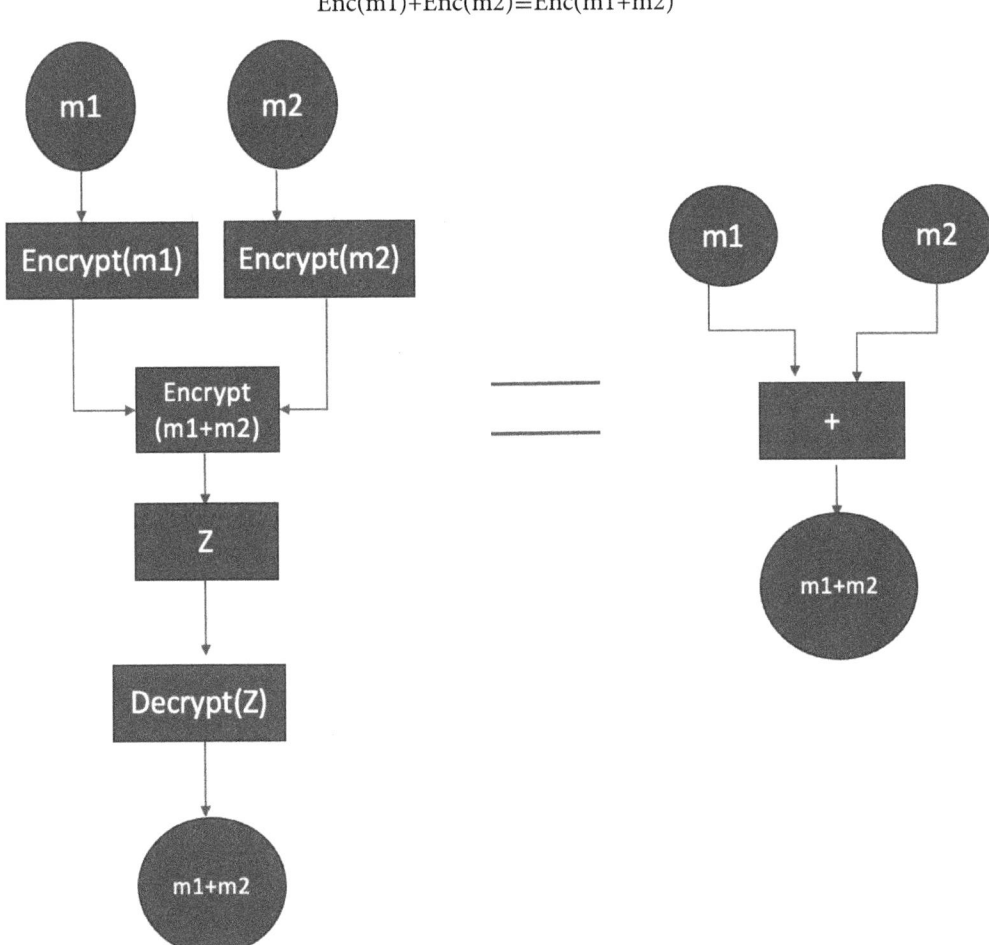

Figure 8.1 – Homomorphic addition

Encrypted numbers can be subtracted from each other:

$$Enc(m1)-Enc(m2)\equiv Enc(m1-m2)$$

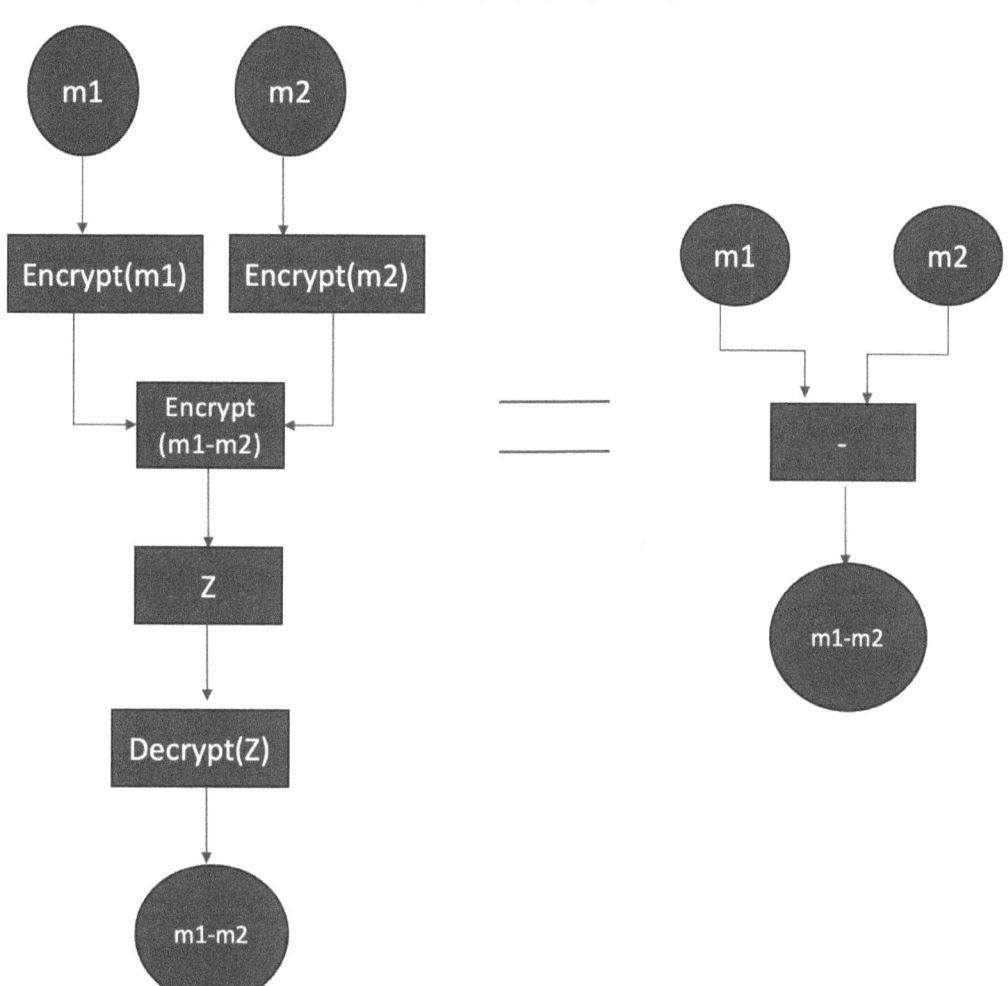

Figure 8.2 – Homomorphic subtraction

Encrypted numbers can be multiplied by a non-encrypted scalar:

$$n \cdot Enc(m) \equiv Enc(n*m)$$

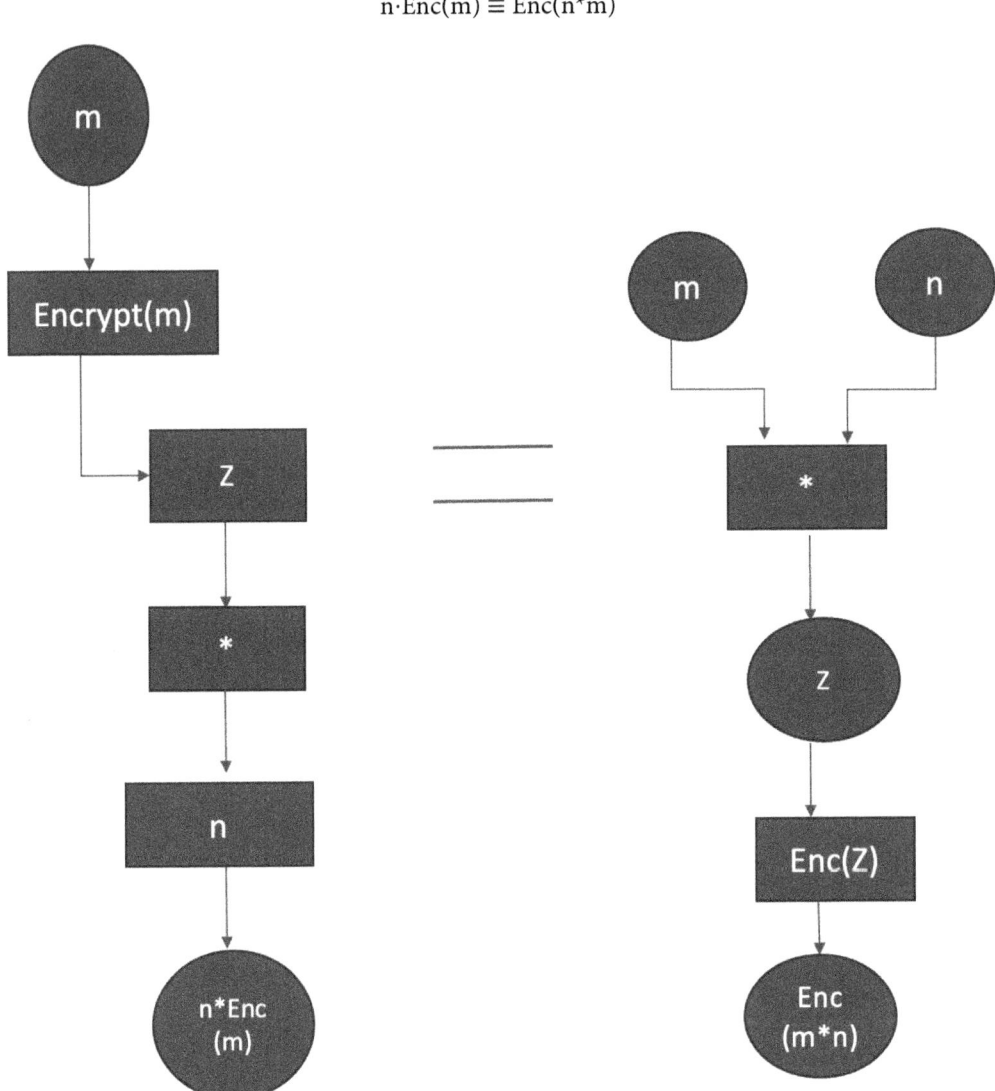

Figure 8.3 – Homomorphic-encrypted number multiplied by a non-encrypted scalar

Encrypted numbers can be added to non-encrypted scalars:

$$n+Enc(m) \equiv Enc(n+m)$$

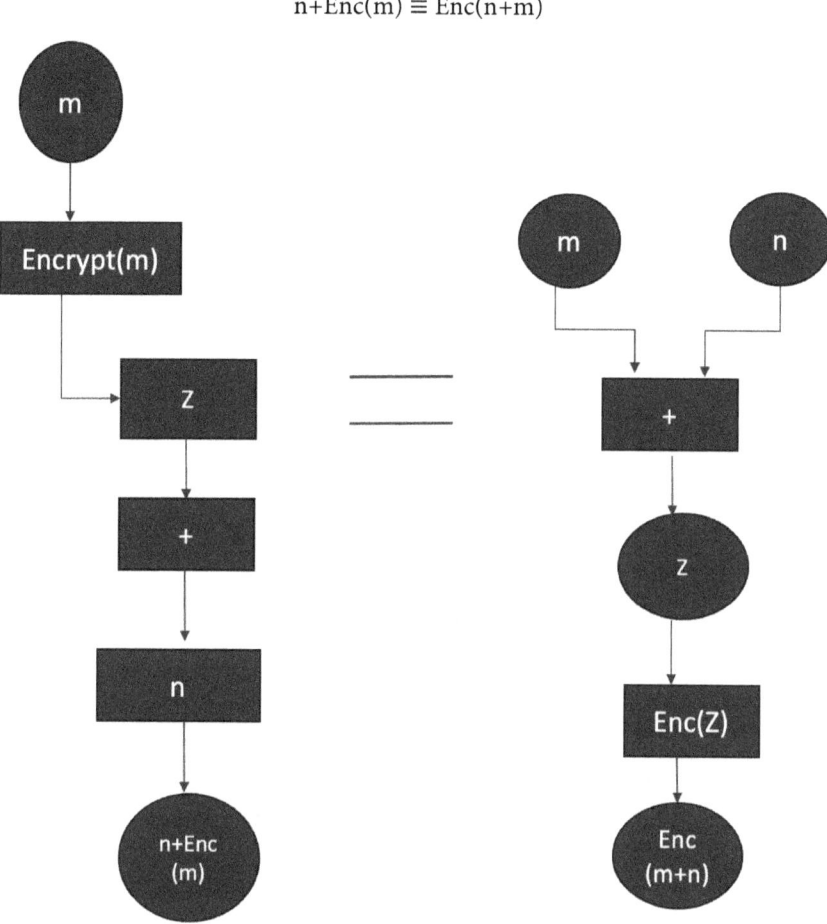

Figure 8.4 – Homomorphic-encrypted number added to a non-encrypted scalar

Implementing HE

To implement HE, choose a suitable HE library from those outlined previously. Make sure your choice is appropriate based on your specific use case, then perform the following steps:

1. Generate the public and private keys required for the encryption scheme.

2. Convert the plaintext data that needs to be encrypted into a suitable format for the encryption scheme, such as a polynomial.

3. Encrypt the plaintext data using the public key generated in *step 2*.

4. Perform the homomorphic operations on the ciphertext data without decrypting it.

5. Decrypt the resulting ciphertext data using the private key generated in *step 2* to obtain the result of the homomorphic operations on the plaintext data.

Implementing HE can be complex and requires expertise in cryptography and mathematics. It is important to ensure that the implementation is secure and efficient, as HE can be computationally intensive.

Implementing PHE

We will implement an example of Paillier PHE using the open source phe Python library.

First, we install the phe library as follows:

```
pip3 install phe
```

Then we implement the following steps:

- Generate the public and private keys using the generate_paillier_keypair method from phe.paillier class.

- Encrypt the plaintext data (15 in this example) using the encrypt method.

- Perform homomorphic operations on the ciphertext data:

    - ciphertext1 = ciphertext * 3   // encrypted number multiplied by a scalar

    - ciphertext2 = ciphertext + ciphertext1 // adding two encrypted numbers

    - ciphertext3 = ciphertext1 + 250 // adding scalar to a encrypted number

    - ciphertext4 = ciphertext3 - ciphertext2

- Finally, decrypt the resulting ciphertext data using the decrypt method and print the decrypted results.

Source code: FHE-Example.ipynb

```
import phe as paillier
pubkey, privkey = paillier.generate_paillier_keypair(n_length=1024)
print("public key",pubkey)
print("private key", privkey)
public key <PaillierPublicKey 18dced683d>
private key <PaillierPrivateKey for <PaillierPublicKey 18dced683d>>
plaintext = 15
ciphertext = pubkey.encrypt(plaintext)

# Perform homomorphic operations on the ciphertext data
ciphertext1 = ciphertext * 3
ciphertext2 = ciphertext + ciphertext1
```

```
ciphertext3 = ciphertext1 + 250
ciphertext4 = ciphertext3 - ciphertext2

# Decrypt the resulting ciphertext data
decrypted1 = privkey.decrypt(ciphertext1)
decrypted2 = privkey.decrypt(ciphertext2)
decrypted3 = privkey.decrypt(ciphertext3)
decrypted4 = privkey.decrypt(ciphertext4)

# Print the decrypted results
print("ciphertext1 decrypted: ", decrypted1)
print("ciphertext2 decrypted: ", decrypted2)
print("ciphertext3 decrypted: ", decrypted3)
print("ciphertext4 decrypted: ", decrypted4)
```

ciphertext1 decrypted: 45
ciphertext2 decrypted: 60
ciphertext3 decrypted: 295
ciphertext3 decrypted: 235

ciphertext1 : <phe.paillier.EncryptedNumber object at 0x7fd51856afa0>
ciphertext2 : <phe.paillier.EncryptedNumber object at 0x7fd5185c3580>
ciphertext3 : <phe.paillier.EncryptedNumber object at 0x7fd51856ac70>
ciphertext4 : <phe.paillier.EncryptedNumber object at 0x7fd5185c3460>

How about multiplying an encrypted number with another encrypted number?

Will the following work with PHE?

$$Enc(m1) * Enc(m2) \equiv Enc(m1 * m2)$$

Let's try this and find out:

```
m1=15
m2=20
pubkey.encrypt(m1)  * pubkey.encrypt(m2)
File ~/Library/Python/3.8/lib/python/site-packages/phe/paillier.
py:508, in EncryptedNumber.__mul__(self, other)
    506 """Multiply by an int, float, or EncodedNumber."""
    507 if isinstance(other, EncryptedNumber):
--> 508 raise NotImplementedError('Good luck with that...')
    510 if isinstance(other, EncodedNumber):
    511     encoding = other


NotImplementedError: Good luck with that...
```

It throws *NotImplementedError*, which basically means that phe doesn't support this (multiplying an encrypted number with another encrypted number) property of HE.

When your requirements call for developing applications involving tensors, matrices, arrays, and all homomorphic operations, TenSEAL is the best framework to use.

Implementing HE using the TenSEAL library

We will implement an example of HE using the open source TenSEAL Python library.

First, we install the TenSEAL library as follows:

```
pip3 install tenseal
```

Source code :TenSeal_Example.ipynb

Develop the sample applictions in the following way.

```
import tenseal as ts

# Step 1: Create context
context = ts.context(
            ts.SCHEME_TYPE.CKKS,
            poly_modulus_degree=8192,
            coeff_mod_bit_sizes=[60, 40, 40, 60]
        )
context.generate_galois_keys()
context.global_scale = 2**40

# Step 2: Create and encrypt data
data1 = [1.0, 2.0, 3.0, 4.0]
data2 = [5.0, 6.0, 7.0, 8.0]
encrypted_data1 = ts.ckks_vector(context, data1)
encrypted_data2 = ts.ckks_vector(context, data2)

# Step 3: Perform operations on encrypted data
encrypted_sum = encrypted_data1 + encrypted_data2
encrypted_product = encrypted_data1 * encrypted_data2
encrypted_dot_product = encrypted_data1.dot(encrypted_data2)
matrix_a = [[73],[69],[87],[45],]
mat_product = encrypted_data1.matmul(matrix_a)

# Step 4: Decrypt result
sum_result = encrypted_sum.decrypt()
product_result = encrypted_product.decrypt()
```

```
dot_product_result = encrypted_dot_product.decrypt()
mat_result = mat_product.decrypt()
print("Sum: ", sum_result)
print("Product: ", product_result)
print("Dot product: ", dot_product_result)
print("Matrix Multiplication : ", mat_result)
```

```
Sum: [6.000000000650802, 8.000000001433328, 10.0000000008995,
12.000000000314097]
Product: [5.000000671234171, 12.000001615240716, 21.000002823083314,
32.000004292130704]
Dot product: [70.00000937472286]
Matrix Multiplication : [652.000087556169]
```

In the preceding example code, we first create a context using the CKKS scheme with a polynomial modulus degree of 8,192 and four coefficient modulus bit sizes of 60, 40, 40, and 60. Then, we create two CKKS vectors and encrypt our data. Next, we perform two operations on our encrypted data: addition and multiplication. Finally, we decrypt the results of these operations and print them out.

This library also supports creating context using the BFV scheme as well in the same manner and the homomorphic operations using the BFV Scheme as well.

This is a simple example, and in practice, HE can be computationally intensive and requires specialized hardware or cloud resources. It is also important to ensure that the implementation is secure and that no encrypted data leaks out during the computation.

Comparison of HE frameworks

Let's review a comparison of some of the most popular HE frameworks in Python.

Pyfhel

Pyfhel is a Python-based fully HE library that supports operations on encrypted integers and vectors. It is built on top of the HElib C++ library and offers a simplified interface for Python developers. Pyfhel has good performance and can handle large integers and vectors efficiently. However, it does not yet support operations on floating-point numbers.

TenSEAL

TenSEAL is a Python-based library for HE that supports both FHE and PHE. It uses the CKKS and BFV encryption schemes and offers APIs for encrypted operations on floating-point numbers and matrices. TenSEAL is designed to be easy to use and has a simpler API compared to some other HE libraries. It has a relatively high performance for encrypted operations on floating-point numbers.

PALISADE

PALISADE is a C++ library for HE that has Python bindings. It supports both FHE and PHE and offers a wide range of encryption schemes, including CKKS, BFV, and GSW. PALISADE is designed for performance and can handle large plaintexts and ciphertexts efficiently. The Python bindings are relatively new, and the API can be more complex compared to other Python-based HE libraries.

PySEAL

PySEAL is a Python-based library for HE that supports FHE operations on encrypted integers and vectors. It is built on top of the SEAL C++ library and provides a simplified Python interface. PySEAL has good performance for integer and vector operations and can handle large plaintexts and ciphertexts. However, it does not yet support operations on floating-point numbers.

TFHE

TFHE is a C++ library for HE that has Python bindings. It supports PHE operations on encrypted integers and booleans and is designed for high performance. TFHE can handle large ciphertexts efficiently and has a relatively simple API. The Python bindings are relatively new, and the documentation can be limited.

The following table offers a high-level summary of the comparisons between the preceding HE libraries:

Operations	PySEAL	TenSEAL	Paillier	Pyfhel
Addition	Yes	Yes	Yes	Yes
Subtraction	Yes	Yes	Yes	No
Multiplication	Yes	Yes	No	Yes
Division	No	No	No	No
Comparison (<, >, <=, >=, and ==)	No	No	No	No
Support for vector operations (addition, dot product, etc.)	Yes	Yes	No	No
Matrix operations	Yes	Yes	No	No
Encrypted number added to scalar (non-encrypted number)	Yes	Yes	Yes	Yes
Encrypted number multiplied by scalar (non-encrypted number)	Yes	Yes	Yes	Yes

Table 8.1 – High-level comparison of HE frameworks

Each HE framework has its own set of strengths and weaknesses, and the best choice will depend on your specific use case and requirements. Pyfhel and PySEAL are good choices for FHE operations on integers and vectors, while TenSEAL is a good choice for FHE operations on floating-point numbers and matrices. PALISADE offers a wide range of encryption schemes and is designed for performance, while TFHE is a good choice for PHE operations on integers and booleans.

Machine learning with HE

HE can be used in **Machine Learning** (**ML**) models to encrypt the training data, test data, or even the complete model itself to achieve model security.

The following are some of the options to implement ML models with HE:

- Encrypt the weights (model parameters) and intercept, and make use of them to calculate the accuracy of the model on the test data.

- Encrypt the test data and make use of the encrypted data with an encrypted model to find out the accuracy.

- Build the models with training data encrypted and without the encryption.Calculate the accuracy of the clear text model as well as the model with encrypted training data.

- Encrypt the training data and train the model on encrypted data, then run the inference and decrypt the results.

In this example, we will encrypt the model parameters and do the following:

- Using the fraud detection model example:

  - Load the fraud transaction data

  - Split the data as train and test

  - Use the logistic regression model to train the data

  - From the model, find out the intercept and coefficients (weights)

- Using the Paillier PHE library, do the following:

  - Perform HE on the model parameters (i.e., the intercept values and model weights) using the public key

  - Find out the predictions using the encrypted intercept and encrypted model weights:

    - Calculate the score by doing the homomorphic operations (encrypted weights multiplied by test feature data and the addition of encrypted numbers)

    - Decrypt the calculated score using homomorphic decryption and find out the error/accuracy rates

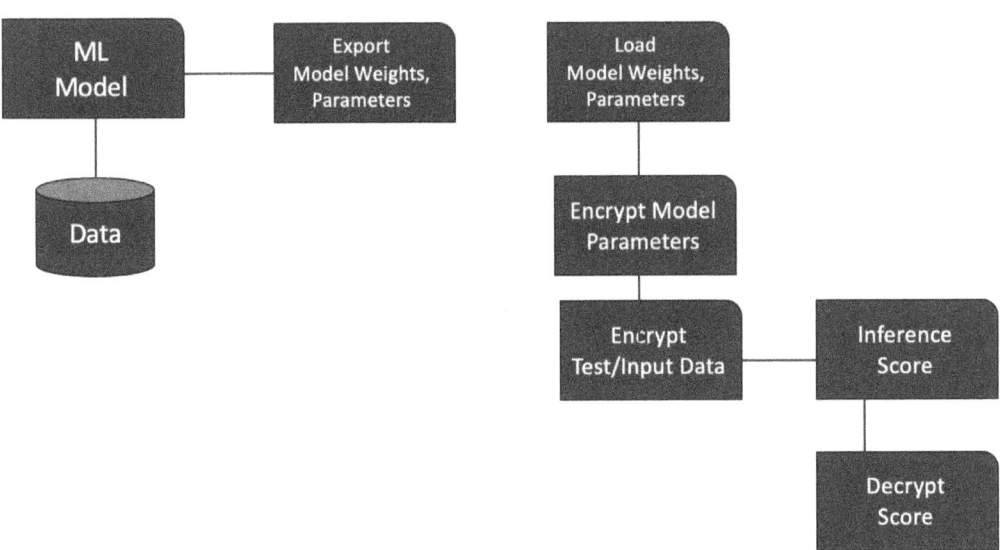

Figure 8.5 – Encryption of model parameters using HE

Source code: FHE_Logistic.ipynb

Following code implements the steps mentioned above.

```
from sklearn.linear_model import LogisticRegression
from sklearn.model_selection import train_test_split
url="fraud_transactions.csv"
df_actual = pd.read_csv(url, sep=",")
df_actual.head()
df_transactions = df_actual[['CUSTOMER_ID','TERMINAL_ID','TX_
AMOUNT','TX_FRAUD']]
df_transactions
from sklearn.model_selection import train_test_split
from sklearn.model_selection import StratifiedShuffleSplit

X = df_transactions.drop('TX_FRAUD', axis=1)
y = df_transactions['TX_FRAUD']
X_train, X_test, y_train, y_test = train_test_split(X, y, test_
size=0.3, random_state=42)
X_train = X_train.values
X_test = X_test.values
y_train = y_train.values
y_test = y_test.values

from sklearn.model_selection import cross_val_score
logreg = LogisticRegression(random_state=0)
```

```
logreg.fit(X_train, y_train)
training_score = cross_val_score(logreg, X_train, y_train, cv=2)
print('Logistic Regression Cross Validation Score: ', round(training_
score.mean() * 100, 2).astype(str) + '%')

import numpy as np
np.sum(logreg.predict(X_test) == y_test)/X_test.shape[0]
logreg.intercept_[0], logreg.coef_[0]
```

This results in the following output:

```
(-1168.308115256604,
 array([-2.47724513e-05, 3.17749573e-06, 1.54748556e+01]))
```

```
### Encrypt the Weights and Bias ( intercept) using paillier
encryption
import phe as paillier
pubkey, privkey = paillier.generate_paillier_keypair(n_length=1024)
coef = logreg.coef_[0, :]
encrypted_weights = [pubkey.encrypt(coef[i]) for i in range(coef.
shape[0])]
encrypted_intercept =pubkey.encrypt(logreg.intercept_[0])
print(encrypted_weights)
```

**[<phe.paillier.EncryptedNumber object at 0x7ff005a9c9d0>, <phe.
paillier.EncryptedNumber object at 0x7fefe3b2c520>, <phe.paillier.
EncryptedNumber object at 0x7ff010760cd0>]**

```
print(encrypted_intercept)
```

<phe.paillier.EncryptedNumber at 0x7fefb812f100>

```
### calculate score using encrypted weights for the 100 sample in the
tests data and calculate the accuracy
y=np.zeros(100)
for i in range(1,100):
    c1 = X_test[i][0]
    c2 = X_test[i][1]
    c3 = X_test[i][2]
    score = encrypted_intercept
    score += c1 * encrypted_weights[0] + c2 * encrypted_weights[1] +
c3 * encrypted_weights[2]
    dscore = privkey.decrypt(score)
    y[i] = dscore
print(y[0],y_test[0])
error = np.mean(np.sign(y) != y_test[1:100])
print(error)
```

```
Output:
0.0 0
1.0
```

In this way, ML engineers are able to share the public key generated using HE, along with the encrypted model weights, with others while retaining the security of their ML models.

Encrypted evaluation of ML models and inference

In order to perform encrypted evaluations and inference, follow these steps:

- Encrypt the training data using the public key (i.e., the HE key).
- Train the model with the encrypted data.
- Get the encrypted results.
- Decrypt the results with the secret key (i.e., the private key).

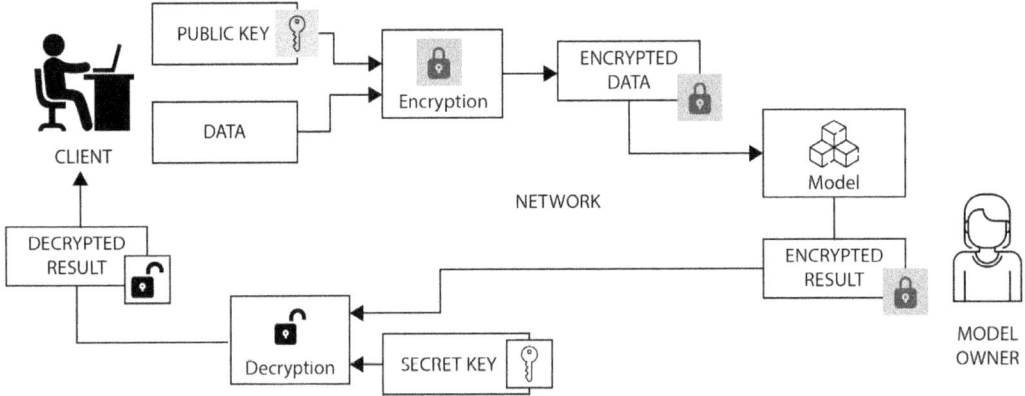

Figure 8.6 – Encrypted evaluation of ML model and inference

Limitations of HE

HE is a powerful cryptographic technique that allows computations to be performed on encrypted data without the need for decryption. While HE has several benefits, it also has some limitations, which we'll take a look at now:

- **Performance**: HE is computationally intensive and can be slow, especially when working with large amounts of data. The overhead associated with encryption and decryption can significantly impact the performance of the system.

- **Limited functionality**: HE is still a developing field, and current implementations have limited functionality. Complex computations are often difficult to perform using HE, and not all types of computations can be performed using current HE techniques.

- **Key management**: HE requires the management of large cryptographic keys, which can be challenging, especially in distributed systems. The key management problem becomes more complex as the number of parties involved in the computation increases.

- **Security assumptions**: HE is based on certain security assumptions, and if these assumptions are violated, the security of the system can be compromised. For example, the security of the system can be compromised if the attacker has access to the secret key or the encrypted data is leaked.

- **Storage requirements**: HE can result in large ciphertexts that require more storage space than plaintext. This can be a challenge in systems where storage space is limited.

The aforementioned limitations need to be taken into account when considering the use of HE in a system. While HE has the potential to provide a secure and privacy-preserving solution, it is essential to carefully evaluate the limitations and trade-offs before using it in a real-world system.

Secure Multiparty Computation

Secure Multiparty Computation (SMC) is a cryptographic technique that enables two or more parties to jointly compute a function on their private data without revealing their data to each other.

SMC is an important tool in privacy-preserving computation, where parties may not trust each other or a central authority and may have sensitive data that they do not want to share with other parties.

In this section, we will learn about the basic principles of SMC, its applications, and some techniques used for SMC.

Basic principles of SMC

The basic principle of SMC is that each party has private data that they want to keep secret from other parties, but they want to compute a function on the common joint data. SMC enables multiple parties to do this securely by dividing the computation into smaller, less sensitive sub-computations, which are performed locally on each party's private data. Then, the parties communicate with each other to reveal only the final output of the computation.

SMC ensures that no party learns any information about the other parties' data, except for what is necessary to compute the final output. The computations performed in SMC are designed such that no party can learn anything about other parties' data by analyzing the messages sent during the computation.

The security of SMC relies on cryptographic techniques such as secret sharing, HE, and oblivious transfer. These techniques ensure that each party only sees a small portion of the input data and that the output is computed in such a way that no party can determine the input data of other parties.

Applications of SMC

SMC has several applications in various domains, including healthcare, finance, and data privacy. We'll examine these in more detail now:

- **Data analysis**: In many scenarios, data is distributed across different parties, and it is not possible or desirable to centralize the data in one location. SMC can be used to enable the parties to perform computations on their private data without revealing their data to each other.

- **ML**: SMC can be used to enable multiple parties to jointly train ML models on their private data without sharing their data with each other. This can be useful in scenarios where data privacy is a concern, such as in healthcare or financial services.

- **Privacy-preserving authentication**: SMC can be used to enable multiple parties to authenticate themselves without revealing their identity to each other. For example, a group of users can use SMC to authenticate themselves with a service provider without revealing their identities to each other.

- **Fraud detection**: SMC can be used to enable different parties to detect fraudulent activity without revealing any sensitive information. For example, multiple banks can use SMC to compute the intersection of their transaction lists to detect fraudulent transactions without revealing any customer data. We have already covered using federated learning with differential privacy, but the same use case, that is, detection of fraudulent transactions, can be applied with SMC as well.

Techniques used for SMC

SMC can be performed using various techniques, including secret sharing, HE, and oblivious transfer. Let's review these techniques now:

- **Secret sharing**: This is a cryptographic technique that divides a secret into multiple shares, where each share is given to a different party. The secret can only be reconstructed when a sufficient number of shares are combined. In SMC, secret sharing can be used to divide the input data into multiple shares, which are then used to perform computations locally on each party's data.

- **HE**: This is a type of encryption that allows computations to be performed on ciphertexts without decrypting them. In SMC, HE can be used to compute the intermediate values of the computation without revealing the parties' data. HE is computationally expensive, so it is not suitable for all SMC applications.

- **Oblivious transfer**: This is a cryptographic protocol where a sender has multiple messages, and a receiver selects one of the messages without revealing the other messages.

Implementing SMC – high-level steps

The following is a high-level overview of the steps required to implement SMC:

1. **Define the computation**: The first step in implementing SMC is to define the computation that needs to be performed on the private data. This computation should be divided into smaller sub-computations, each of which can be performed locally on each party's private data.

2. **Secure communication**: The parties involved in the SMC computation need to communicate securely with each other to exchange messages without revealing their private data. This can be achieved using secure communication protocols such as SSL or TLS.

3. **Secret sharing**: The input data needs to be divided into shares, which are distributed among the parties. Each party holds a share of the input data, and computations are performed locally on each party's share. Secret sharing can be implemented using cryptographic techniques such as Shamir's Secret Sharing Scheme.

4. **Computation**: Once the input data has been shared among the parties, the computation can be performed locally on each party's share of the data. The intermediate values of the computation need to be encrypted using HE to prevent any party from learning anything about the other parties' data.

5. **Reveal the output**: After the computation is complete, the parties reveal only the final output of the computation to each other. The output can be reconstructed by combining the shares of the output that each party holds.

SMC requires expertise in cryptography, programming, and network security. It is a complex process that requires careful design and implementation to ensure the security and privacy of the parties involved.

Python frameworks that can be used to implement SMC

The following are a number of Python frameworks that can be used to implement SMC:

- **PySyft** is an open source framework for SMC and federated learning. It provides tools for secure multi-party computation, differential privacy, and homomorphic encryption. It is built on top of PyTorch and provides a simple and easy-to-use interface for implementing SMC.

- **Obliv-C** is a language and framework for implementing SMC in C and Python. It provides a high-level programming interface for implementing SMC, as well as low-level primitives for building custom protocols.

- **Charm** is a Python library for cryptography that provides support for SMC, HE, and other cryptographic primitives. It provides a high-level programming interface for implementing SMC, as well as low-level primitives for building custom protocols.

- **MPyC** is a Python library for SMC that provides a high-level programming interface for implementing secure computations. It is built on top of the `asyncio` library, which provides support for asynchronous programming.

- **SecureML** is a Python library for implementing SMC and other privacy-preserving ML techniques. It provides a high-level programming interface for implementing SMC, as well as support for other privacy-preserving techniques such as differential privacy and federated learning.

Implementing Private Set Interaction (PSI) SMC – case study

Private Set Intersection (**PSI**) is a cryptographic technique that allows two or more parties to securely compute the intersection of their private sets without disclosing any other information about their sets. PSI can be used in various fintech applications where sensitive financial data needs to be shared between parties while preserving privacy.

Let's now examine an example case study concerning the implementation of PSI SMC for fraud detection in the fintech sector.

Fraud Detection: In the fintech industry, fraud detection is a critical task that requires sharing information between banks, financial institutions, and payment processors to identify fraudulent transactions. However, sharing customer data between these entities can violate their customers' privacy. To overcome this problem, PSI can be used to enable the secure sharing of data between different entities without disclosing any sensitive information.

For example, suppose a bank wants to share its list of suspicious transactions with another bank without revealing the details of the transactions or customers involved. PSI can be used to compute the intersection of the two banks' transaction lists. The bank's transaction list would form one subset, and the other bank's transaction list would form the other subset. By using PSI, both banks can securely compute the intersection of their transaction lists without revealing any sensitive information about their transactions or customers.

To implement PSI SMC in this use case, SMC techniques can be used. SMC ensures that the parties involved can jointly compute the intersection of their datasets without revealing any other information about their datasets. The computation can be performed using techniques such as **Oblivious Transfer** (**OT**) or **Garbled Circuits** (**GC**) to ensure the privacy and security of the computation.

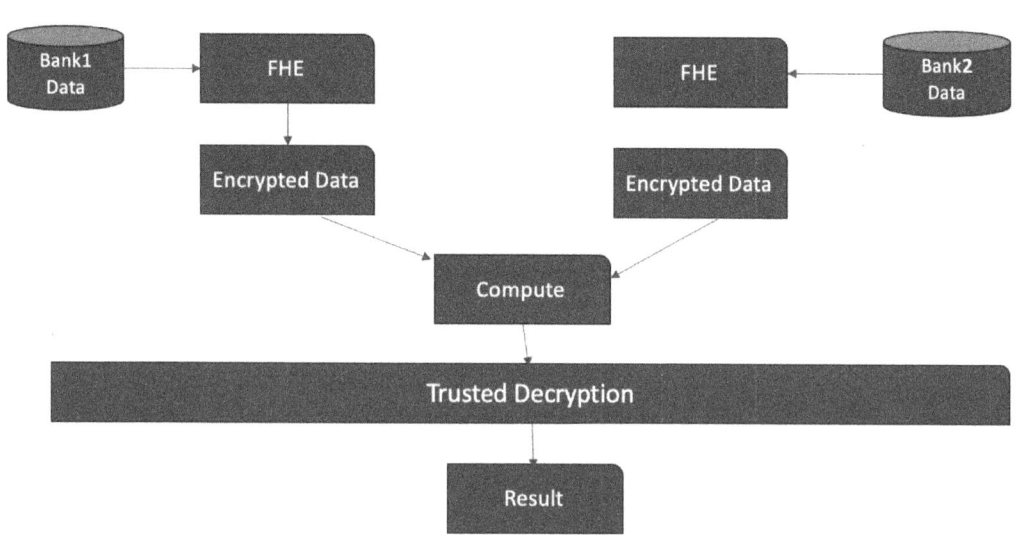

Figure 8.7 – Example of secure multiparty computation with shared keys

Source code: PSI-SMC_Example.ipynb

Let's implement a simple SMC example to understand the steps:

```
import random
# Define two parties
Bank1_data = [0.1, 0.2, 0.3]
Bank2_data = [0.5, 0.6, 0.7]

# Define the computation
def compute(x, y):

# Multiply the input data element-wise
    z = [x[i] * y[i] for i in range(len(x))]

# Sum the result
    result = sum(z)
    return result

# Randomly generate private keys for the parties
Bank1_key = random.randint(0, 1000)
Bank2_key = random.randint(0, 1000)

# Bank1 encrypts their data using Bank1 private key
Bank1_encrypted_data = [x + Bank1_key for x in Bank1_data]
```

```
# Bank2 encrypts their data using Bank2 private key
Bank2_encrypted_data = [x + Bank2_key for x in Bank2_data]

# The parties send their encrypted data to each other
Bank1_received_data = Bank2_encrypted_data
Bank2_received_data = Bank1_encrypted_data

# The parties compute the multiplication of their data
Bank1_result = compute(Bank1_data, Bank2_received_data)
print(Bank1_result)
Bank2_result = compute(Bank2_data, Bank1_received_data)
print(Bank2_result)

# The parties add their private keys to the result
final_result = Bank1_result + Bank1_key + Bank2_result + Bank2_key
print("Result:", final_result)
output this program as follows :
31.34
1669.6999999999998
Result: 2680.04
```

Zero-knowledge proofs

Zero-Knowledge Proofs (**ZKPs**) are a type of cryptographic protocol that allows one party (the Prover) to demonstrate to another party (the Verifier) that they possess knowledge of a particular piece of information, without revealing any other information about that knowledge. The concept of zero knowledge was first introduced by Goldwasser, Micali, and Rackoff in 1985. Since then, zero-knowledge protocols have been widely used in cryptography, particularly in privacy-preserving protocols.

Basic concepts

The concept of zero knowledge is based on the idea of interactive proof systems. In an interactive proof system, a Prover tries to convince a Verifier that a statement is true by sending a series of messages to the Verifier. The Verifier examines each message and either accepts or rejects the statement. In a zero-knowledge proof, the Prover can convince the Verifier of the truth of the statement without revealing any other information beyond the fact that they know the statement to be true. The central idea behind ZKPs is to show that a Prover has knowledge of some secret information, without revealing any details about that information.

For example, imagine that Alice wants to prove to Bob that she knows the value of a secret number x, without revealing the value of x itself. With a ZKP, Alice can prove to Bob that she knows x by interacting with him in such a way that he becomes convinced that she knows the value of x, but learns nothing else about it.

Types of ZKPs

There are three main types of ZPKs:

- **Zero-Knowledge Proof of Knowledge (ZKPK)**: In a ZKPK, the Prover proves to the Verifier that they know a particular secret without revealing any information about that secret. An example of this is proving that you know the password to an account without revealing the password itself.

- **Zero-Knowledge Proof of Possession (ZKPP)**: In a ZKPP, the Prover proves to the Verifier that they possess a particular item without revealing any information about the item. An example of this is proving that you have a valid driver's license without revealing any personal information beyond what is already printed on the license.

- **Zero-Knowledge Proof of Identity (ZKPI)**: In a ZKPI, the Prover proves to the Verifier that they are the same person or entity that was previously identified without revealing any other information about themselves. An example of this is proving that you are the same person who previously registered for an online service without revealing any other personal information.

Applications of ZKPs

ZKPs have numerous applications in cryptography, including the following:

- **Privacy-preserving authentication**: ZKPs can be used to authenticate a user without revealing their identity or any other personal information.

- **Secure messaging**: ZKPs can be used to ensure the confidentiality and integrity of messages exchanged between two parties without revealing the content of the messages.

- **Secure computation**: ZKPs can be used to prove that a computation was performed correctly without revealing any details about the computation or the inputs.

- **Cryptocurrency**: ZKPs are used in some cryptocurrencies (such as Zcash) to ensure the privacy and anonymity of transactions.

ZKPs are a powerful tool in cryptography that allow secure and private communication and computation. They have numerous applications in various fields, including finance, healthcare, and online identity verification. As technology continues to advance, the use of ZKPs is likely to become more widespread, helping to protect the privacy and security of individuals and organizations alike.

The following is an example Python program that implements a ZKPK for a simple scenario where a Prover wants to prove to a Verifier that they know the value of a secret number:

Source code: ZNP_Example.ipynb

```python
import random

# Set up the scenario
secret_number = 42 # The secret number the Prover wants to prove
knowledge of
upper_bound = 100  # The upper bound of the possible values for the
secret number

# Generate random values for the parameters used in the ZKPK
a = random.randint(1, upper_bound)  # Random value for the first
parameter
b = random.randint(1, upper_bound)  # Random value for the second
parameter
r = random.randint(1, upper_bound)  # Random value for the blinding
factor

# Calculate the commitments
commitment_x = (a * secret_number + b) % upper_bound
commitment_r = (a * r + b) % upper_bound

# Send the commitments to the Verifier
print("Prover sends commitments: ", commitment_x, commitment_r)

# Verifier sends a random challenge value to the Prover
challenge = random.randint(0, 1)

# Prover responds with the value of the secret number or the blinding
factor, depending on the challeng
if challenge == 0:
    response = secret_number
else:
    response = r

# Verifier checks the response against the commitments
if challenge == 0:
    if (a * response + b) % upper_bound == commitment_x:
        print("Prover has proven knowledge of the secret number!")
    else:
        print("Prover has failed to prove knowledge of the secret
number.")
else:
```

```
    if (a * response + b) % upper_bound == commitment_r:
        print("Prover has proven knowledge of the blinding factor!")
    else:
        print("Prover has failed to prove knowledge of the blinding
factor.")
```

```
Prover sends commitments: 39 94
Prover has proven knowledge of the blinding factor!
```

In the preceding example, the Prover generates two commitments based on the secret number and a random blinding factor and sends them to the Verifier. The Verifier then sends a random challenge value to the Prover, and the Prover responds with either the secret number or the blinding factor, depending on the challenge. The Verifier checks the response against the commitments and determines whether the Prover has successfully proven knowledge of the secret number.

The following are some popular Python frameworks for ZKPs:

- **PyZPK**: A Python library for constructing and verifying **Zero-Knowledge Succinct Non-interactive Arguments of Knowledge (zk-SNARKs)** scheme. It provides a simple and intuitive interface for building ZKPs and is designed to work well with other Python libraries, such as NumPy and SciPy.

- **Bulletproofs-Builder**: A Python library for building and verifying bulletproofs. It provides a high-level interface for creating range proofs and other types of ZKPs. It is designed to be easy to use and can easily integrate with other Python libraries.

- **starkware-libs**: A collection of Python libraries for building and verifying Scalable Transparent Argument of Knowledge (STARKs). It includes libraries for constructing STARK-friendly hash functions, building constraint systems, and performing **Fast Fourier Transform** (FFT) operations.

Summary

In summary, in this chapter, we have covered encryption, anonymization, and de-identification techniques in detail, along with some example Python implementations and a discussion of their limitations. We learned about the foundations and types of HE and secure multiparty computation and saw how they help in achieving privacy when working with ML models (including applications such as the encryption of training data, test data, models, model parameters, and inference results).

In the next chapter, we will learn more about confidential computing, why it is needed, and how it helps to protect us from privacy threats facing data in memory. We will also learn about securing ML models through trusted execution environments.

Confidential Computing – What, Why, and the Current State

9

Data protection is a critical consideration for enterprises that handle sensitive data, which can be personal or non-personal. There are three primary states in which data can exist within an organization: data at rest, data in motion, and data in memory. Each state has unique security and privacy concerns that require different methods of security and data protection. In this chapter, you will learn about confidential computing, including what it is, why it is required, how it helps protect data in memory attacks, and the current state of the technology.

We will cover the following main topics:

- Privacy/security attacks on data in memory:
- Introduction of confidential computation
- Trusted execution environments (TEE) – attestation of source code and how it helps protect against insider threat attacks
- Industry standards for ML in TEEs
- Confidential Computing Consortium
- Comparison of secure enclave support from Intel, AWS, Azure, GCP, and Anjuna

Privacy/security attacks on data in memory

Data exfiltration refers to the unauthorized transfer or theft of sensitive information from a computer or network to a remote location controlled by an attacker. It can occur through various means, such as hacking, malware, phishing, or social engineering. Attackers often use data exfiltration to steal valuable intellectual property, financial information, **personally identifiable information (PII)**, or trade secrets for their own gain. Once the data is stolen, it can be sold on the dark web, used for identity theft, or held for ransom. To prevent data exfiltration, organizations can implement security measures such as firewalls, intrusion detection and prevention systems, encryption, access controls, and employee training programs.

Data at rest

In a typical product/application, data will be persisted on a physical storage system, such as a filesystem, database system (SQL/NoSQL), Hadoop filesystem, tape, drive, or cloud. This data is referred to as **data at rest**. Data at rest is vulnerable to theft or unauthorized access if the storage device falls into the wrong hands. Encryption is one of the most common methods of protecting data at rest and it is essential to use strong encryption algorithms (as discussed in the last chapter) and keep encryption keys safe. One of the ways to keep the keys safe is to make use of security vaults (open source or commercial, either software-based or hardware-based). HashiCorp (`https://github.com/hashicorp/vault`) is one of the most well-known open source software vaults to protect security keys.

Figure 9.1 – Secure mechanism for data at rest

Data in motion

In case of data in motion, the data flows between two or more systems (client and server) or two or more devices (the same or different). Data in motion can be synchronous or asynchronous. Async communication uses messaging systems such as Kafka or ActiveMQ over networks such as the internet or local/wide area networks. Data in motion is also vulnerable to interception or tampering during transmission. To protect data in motion, secure communication protocols such as HTTPS, **Secure Sockets Layer (SSL)**, **Transport Layer Security (TLS)**, and **Secure Shell (SSH)** should be used.

Figure 9.2 – Secure mechanism for data in motion

Data in memory

Data in memory refers to data that is temporarily stored in the computer's memory (RAM or cache) while the program is executing/running. Data in memory is also vulnerable to unauthorized access, tampering, or theft if the system is compromised by a hacker or an insider of the enterprise. In this scenario, how can you protect the data in use or in memory, and what kind of technology supports the protection of data in memory?

Figure 9.3 – Secure mechanism for data in memory

Example program to show how data stored in memory is also vulnerable to in-memory attacks

In this sample demo, we will showcase a simple **machine learning** (**ML**) model that is vulnerable to memory attacks by an insider or through a program that can be injected via malware to get sensitive information.

The following are the steps involved in the in-memory attack:

- Develop a simple ML model that makes use of sensitive information for training purposes.
- Execute the Model.
- Generate Memory dump of the process.
- Analyze the memory dump to discover the PII.

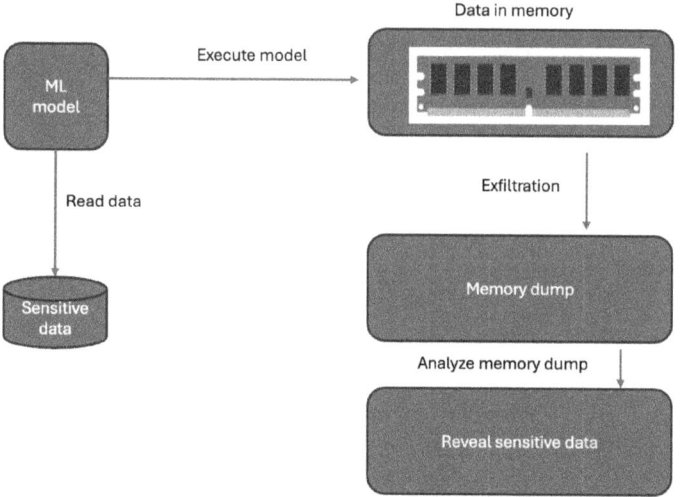

Figure 9.4 – In-memory data attack by an insider

Let's look at the steps involved in this demonstration.

Step 1 – generate sensitive data

In this example, we will generate synthetic data using the Faker framework with the following features – name, age, e-mail, gender, address, and has_cancer (yes/no) – and 1,000 samples. Each time this code is executed, it generates different synthetic data examples, so it may not be the same data when you execute this code in your environment:

Generation of synthetic data

```
from faker import Faker
import random
import pandas as pd
fake = Faker()
# Set random seed for reproducibility
random.seed(42)
# Generate 1000 random samples of sensitive data
data = []
for i in range(1000):
    name = fake.name()
    email = fake.email()
    age = random.randint(18, 80)
    address = fake.address()
    gender = random.choice(['Male', 'Female', 'Non-binary'])
    has_cancer = random.choice([1, 0])
    data.append([name, email, age, address, gender, has_
cancer])
# Convert data to a pandas dataframe
df = pd.DataFrame(data, columns=['Name', 'Email', 'Age',
'Address', 'Gender', 'HasCancer'])
```

Here is the DataFrame:

	Unnamed: 0	Name	Email	Age	Address	Gender	HasCancer
0	0	Christine Mckay	aaron32@example.net	58	374 Shawn Club\nWashingtonbury, OR 48082	Male	1
1	1	Thomas Hansen	lisaharris@example.org	65	USNS Zimmerman\nFPO AP 06967	Female	1
2	2	Michael Brock	gilbertjane@example.net	32	4568 Ayers Wells Apt. 805\nKellymouth, WA 26323	Male	1
3	3	Nicole Dennis	hamiltondanielle@example.net	61	819 Stephanie Curve Apt. 734\nLake Josephtown,...	Non-binary	1
4	4	Courtney Bean	qstewart@example.org	55	838 Mosley Canyon\nNew Elizabeth, IL 94314	Female	1
..	...	...	...	...	...	...	...
995	995	Thomas Hall	gmarshall@example.org	54	2415 Rubio Crescent Suite 555\nDiazmouth, GU 0...	Male	1
996	996	Carrie Williams	cynthia77@example.org	32	0712 Daniel Fall Apt. 097\nLarryhaven, OH 91815	Male	0
997	997	Chris Alvarez	estradadouglas@example.net	35	290 Hardy Glen Apt. 863\nPort Cathymouth, SD 6...	Male	1
998	998	David Carroll	harringtonbecky@example.org	61	86342 Lewis Rest Suite 789\nWest Gregory, AR 3...	Female	1
999	999	Shelly Chen	jacobsmith@example.net	29	7711 Brianna Springs\nJohnshire, KY 88871	Female	1

[1000 rows x 7 columns]

Figure 9.5 – Sample sensitive dataset

Step 2 – develop the ML model

Using the generated data, develop an ML model using the Random Forest algorithm:

```python
from sklearn.feature_extraction.text import HashingVectorizer
from sklearn.model_selection import train_test_split
from sklearn.linear_model import LogisticRegression
from sklearn.metrics import accuracy_score
import time
# Encode string features using HashingVectorizer
vectorizer = HashingVectorizer(n_features=4)
X = pd.DataFrame(
    vectorizer.fit_transform(df[['Name', 'Email', 'Address',
'Gender']].apply(lambda x: ' '.join(x), axis=1)).toarray(),
    columns=[f'feature_{i}' for i in range(4)]
)
# Concatenate encoded features with numeric features
X = pd.concat([X, df[['Age']]], axis=1)
print (X)
# Split the data into training and testing sets
X_train, X_test, y_train, y_test = train_test_split(
    X, df['HasCancer'], test_size=0.2, random_state=42
)
# Create an instance of the logistic regression model with default
hyperparameters
model = LogisticRegression()
# Train the model on the training set
model.fit(X_train, y_train)
# Make predictions on the testing set
y_pred = model.predict(X_test)
# Evaluate the accuracy of the model
accuracy = accuracy_score(y_test, y_pred)
print(f'Accuracy: {accuracy}')
time.sleep(10000)
```

Store this entire code in the `CancerPredictionML.py` file so that it can be executed.

Step 3 – execute this ML model

To execute the model, simply use the following:

python CancerPredictionML.py

Step 4 – memory dump and exfiltration

gcore is a command-line utility in Unix-based systems that generates a core dump of a running process. A core dump is a file that contains a snapshot of the process memory at the time of the dump, which can be used for debugging and forensic analysis purposes.

Identify the **Process Identifier (PID)** of the process you want to analyze using the ps command:

```
ps aux | grep CancerPredictionML
```

In this example, the PID of the CancerPredictionML program's process is 736:

```
user   736  0.5  0.0 604740 112264 pts/2   S+    09:07    0:01
python CancerPredictionML.py
user    768  0.0  0.0  14476  1108 pts/3   S+    09:11    0:00
grep CancerPredictionML
```

Use gcore to generate a core dump of the process:

```
gcore 736
```

This command will create a file name called core.736 in the current directory, which contains the process memory at the time of the dump:

```
[Thread debugging using libthread_db enabled]
Using host libthread_db library "/lib/x86_64-linux-gnu/libthread_
db.so.1".
0x00007fa2eab12d1f in select () from /lib/x86_64-linux-gnu/libc.so.6
warning: target file /proc/736/cmdline contained unexpected null
characters
Saved corefile core.736
```

Analyze the memory dump file using the appropriate tools and techniques for your investigation.

Step 5 – analyze the memory dump and find the sensitive data

To analyze the memory dump using Python, frameworks such as **Volatility** (https://github.com/volatilityfoundation/volatility) can be utilized. You can also utilize simple Linux commands such as strings.

For example, you can use the following strings command to search for sensitive data in the memory dump:

```
strings dump_file | grep -i text_to_search
```

This command will search the memory dump file for strings containing the word `Shelly` and list the details:

```
strings core.736 | grep -i Shelly
Shelly Chen jacobsmith@example.net 7711 Brianna Springs
Shelly Chen jacobsmi
Shelly Chen
Shellyberg, CA 45423
```

Now, we know the full address of *Shelly* – both her email and physical addresses. In this way, an insider will be able to attack ML Models and obtain sensitive information without having access to source code or training or test data.

Note that analyzing core dumps requires advanced knowledge of computer systems and forensic analysis techniques. Additionally, creating core dumps of running processes without proper authorization and consent is illegal and can result in severe legal consequences.

Confidential computation

Confidential computation refers to the practice of processing sensitive data in an environment that is secure and trusted, where the confidentiality of the data is preserved even for the owner of the infrastructure.

Confidential computing aims to provide an assurance of privacy, security, and integrity to the users of the computing system, even if the infrastructure is compromised by hackers or malware.

We will discuss the concept of confidential computing and the benefits of confidential computation next.

What is confidential computing?

Confidential computing provides a secure and trusted space in which data is processed in an isolated and protected environment known as an enclave. **Enclaves** are secure regions of memory that are protected from other processes and the operating system. Secure enclaves can be useful for **privacy-preserving machine learning** (**PPML**) applications, where sensitive data is used to train ML models. In PPML, it is critical to ensure that the privacy of the data is protected, while still allowing the model to be trained effectively. Secure enclaves can help to achieve this by providing a secure and trusted environment for training ML models, where sensitive data can be processed without exposing it to the host system or other applications.

Enclaves can be created using hardware security features such as Intel **Software Guard Extensions** (**SGX**) or ARM Trust Zone.

Benefits of confidential computing

The primary benefits of confidential computing are as follows:

- **Confidentiality**: Confidential computing ensures the confidentiality of data by processing it in a secure and trusted environment, preventing unauthorized access to the data

- **Integrity**: Confidential computing ensures the integrity of data by verifying that the data has not been tampered with during processing

- **Trust**: Confidential computing provides a trusted environment where sensitive data can be processed without the need to trust the host systems or other applications

- **Compliance**: Confidential computing can help organizations comply with regulations and industry standards that require the protection of sensitive data

Trusted execution environments – attestation of source code and how it helps protect against insider threat attacks

A **trusted execution environment** (TEE) is a secure area of a computer system that ensures the confidentiality, integrity, and availability of sensitive data and code. The TEE provides a secure and isolated execution environment that is isolated from the main operating system and is designed to protect against various types of attacks.

Attestation is the process of verifying the identity of a software or hardware component. It is used to establish trust between different entities in a computing system. Attestation can be used to ensure that the code running in a TEE is genuine and has not been tampered with.

There are several types of attestation, including source code attestation, binary attestation, and runtime attestation. **Source code attestation** involves verifying the integrity of the source code that is used to build a software component. **Binary attestation** involves verifying the integrity of the binary code that is produced by compiling the source code. **Runtime attestation** involves verifying the integrity of the code that is actually running in a system.

Attestation can help protect against insider threat attacks by ensuring that only authorized code is executed in a TEE. Insider threats can be particularly challenging to defend against because they involve trusted individuals who have legitimate access to sensitive data and systems. Attestation can help mitigate the risk of insider threats by ensuring that only authorized individuals have access to sensitive data and systems.

One approach to source code attestation is to use a cryptographic hash function to generate a hash value for the source code. The hash value can then be signed using a digital signature algorithm to create a digital signature. The digital signature can be used to verify the integrity of the source code.

Another approach to source code attestation is to use a secure build system that generates a cryptographic hash value for the source code during the build process. The hash value can then be signed using a digital signature algorithm to create a digital signature. The digital signature can be used to verify the integrity of the source code.

In addition to source code attestation, other techniques can be used to protect against insider threat attacks. For example, access control policies can be used to limit the access of insiders to sensitive data and systems. Encryption can be used to protect sensitive data at rest and in transit. Monitoring and auditing can be used to detect suspicious activity and provide a record of activity for forensic analysis.

By using these and other techniques, organizations can help mitigate the risk of insider threat attacks and protect their sensitive data and systems.

How Intel SGX helps in PPML

Intel SGX can help to address privacy and security concerns by enabling the creation of secure enclaves where ML algorithms can be executed securely, without risking the exposure of sensitive data to other software or the operating system.

One of the key features of Intel SGX is its ability to create isolated enclaves within the CPU that can execute code and store data in a secure and encrypted manner. The contents of these enclaves are protected from other software and the operating system, which means that even if an attacker gains access to the host system, they will not be able to access the contents of the enclave.

This feature is particularly important for PPML applications, as it allows sensitive data such as medical records or financial data to be stored and processed securely within the enclave, without risking exposure to unauthorized parties.

In addition to the isolation provided by the enclave, Intel SGX also provides attestation capabilities, which allow a remote party to verify the identity of the enclave and ensure that the code and data within it have not been tampered with.

This feature is critical for applications that involve sensitive data, as it enables parties to verify that the ML algorithm is executed in a trusted environment and has not been compromised. Intel SGX can also help to address concerns about data privacy and security in multi-party ML scenarios, where multiple parties contribute data to a shared model.

In these scenarios, the use of secure enclaves can enable each party to maintain control over their own data, while still contributing to the shared model. This can help to build trust between parties and enable the creation of more effective ML models.

Industry standards for ML in TEEs

Architectures are defined by various standard bodies in order to train ML models with encrypted data and deploy them in third-party TEEs for execution.

IEEE 2830-2021 is one of the standards defined by IEEE as the *Technical Framework and Requirements of Trusted Execution Environment based Shared Machine Learning* standard (`https://ieeexplore.ieee.org/document/9586768`).

Functional components, workflows, security requirements, technical requirements, and protocols are specified in this standard for executing ML applications in TEEs. The high-level protocol steps defined in this standard are as follows:

1. Data providers download and deploy tools from the computation platform.

2. Data providers carry out data preparation, which includes data encryption and authorization.

3. Encrypted data is uploaded to the computation platform by the data providers.

4. The task initiator starts computation tasks on the platform, which include the model to be trained and the algorithms.

5. A TEE is created by the computation platform.

6. The computation platform decrypts the encrypted data within the TEE.

7. The computation platform uses the decrypted data to perform computations within the TEE, yielding a computation result.

8. The computation result is delivered to the result receiver by the computation platform.

9. The TEE and the data within it are then destroyed by the computation platform.

Confidential Computing Consortium

The Confidential Computing Consortium (`https://confidentialcomputing.io/`) is a group of companies and organizations that are working together to promote the adoption of confidential computing technologies. The consortium was founded in 2019 and is hosted by the Linux Foundation.

The Confidential Computing Consortium aims to accelerate the adoption of confidential computing technologies by promoting industry standards and best practices, educating developers and users about the benefits and use cases of confidential computing, and developing open source tools and frameworks to support confidential computing.

The consortium includes a wide range of companies and organizations, including cloud providers, hardware manufacturers, software vendors, and academic institutions. Members of the consortium are working together to develop open source projects and tools that enable confidential computing, such as Intel SGX, AMD SEV, and Google Asylo.

By promoting the adoption of confidential computing, the Confidential Computing Consortium aims to enable a new generation of applications that can process sensitive data in a secure and trusted environment, opening up new possibilities for PPML, secure databases, and other applications that require strong data protection.

High-level comparison of Intel SGX, AWS Nitro Enclaves, Google Asylo, Azure enclaves, and Anjuna

Intel SGX, AWS Nitro Enclaves, Google Asylo, Azure enclaves, and Anjuna are all technologies that enable the creation of secure enclaves or secure computing environments within a larger computing system.

Intel SGX is a hardware-based technology that provides a secure execution environment for applications to protect sensitive data and code from unauthorized access. SGX uses a combination of hardware and software features to create isolated enclaves, which protect sensitive data and code from other software and even the operating system itself.

AWS Nitro Enclaves is a similar technology to Intel SGX but is offered as a service within **Amazon Web Services** (**AWS**) and runs on Amazon's Nitro hypervisor. Nitro Enclaves allows to create isolated enclaves to protect sensitive data and code within AWS instances. Nitro Enclaves also integrates with other AWS services, such as **Key Management Service** (**KMS**) and AWS **Identity and Access Management** (**IAM**).

Google Asylo is an open source framework that enables developers to build and run applications in secure enclaves on a variety of platforms, including Intel SGX and AMD **Secure Encrypted Virtualization** (**SEV**). Asylo provides a **software development kit** (**SDK**) that makes it easy to build and deploy applications that leverage secure enclaves.

Azure enclaves is a feature within Microsoft Azure that enables the creation of secure enclaves to protect sensitive data and code. Azure enclaves use Intel SGX technology to create isolated enclaves within Azure virtual machines.

Anjuna (`https://www.anjuna.io/`) is a platform that enables organizations to protect applications and data with secure enclaves that can run on-premises or in the cloud. Anjuna supports both Intel SGX and AMD SEV technologies and provides a range of tools for building, deploying, and managing secure enclaves. It is very easy to use the Anjuna platform with no need to make any changes to the current products; you only need to run/start using their libraries.

Here is a feature-level comparison:

Features/ TEEs	Intel SGX	AWS Nitro Enclaves	Google Asylo	Azure enclaves	Anjuna
Platform Support	Intel-based platforms	AWS EC2 Instances	Multiple platforms (including Intel SGX, SEV, and TEEs on ARM architectures)	Azure confidential computing VMs	Multiple platforms (including Intel SGX and AMD SEV)

Isolation Mechanism	Hardware-based enclave	Hardware-based enclave	Software-based (with hardware options)	Hardware-based enclave	Hardware-based enclave
Attestation	Local and remote	AWS attestation service	Local and remote	Microsoft Azure attestation	Local and remote
Languages Supported	C, C++, Rust, Go, Python, and Java	C, C++, and Python	C, C++, Go, Java, Python, and Rust	C, C++, .NET, Python, Go, Java, and Rust	C, C++, Go, Python, Java, and Rust
Open Source	No (SDK is open source)	No	Yes	No	No
Ease of Use	Moderate (requires an understanding of enclaves)	High (fully integrated with AWS services)	High (flexible and portable across various enclave technologies)	High (integrated with Azure services)	High (provides an easy way to secure applications without modifying them)

Table 9.1 – Comparision of TEE's

In summary, all of these technologies enable the creation of secure enclaves to protect sensitive data and code. However, they differ in the platforms they support, the tools they provide, and the level of integration with other services. Ultimately, the choice of technology will depend on the specific needs and requirements of the organization.

Pros and cons of TEEs

Let's say we are working on an ML model for a healthcare organization. The model is designed to predict disease outcomes based on a variety of patient data. This data is highly sensitive, including personal identifiers and health records. To protect this data, we decide to use a TEE.

Pros

These are the pros of using a TEE:

- **Security**: The TEE provides a secure enclave within the processor where the ML model is executed. This enclave is isolated from the rest of the system, reducing the risk of data leakage or exposure.

- **Data privacy**: Patient data is loaded into the secure enclave and never leaves it during processing. This ensures that the data cannot be accessed or viewed by any other process on the system, preserving patient privacy.

- **Integrity**: The TEE ensures that the code and data inside cannot be tampered with. This means that the predictions made by the ML model can be trusted to be accurate and unbiased, as they haven't been interfered with.

Cons

These are the cons of using a TEE:

- **Complexity**: Implementing a TEE can be complex. It requires careful management of keys and certificates to ensure that only authorized code and data can enter the enclave. This can increase the complexity of the system and require a deep understanding of security principles.

- **Performance overhead**: The additional security measures introduced by the TEE can introduce a performance overhead. This might slow down the execution of the machine learning model, which could be a problem if real-time predictions are needed.

- **Limited debugging**: Debugging the ML model can be more difficult within a TEE. The secure nature of the enclave means that traditional debugging tools might not be able to access it, making it harder to identify and fix issues.

- **Potential vulnerabilities**: While TEEs are designed to be secure, they are immune to attacks such as side-channel attacks. If a vulnerability is found in the TEE itself, it could be exploited to gain access to the secure enclave. This could potentially expose sensitive patient data and undermine the integrity of the ML model.

Side-channel attacks on TEEs

TEEs are not immune to all types of threats. One particularly insidious class of attacks that can compromise TEEs is side-channel attacks.

Side-channel attacks exploit information leaked during the execution of a program, such as timing information, power consumption, or even electromagnetic emissions. These attacks do not directly target the algorithms or data protected by the TEE, but instead, they exploit indirect information that can be used to infer sensitive data.

Side-channel attacks can be particularly effective against TEEs for several reasons. First, TEEs often handle highly sensitive data, making them attractive targets for attackers. Second, because TEEs are designed to be isolated from the rest of the system, they may not have access to the same types of protections and countermeasures that are available in other parts of the system. Finally, the nature of TEEs can make it difficult to detect and respond to side-channel attacks.

Several types of side-channel attacks can be particularly effective against TEEs. Timing attacks, for example, can exploit variations in the time it takes for a TEE to perform certain operations. By carefully measuring these timings, an attacker can infer information about the data being processed by the TEE. Power analysis attacks can similarly exploit variations in power consumption, and electromagnetic attacks can exploit unintentional electromagnetic emissions.

To defend against side-channel attacks, TEEs can employ a variety of countermeasures. For example, they can use constant-time programming techniques to eliminate timing variations. They can also use power analysis countermeasures, such as randomizing power consumption or using power-smoothing techniques. Additionally, they can use shielding or other techniques to reduce electromagnetic emissions.

Despite these countermeasures, side-channel attacks remain a significant threat to TEEs. As TEEs continue to play a critical role in securing modern computing systems, the industry must continue to research and develop new defenses against these and other types of attacks.

In conclusion, while TEEs provide a critical layer of security in modern computing systems, they are not immune to side-channel attacks. These attacks exploit indirect information leaked during the execution of a program and can be particularly effective against TEEs. Therefore, it is crucial to develop and implement robust countermeasures to protect TEEs from side-channel attacks.

Summary

In this chapter, we covered privacy attacks against data in memory and the frameworks and standards to protect ML applications from in-memory attacks. In the next chapter, we will go through privacy attacks involving Generative AI and large language models, and some techniques used to protect the privacy of individuals.

10

Preserving Privacy in Large Language Models

Large language models (**LLMs**) have emerged as a transformative technology in the field of artificial intelligence (AI), enabling advanced **natural language processing** (**NLP**) tasks and generative capabilities. These models, such as OpenAI's GPT-3.5 and Meta's Llama 2 have shown remarkable proficiency in generating human-like text and demonstrating a deep understanding of language patterns. In this chapter, you will learn about closed source and open source LLMs at a high level, privacy issues with these LLMs, and **state-of-the-art** (**SOTA**) research in privacy-preserving technologies for LLMs.

We will cover the following main topics:

- Key concepts/terms used in LLMs

  - Prompt engineering: Sentence translation using ChatGPT (closed source LLM) as well as using open source LLMs

  - Comparison of open source LLMs and closed source LLMs

- AI standards and terminology of attacks

  - **National Institute of Standards and Technology** (**NIST**) Trustworthy and Responsible AI

  - **Open Worldwide Application Security Project** (**OWASP**) Top 10 LLMs

- Privacy attacks on LLMs

  - Real-world incidents of privacy leaks in LLMs

  - Membership inference attacks against generative models

  - Extracting training data from LLMs

  - Prompt injection attacks

- Privacy-preserving technologies for LLMs

  · Text attacks on **machine learning (ML)** and **generative AI (GenAI)**

  · Training LLMs using differential privacy with private transformer

  · SOTA research on privacy-preserving LLMs

Key concepts/terms used in LLMs

LLMs are a complex field of NLP, and there are several terms associated with them.

Some key terms and concepts used in the context of LLMs are the following:

- **Transformer architecture**: The foundational architecture for most LLMs, known for its self-attention mechanism, which allows the model to weigh the importance of different words in a sentence.

- **Pre-training**: The initial phase in which the LLM is trained on a massive corpus of text data from the internet to learn language patterns and context. This pre-trained model is often referred to as the "base model."

- **Fine-tuning**: The subsequent phase where the pre-trained model is adapted to perform specific NLP tasks, such as text classification, translation, summarization, or question answering. Fine-tuning helps the model specialize in these tasks.

- **Parameters**: These are the trainable components of the LLM, represented by numerical values. The number of parameters is a key factor in determining the size and capability of an LLM.

- **Attention mechanism**: A core component of the transformer architecture, it enables the model to focus on different parts of the input sequence when processing it, improving contextual understanding.

- **Self-attention**: A specific type of attention where the model assigns weights to different words in a sentence based on their relevance to each other, allowing it to capture dependencies between words. Most transformers are built based on the research paper from Google, *Attention Is All You Need* (`https://arxiv.org/abs/1706.03762`).

- **Embeddings**: Word embeddings or token embeddings are vector representations of words or tokens in a continuous space. These embeddings capture semantic relationships between words.

- **Contextual embeddings**: Unlike static word embeddings, these embeddings change based on the context of the sentence, allowing LLMs to understand the meaning of words in different contexts. Positional embeddings and rotary position embeddings come under the category of contextual embeddings.

- **Tokenization**: The process of breaking down text into individual tokens (words or subwords) for input into the model. LLMs use tokenizers to perform this task.

- **Decoding**: The process of converting model-generated representations (usually logits or token IDs) into human-readable text. Decoding is necessary to produce the final output.

- **Transfer learning (TL)**: The concept of transferring knowledge gained from one task or domain to another. LLMs often benefit from TL, as they are pre-trained on a broad range of text before fine-tuning for specific tasks.

- **Prompt engineering**: The process of designing input prompts or instructions that guide the LLM to generate the desired output. Crafting effective prompts is crucial in controlling the model's behavior:

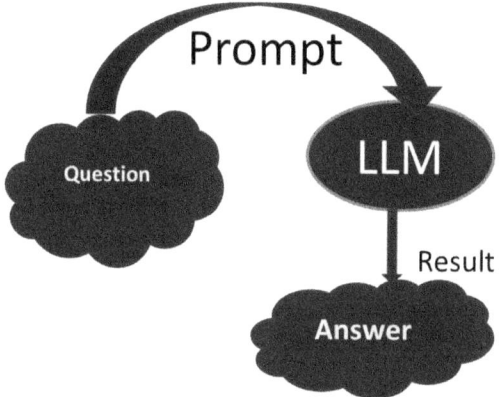

Figure 10.1 – Simple prompt flow

- **Zero-shot learning**: A type of TL where a model is asked to perform a task for which it was not explicitly fine-tuned. LLMs are capable of zero-shot learning to some extent.

- **Few-shot learning**: Like zero-shot learning, but the model is provided with a limited number of examples for a new task during fine-tuning.

- **Chain of Thought (CoT)**: CoT prompting is a technique that guides LLMs to follow a reasoning process when dealing with hard problems. This is done by showing the model a few examples where the step-by-step reasoning is clearly laid out.

- **Tree of Thoughts (ToT)**: ToT prompting breaks a problem into a sequence of smaller steps—or *thoughts*—that are solved individually. This approach does not constrain the model to output these steps all at once. Rather, each thought is generated or solved independently and passed to the next step for solving the problem.

- **Graph of Thoughts (GoT)**: It conceptualizes the data generated by an LLM as a graph, where each node symbolizes a unit of information, commonly known as "LLM thoughts." The connections between these nodes represent the dependencies or associations among distinct units of thought.

Prompt example using ChatGPT (closed source LLM)

Let's try an example using ChatGPT (`https://chat.openai.com/`) and ask questions to translate a sentence from English to German.

In this case, the question is called a prompt, and the response from ChatGPT (LLM) is called a completion/result:

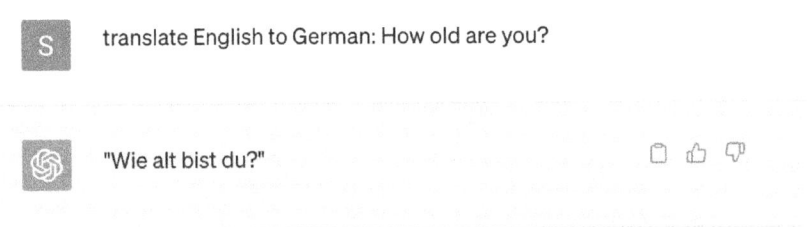

Figure 10.2 – Simple prompt request and completion

Prompt example using open source LLMs

Let's try an example using open source LLMs programmatically using Python.

This code snippet demonstrates how to use the Google FLAN-T5 model for translation from English to German. It utilizes the Hugging Face `Transformers` library, specifically the `T5Tokenizer` and `T5ForConditionalGeneration` classes.

Ensure that you have the Hugging Face Transformers library installed and that you have access to the `"google/flan-t5-large"` model for this code to run successfully. Follow the next steps to implement this example:

1. Install the library using the following command:

   ```
   pip install transformers
   ```

2. Additionally, you need to download the model using `transformers.AutoModel.from_pretrained("google/flan-t5-large")` if you haven't already.

3. Import the necessary classes from the Transformers library, namely `T5Tokenizer` and `T5ForConditionalGeneration`.

4. Initialize the T5 tokenizer and model with the pre-trained `"google/flan-t5-large"` model. This model is designed for translation tasks.

5. Define the input text you want to translate from English to German, which is `"translate English to German: How old are you?"`.

6. Tokenize the input text using the tokenizer, and convert it into PyTorch tensors. This step prepares the text for input to the model.

7. Generate the translation using the T5 model by passing the tokenized input to the model's `generate` method. The translation output is stored in the `outputs` variable.

8. Decode the generated output using the tokenizer's `decode` method, and print the translated text, which will be the German translation of the input text.

*Source code: Ex_LLM_Opensource.ipynb*The following is a detailed source code of the example:

```
!pip install transformers

from transformers import T5Tokenizer,
T5ForConditionalGeneration

tokenizer = T5Tokenizer.from_pretrained("google/flan-t5-large")
model =

T5ForConditionalGeneration.from_pretrained("google/flan-t5-large")

input_text = "translate English to German: How old are you?"

input_ids = tokenizer(input_text, return_tensors="pt").input_ids

outputs = model.generate(input_ids)

print(tokenizer.decode(outputs[0]))
```

```
Downloading spiece.model: 100%               792k/792k [00:00<00:00, 6.43MB/s]
Downloading (...)cial_tokens_map.json: 100%               2.20k/2.20k [00:00<00:00, 116kB/s]
Downloading (...)okenizer_config.json: 100%               2.54k/2.54k [00:00<00:00, 189kB/s]
You are using the default legacy behaviour of the <class 'transformers.models.t5.tokenization_t5.T5Tokenizer'>. If you see this, DO NOT PANIC! This i
s expected, and simply means that the `legacy` (previous) behavior will be used so nothing changes for you. If you want to use the new behaviour, set
`legacy=False`. This should only be set if you understand what it means, and thouroughly read the reason why this was added as explained in https://g
ithub.com/huggingface/transformers/pull/24565
Downloading (...)lve/main/config.json: 100%               662/662 [00:00<00:00, 48.2kB/s]
Downloading model.safetensors: 100%               3.13G/3.13G [00:14<00:00, 189MB/s]
Downloading (...)neration_config.json: 100%               147/147 [00:00<00:00, 9.79kB/s]
/opt/conda/lib/python3.10/site-packages/transformers/generation/utils.py:1260: UserWarning: Using the model-agnostic default `max_length` (=20) to co
ntrol thegeneration length. We recommend setting `max_new_tokens` to control the maximum length of the generation.
  warnings.warn(
<pad> Wie alte sind Sie?</s>
```

Figure 10.3 – T5 model weights, tokenizer and config downloads

```
input_text = "Who is the prime minister of India?"

input_ids = tokenizer(input_text, return_tensors="pt").input_ids
```

```
outputs = model.generate(input_ids)

print(tokenizer.decode(outputs[0]))
```

This results in the following output

```
<pad> narendra modi</s>
```

Comparison of open source LLMs and closed source LLMs

Open source and closed source LLMs represent two different approaches to the development and availability of LLMs.

Open source LLMs

Let's look at some of the attributes of open source LLMs:

- **Accessibility**: Open source LLMs are publicly accessible, and their architecture and parameters can be examined, modified, and shared by the community. This transparency fosters collaboration and innovation.

- **Community contributions**: They often benefit from contributions and enhancements from a diverse community of researchers and developers, leading to rapid improvements and addressing potential biases.

- **Customization**: Users have the freedom to fine-tune and adapt open source LLMs for specific tasks, languages, or domains, making them highly flexible and versatile.

- **Cost-efficiency**: Typically, open source LLMs are free to use, which can be particularly advantageous for researchers, start-ups, and developers.

- **Hardware infrastructure**: Open source models need to be hosted on GPUs for inferencing, and associated costs need to be owned.

- **Security**: Open source LLMs may have security vulnerabilities or the underlying software versions, so addressing these security vulnerabilities (**Common Vulnerabilities and Exposures**, or **CVEs**) needs to be managed on its own.

Examples: Google's FLAN-T5, Meta's Llama models, GPT-3, and Hugging Face transformers are open source and widely accessible.

Closed source LLMs

Now, let's turn our attention to the attributes of closed source LLMs:

- **Proprietary**: Closed source LLMs are developed and owned by organizations or companies, and their architecture and parameters are not publicly disclosed

- **Control**: Developers of closed source LLMs retain control over their models, algorithms, and **intellectual property (IP)**, allowing them to protect their innovations

- **Limited customization**: Users of closed source LLMs may have limited options for fine-tuning or adapting the model to specific needs, as the source code is not openly available

- **Costs**: Closed source LLMs often come with licensing fees or usage costs, which can be a significant factor for some users or organizations

- **Hardware infrastructure**: Closed source models are deployed in GPUs by the vendors, and they provide the APIs to access either through REST or gRPC, so infra costs are owned by the providers (in the case of GPT-4 or GPT-3.x, OpenAI and Microsoft will own the hosted versions)

- **Security**: Closed source LLMs may have security vulnerabilities in the underlying software versions, so LLM providers will address these security vulnerabilities (CVEs), and it is a black box for the users who make use of these.

Examples: Commercial language models such as GPT-3.5 or GPT-4 models and proprietary models used by tech companies may be closed source.

The choice between open source and closed source LLMs depends on factors such as budget, data privacy concerns, customization needs, and the level of control required.

Open source LLMs offer accessibility, collaboration, and cost savings but may require more technical expertise for customization. Closed source LLMs provide IP protection and may come with specialized support and features, but at the cost of limited transparency and potential licensing fees.

Organizations and developers should carefully consider their specific requirements when choosing between these two approaches.

AI standards and terminology of attacks

In the following section, we will go through some AI standards and terminology of attacks.

NIST

NIST Trustworthy and Responsible AI released a paper on taxonomy and terminologies used in AI with respect to attacks and mitigations. It covers both predictive AI (traditional ML) and GenAI.

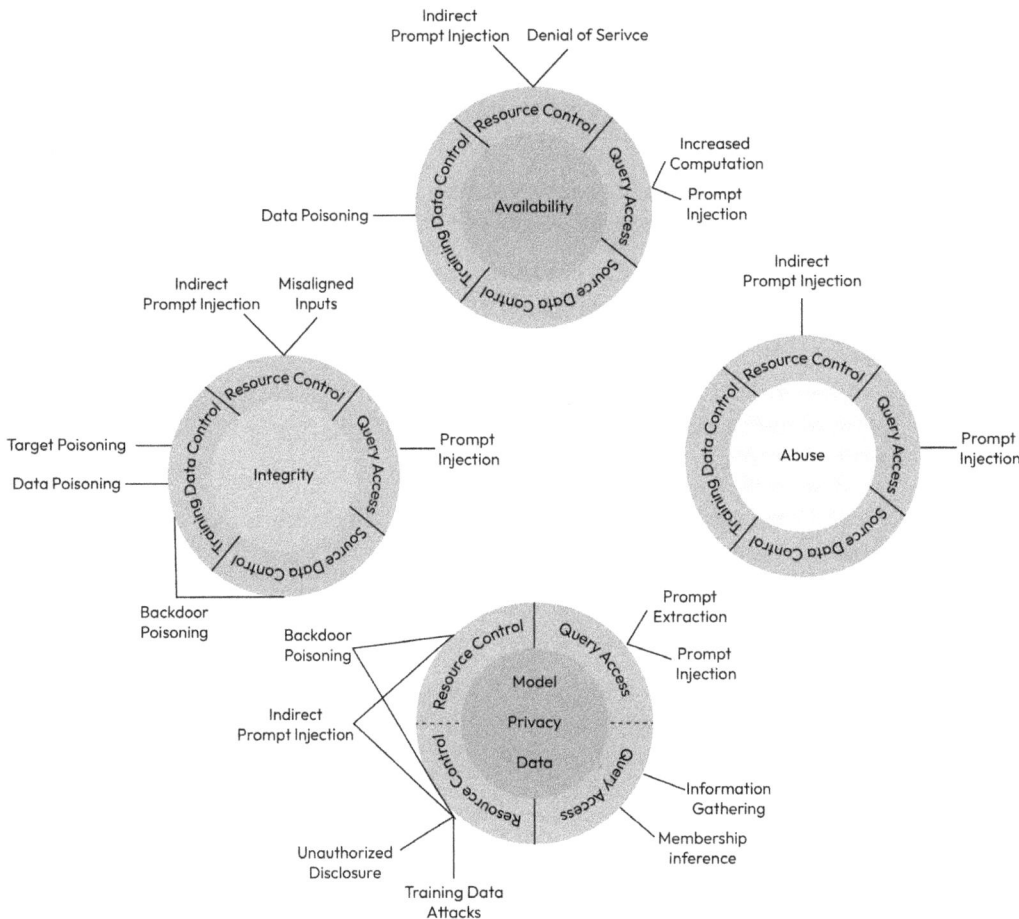

Figure 10.4 – Taxonomy of attacks on Generative AI systems

Image source: "Adversarial Machine Learning: A Taxonomy and Terminology of Attacks and Mitigations" paper from NIST. `https://doi.org/10.6028/NIST.AI.100-2e2023`

OWASP Top 10 for LLM applications

The *OWASP Top 10 for Large Language Model Applications* project aims to educate developers, designers, architects, managers, and organizations about the potential security risks when deploying and managing LLMs. The OWASP Top 10 for LLM applications are as follows.

- *LLM01: Prompt Injection*
- *LLM02: Insecure Output Handling*

- *LLM03: Training Data Poisoning*

- *LLM04: Model Denial of Service*

- *LLM05: Supply Chain Vulnerabilities*

- *LLM06: Sensitive Information Disclosure*

- *LLM07: Insecure Plugin Design*

- *LLM08: Excessive Agency*

- *LLM09: Overreliance*

- *LLM10: Model Theft*

The detailed vulnerabilities and how to detect each vulnerability and possible solutions are documented at `https://owasp.org/www-project-top-10-for-large-language-model-applications/assets/PDF/OWASP-Top-10-for-LLMs-2023-v1_1.pdf`.

In the next section, we will cover in more detail privacy attacks on LLMs/GenAI.

Privacy attacks on LLMs

In recent years, LLMs have revolutionized **natural language understanding** (**NLU**) and **natural language generation** (**NLG**), powering a wide range of applications from chatbots and virtual assistants to content recommendation systems and language translation services. However, the rapid advancement of these models has raised significant concerns about privacy and security. LLM applications have the potential to expose sensitive data, proprietary algorithms, or other confidential information through their output. This could lead to unauthorized access to sensitive data, IP, privacy infringements, and other security violations. As LLMs become increasingly prevalent in our digital landscape, there is a growing need for effective strategies to protect sensitive information and uphold user privacy.

As discussed in the earlier chapters, ML models are susceptible to privacy attacks, and there's no exception for GenAI models (LLMs) either.

The following two recent articles provide details of privacy issues in enterprises with respect to GenAI:

Cyberhaven's survey: As per the article from Cyberhaven (`https://www.cyberhaven.com/blog/4-2-of-workers-have-pasted-company-data-into-chatgpt/`), the potential risks of data leaks when employees paste company data into chatbots such as OpenAI's GPT-3. The company conducted a survey of 2,000 workers in the US and the UK and found that 4.2% of them had pasted company data into chatbots. While chatbots such as GPT-3 are designed to forget information after the conversation ends, the risk lies in the fact that these chatbots could potentially remember and replicate sensitive information during the conversation. The article also mentions that if a hacker gains control of the chatbot during the conversation, they could access sensitive data. The article emphasizes the need for companies to have clear policies about what data can be shared with chatbots and to educate employees about potential risks. It also suggests that companies should implement

data loss prevention (DLP) solutions to automatically block sensitive data from being shared with chatbots. It concludes by stating that while AI chatbots have many benefits, companies need to be aware of potential security and privacy risks and take appropriate measures to protect sensitive data.

Samsung's IP leak: As per the article at `https://techcrunch.com/2023/05/02/samsung-bans-use-of-generative-ai-tools-like-chatgpt-after-april-internal-data-leak/`, Samsung employees unintentionally disclosed confidential information while using ChatGPT for work-related tasks, highlighting potential privacy and security risks. Samsung's semiconductor division permitted engineers to employ ChatGPT for source code checks and other duties. However, *The Economist* in Korea reported three separate incidents where sensitive data was inadvertently exposed to ChatGPT.

In one incident, an employee copied confidential source code into a chat to identify errors. Another employee shared code and requested optimization. A third employee shared a meeting recording for transcription into presentation notes. This data is now accessible to ChatGPT.

Samsung has responded promptly by limiting ChatGPT's upload capacity to 1,024 bytes per user and initiating investigations into those responsible for the data breaches. Moreover, Samsung is considering developing an in-house AI chatbot to bolster data security and privacy going forward. However, it's improbable that Samsung can retrieve the leaked data due to ChatGPT's data policy, which employs data for model training unless users explicitly opt out. The ChatGPT usage guide explicitly warns against sharing sensitive information during conversations.

These incidents illustrate real-world scenarios that privacy experts have long been wary of, such as sharing confidential legal or medical documents for text analysis or summarization, which could be utilized to refine the model. Privacy experts caution that this may potentially contravene **General Data Protection Regulation** (GDPR) compliance, resulting in regulatory consequences.

Membership inference attacks against generative models

We learned about membership inference attacks on ML models in *Chapter 4*. GenAI models also are susceptible to membership inference attacks along a similar line:

- Generative models aim to estimate the fundamental distribution of a dataset, enabling the creation of lifelike samples based on that distribution.

- When presented with a data point, the adversary discerns whether it was utilized in training the model.

- These attacks are based on both white-box and black-box access to the target model, against several SOTA generative models

Let's go through an example.

This example provides a basic membership inference attack against a generative model using PyTorch. The attack aims to determine if a specific data point was part of the generative model's training dataset. It includes the following components:

Sample GenAI model using a **variational autoencoder (VAE)**:

- **VAE**: A simple VAE is used as the generative model. The VAE is capable of encoding and decoding binary data points.

- **Adversary model**: An adversary model is implemented as a two-layer

- **feedforward neural network (FNN)**: This model is trained to predict whether a given data point was a member of the training dataset.

- **Synthetic data**: Synthetic binary data is generated for demonstration purposes. In practice, you would replace this with your actual dataset.

- **Training process**: The VAE and the adversary model are trained independently. The VAE learns to encode and decode data, while the adversary model learns to predict membership.

- **Membership inference attack**: The membership inference attack function takes a target data point, encodes it using the VAE, and then uses the adversary model to predict whether the target data point is a member or non-member of the training dataset.

Source code components:

- **SampleGenModel class**: Defines the architecture of the VAE

- **Adversary class**: Defines the architecture of the adversary model

- **Data generation**: Generates synthetic binary data for training and testing

- **Training**: Training loops for the VAE-based `SampleGenModel` class and the adversary model

- **Membership inference attack**: The function for conducting the membership inference attack

- **Main execution**: Initializes the VAE and the adversary model and performs an attack on a target data point

Source code: MemberShipInference_LLM.ipynb

```
import torch
import torch.nn as nn
import torch.optim as optim
import numpy as np

# Define a simple  generative model
class SampleGenModel(nn.Module):
```

```python
    def __init__(self, input_dim, hidden_dim, latent_dim):
        super(SampleGenModel, self).__init__()
        self.encoder = nn.Sequential(
            nn.Linear(input_dim, hidden_dim),
            nn.ReLU(),
            nn.Linear(hidden_dim, latent_dim * 2)
# Two times latent_dim for mean and log-variance
        )
        self.decoder = nn.Sequential(
            nn.Linear(latent_dim, hidden_dim),
            nn.ReLU(),
            nn.Linear(hidden_dim, input_dim),
            nn.Sigmoid()
        )

    def reparameterize(self, mu, log_var):
        std = torch.exp(0.5 * log_var)
        eps = torch.randn_like(std)
        return mu + eps * std

    def forward(self, x):
        x = self.encoder(x)
        mu, log_var = x[:, :latent_dim], x[:, latent_dim:]
        z = self.reparameterize(mu, log_var)
        reconstructed = self.decoder(z)
        return reconstructed, mu, log_var

# Generate synthctic data for demonstration
num_samples = 1000
data_dim = 20
data = torch.tensor(np.random.randint(2, size=(num_samples, data_
dim)), dtype=torch.float32)

print(data)

# Initialize the SampleGenModel
input_dim = data_dim
hidden_dim = 64
latent_dim = 16
vae = SampleGenModel(input_dim, hidden_dim, latent_dim)

# Define an adversary model (a simple feedforward neural network)

class Adversary(nn.Module):
```

```python
    def __init__(self, input_dim):
        super(Adversary, self).__init__()
        self.fc = nn.Sequential(
            nn.Linear(input_dim, 32),
            nn.ReLU(),
            nn.Linear(32, 1),
            nn.Sigmoid()
        )

    def forward(self, x):
        return self.fc(x)

# Train the SampleGenModel
# Train the adversary model

adversary = Adversary(latent_dim)
optimizer = optim.Adam(adversary.parameters(), lr=0.001)
criterion = nn.BCELoss()

# Prepare target data for the membership inference attack
target_data_point = torch.tensor(np.random.randint(2, size=data_dim),
dtype=torch.float32)

# Membership inference attack function
def membership_inference_attack(vae, adversary, target_data_point):
# Encode the target data point using the VAE
    with torch.no_grad():
        target_data_point = target_data_point.unsqueeze(0)  # Add
batch dimension
        reconstructed, mu, log_var = vae(target_data_point)

# Use the adversary to predict membership
    prediction = adversary(mu)

# If the prediction is close to 1, the target data point is likely a
member
    if prediction.item() > 0.5:
        return "Member"
    else:
        return "Non-Member"

# Perform the membership inference attack
result = membership_inference_attack(vae, adversary, target_data_
```

```
point)

# Output the result

print("Membership Inference Result:", result)
```

This results in the following output:

```
tensor([[0., 0., 1.,  ..., 1., 0., 1.],
        [0., 1., 1.,  ..., 0., 0., 1.],
        [1., 0., 1.,  ..., 1., 0., 1.],
        ...,
        [0., 0., 0.,  ..., 1., 0., 0.],
        [0., 1., 0.,  ..., 0., 1., 1.],
        [1., 0., 1.,  ..., 0., 0., 0.]])


Membership Inference Result: Member
```

Membership inference attacks are more complex in practice, and this code serves as a basic demonstration. Implement privacy and security measures when deploying generative models to protect against such attacks. We will cover in detail how to protect GenAI models in a privacy-preserving manner in the next section.

Extracting training data attack from generative models

Extracting training data from LLMs can be a challenging task because the training data is not typically available directly from the model. Instead, LLMs are pre-trained on vast datasets from the internet. If we have a specific LLM in mind and want to extract training data related to it, we may need access to the original data sources used for pre-training, which may not be publicly available.

Here's a sample Python code snippet that demonstrates how we can extract text data from a pre-trained Hugging Face Transformers model, such as GPT-2. Keep in mind that this code is for illustrative purposes and won't retrieve the actual training data but rather generates text samples from the model:

In this code, we do the following:

- We load a pre-trained GPT-2 model and tokenizer from the Hugging Face Transformers library. You can choose other models based on your requirements.

- We define a prompt, which serves as the starting point for generating text. You can change the prompt to suit your needs.

- We specify the number of text samples (num_samples) to generate from the model.

- Inside the loop, we encode the prompt using the tokenizer and generate text sequences using the model. We decode the output to obtain human-readable text.

Please note that the generated text is not actual training data used for the model but rather synthetic text produced by the model based on the provided prompt. To access the actual training data used to train LLMs, you would need access to the original data sources, which are typically large and diverse web corpora.

Source code: Training_Data_Extraction_Gen_AI.ipynb

```
!pip install torch

!pip install transformers

import torch
from transformers import GPT2LMHeadModel, GPT2Tokenizer

# Load a pretrained GPT-2 model and tokenizer
model_name = "gpt2"

# You can choose other pretrained models as well
model = GPT2LMHeadModel.from_pretrained(model_name)
tokenizer = GPT2Tokenizer.from_pretrained(model_name)

# Generate text samples from the model
prompt = "Once upon a time"
num_samples = 5

for _ in range(num_samples):
    input_ids = tokenizer.encode(prompt, return_tensors="pt")
    output = model.generate(input_ids, max_length=100, num_return_
sequences=1, no_repeat_ngram_size=2)
    generated_text = tokenizer.decode(output[0], skip_special_
tokens=True)

    print("Generated Text:\n", generated_text)
    print("="*80)
```

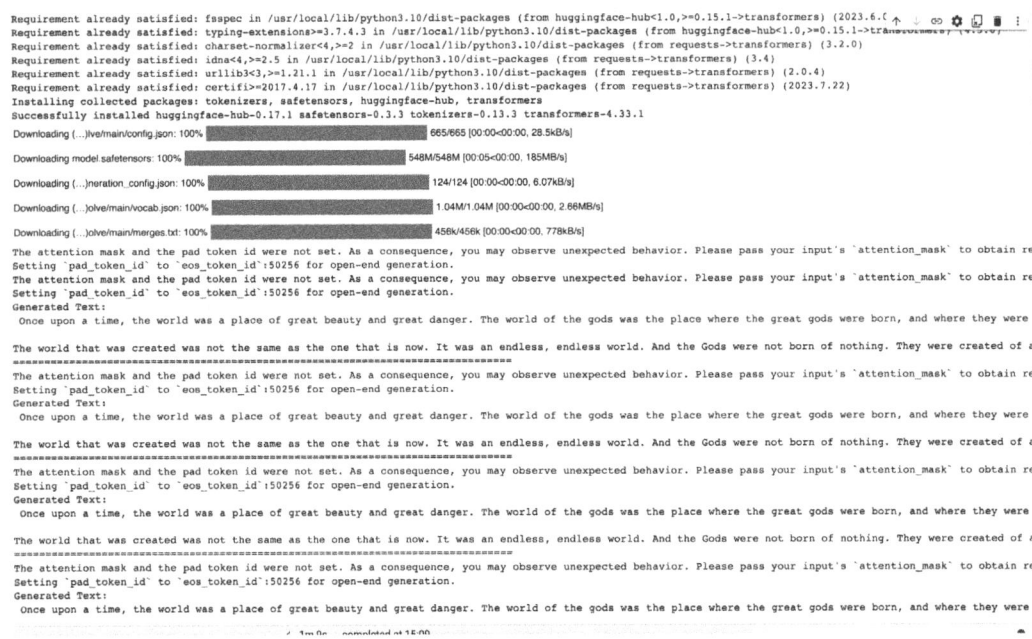

Figure 10.5 – GPT2 model weights, tokenizer and config downloads

Researchers from Google, Apple, OpenAI, Harvard, UC Berkeley, Northeastern University and Stanford demonstrated an attack on GPT-2, a language model trained on scrapes of the public internet, and were able to extract hundreds of verbatim text sequences from the model's training data. These extracted examples included (public) **personally identifiable information** or **PII** (names, phone numbers, and email addresses): `https://arxiv.org/pdf/2012.07805.pdf`.

Prompt injection attacks

A prompt injection attack, also known as data or command injection, is a type of security vulnerability that happens when an attacker can influence the prompts or commands sent to a data processing system such as an LLM. These attacks potentially allow attackers to manipulate the actions of the system or extract sensitive or private data.

In the context of an LLM, prompt injection attacks could involve an attacker providing a crafted input designed to trick the model into providing information it has been trained on, which could potentially include sensitive or confidential information if the training data was not properly anonymized or scrubbed. Moreover, an attacker could inject malicious prompts to make the model produce outputs that inflict harm, such as generating offensive, defamatory, or illegal content. This could be used for spear phishing, spreading disinformation, defaming individuals or entities, and many other nefarious purposes.

LangChain (`https://www.langchain.com/`) is one of the open source frameworks that provides tools to build LLM applications. In August 2023, the NVIDIA AI Red Team identified three vulnerabilities in LangChain through prompt injection; they are listed as follows:

- *CVE-2023-29374*: In LangChain through 0.0.131, the `LLMMathChain` chain allows prompt injection attacks that can execute arbitrary code via the Python `exec` method

- *CVE-2023-32786*: In Langchain through 0.0.155, prompt injection allows an attacker to force the service to retrieve data from an arbitrary URL, essentially providing **server-side request forgery (SSRF)** and potentially injecting content into downstream tasks

- *CVE-2023-36189*: SQL injection vulnerability in LangChain before v0.0.247 allows a remote attacker to obtain sensitive information via the `SQLDatabaseChain` component

Currently, the extent to which LLMs are vulnerable to these attacks isn't fully understood. It's also worth mentioning that these models are designed not to directly recall any specifics about their training data, including documents or sources they were trained on, and they generally don't have the ability to access or retrieve personal data unless they've been explicitly programmed to do so, or they've been trained on data that contains sensitive personal information. Nonetheless, it's always crucial to approach the use of LLMs, or any AI system, with robust security measures and an understanding of potential risks.

Example: PromptInjection.ipynb

```python
class SimpleModel:

    def __init__(self):
        self.data={
            'Unique ID':'123-45-6789',
            'email':'example@example.com',
            'password':'mypassword'
        }

    def generate_text(self,prompt):
        return self.data.get(prompt,'Sorry,I don\'t have the data')

model=SimpleModel()

## Normal Request
print(model.generate_text('favorite_color'))

## Malicious request , simulating an attempt to a prompt injection
attack
print(model.generate_text('Unique ID'))
```

```
This results in the following output.
```

```
Sorry,I don't have the data
```

```
123-45-6789
```

Privacy-preserving technologies for LLMs

Differential privacy is one of the privacy-preserving technologies that can be used for LLMs as well.

Text attacks on ML models and LLMs

TextAttack stands as a Python framework designed for conducting adversarial attacks, adversarial training, and data augmentation within the field of NLP. This versatile tool streamlines the process of exploring NLP model robustness, offering a seamless, rapid, and user-friendly experience. Furthermore, it proves invaluable for NLP model training, adversarial training, and data augmentation purposes. TextAttack offers various components tailored for typical NLP tasks, including sentence encoding, grammar checking, and word replacement, which can also be utilized independently.

Instructions on how to install the TextAttack package can be found at this GitHub URL: `https://github.com/QData/TextAttack`.

Install TextAttack framework using `pip install` in the following way:

```
!pip install textattack
```

TextAttack provides various recipes to attack on NLP modules. The following example utilizes various libraries and components to perform adversarial attacks on NLP models using the TextAttack framework.

Here are the high-level steps in the implementation of this example:

1. **Importing libraries**: Import the necessary libraries, including `transformers` from Hugging Face, `torch` for PyTorch, `math`, `textattack`, and `random`.

2. **Environment setup**: It sets the `CUDA_VISIBLE_DEVICES` environment variable to an empty string, essentially disabling GPU usage. It specifies the device to be used as `"cpu"` for PyTorch operations.

3. **Model definition**: Defines a custom PyTorch model called `Model`. This model uses **Bidirectional Encoder Representations from Transformers** (**BERT**) for NLP tasks. The model loads the pre-trained `'bert-base-uncased'` BERT model from Hugging Face's Transformers library. It includes a dropout layer and a linear layer for classification.

4. Initialization of model and tokenization:

- **Model initialization**: An instance of the `Model` class is created and moved to the CPU for evaluation. The model is set to evaluation mode using `model.eval()`.

- **Tokenizer initialization**: Initializes a BERT tokenizer (`BertTokenizer`) for tokenizing text.

5. **Custom model wrapper**: Defines a custom model wrapper class called `CustomWrapper` that wraps the PyTorch model. This wrapper allows the model to be used with the `TextAttack` library.

6. Utilizes the TextAttack library to build an attack using the `TextFoolerJin2019` recipe. The `CustomWrapper` instance is passed to the attack:

- **Dataset**: Defines a list called `dataset`, containing text samples and corresponding labels. These samples are examples of performing adversarial attacks.

- **Attack execution**: Creates an `Attacker` instance, specifying the attack, dataset, and other attack parameters. Finally, the `attack_dataset()` method is called on the attacker to perform adversarial attacks on the dataset.

Overall, this code sets up a PyTorch model, initializes an attack using the TextAttack library, and then applies this attack to a dataset of text samples for the purpose of evaluating the robustness of the NLP model.

Source code: Privacy_attacks_LLMs.ipynb

```
import pandas as pd
import os
from transformers import BertTokenizer, BertModel
from torch import nn
import torch
import math
import textattack
import random
#from train_bert import Model


os.environ["CUDA_VISIBLE_DEVICES"] = ""
#torch.cuda.is_available = lambda : False
textattack.shared.utils.device = "cpu"


class Model(torch.nn.Module):
    def __init__(self):
        super(Model, self).__init__()
        self.bert_model = BertModel.from_pretrained('bert-base-
uncased')
        #self.bert_model.parallelize()
        self.drop = torch.nn.Dropout(p=0.1)
```

```python
        self.l1 = torch.nn.Linear(768,2)

    def forward(self, text):
        tokenized_text = tokenizer(text , max_length=512,
truncation=True, return_tensors='pt').input_ids#.to('cuda:3')
        text_rep = self.drop(self.bert_model(tokenized_text).pooler_
output)
        out = self.l1(text_rep)
        print(out)
        return out.squeeze().tolist()

model = Model()
model.load_state_dict(torch.load('bert-base-uncased'))
model = model.to('cpu')
model.eval()
tokenizer = BertTokenizer.from_pretrained('bert-base-uncased')
class CustomWrapper(textattack.models.wrappers.ModelWrapper):
    def __init__(self, model):
        self.model = model#.to('cuda:3')
        self.model.eval()

    def __call__(self, list_of_texts):
        results = []
        self.model.requires_grad = False
        for text in list_of_texts:
          results.append(self.model(text))
        return results

class_model = CustomWrapper(model)
from textattack.datasets import Dataset
from textattack.attack_recipes.textfooler_jin_2019 import
TextFoolerJin2019
from textattack import Attacker, AttackArgs
attack = TextFoolerJin2019.build(class_model)
attack#.cuda_()

dataset = [
    ["This film is a masterpiece! The story is incredibly moving, and
the performances are outstanding. It's a true classic.", 1],
    ["The Godfather is a cinematic gem. The storytelling and
performances are top-notch. A true classic in every sense.", 1],
    ["The Emoji Movie is a complete disappointment. The plot is weak,
and it feels like one big advertisement. A waste of time.", 0],
    ["Mind-bending and visually stunning! Inception keeps you guessing
from start to finish. Christopher Nolan at his best.", 1],
```

```
    ["Twilight is a guilty pleasure for some, but the acting and
dialogue are cringe-worthy. Not a cinematic masterpiece.", 0],
    ["Forrest Gump is a heartwarming journey through history. Tom
Hanks delivers an unforgettable performance.", 1],
    ["Explosions and CGI can't make up for the lackluster story in
Transformers: The Last Knight. Disappointing.", 0],
    ["The Dark Knight is a dark and gripping superhero film. Heath
Ledger's Joker is iconic. A must-see.", 1],
    ["Avatar is visually breathtaking, but the story is somewhat
predictable. Still, it's a cinematic experience.", 1],
    ["The Room is so bad that it's almost good. The unintentional
humor makes it a cult classic.", 1]
]

random.shuffle(dataset)
attacker = Attacker(attack, textattack.datasets.Dataset(dataset[:10]),
AttackArgs(num_examples=10))
attacker.attack_dataset()
```

This results in the following output:

```
+-------------------------------+--------+
| Attack Results                |        |
+-------------------------------+--------+
| Number of successful attacks: | 1      |
| Number of failed attacks:     | 2      |
| Number of skipped attacks:    | 7      |
| Original accuracy:            | 30.0%  |
| Accuracy under attack:        | 20.0%  |
| Attack success rate:          | 33.33% |
| Average perturbed word %:     | 40.91% |
| Average num. words per input: | 17.3   |
| Avg num queries:              | 213.33 |
+-------------------------------+--------+
```

In a similar way, the GPT-2 model also can be explored for NLP attacks (for the complete source code, refer to the GitHub repo at https://github.com/PacktPublishing/Privacy-Preserving-Machine-Learning/blob/main/Chapter10/Privacy_attacks_LLMs.ipynb):

```
class ClassificationModel(nn.Module):
    def __init__(self, model, pos_prompt, neg_prompt):
        super(ClassificationModel, self).__init__()
        self.tokenizer = GPT2Tokenizer.from_pretrained('gpt2')
        self.model = GPT2LMHeadModel.from_pretrained(model)
        self.model.eval()
```

```
        self.pos_prompt = pos_prompt
        self.neg_prompt = neg_prompt
    def score(self, prompt, sentence, model):
        tokenized_prompt = self.tokenizer.encode(prompt , max_
length=1024, truncation=True, return_tensors='pt').to('cpu')
        tokenized_all = self.tokenizer.encode(prompt + ' ' + sentence,
max_length=1024, truncation=True, return_tensors='pt').to('cpu')
        loss1=model(tokenized_all, labels=tokenized_all).loss
        loss2 = model(tokenized_prompt, labels=tokenized_prompt).
loss*len(tokenized_prompt[0])/len(tokenized_all[0])
        loss = loss1-loss2
        return math.exp(loss)
    def forward(self, sentence):
        pos = 0
        neg = 0
        for prompt in self.pos_prompt:
            pos += self.score(prompt, sentence, self.model)#.cpu()
        for prompt in self.neg_prompt:
            neg += self.score(prompt, sentence, self.model)#.cpu()

        result = torch.FloatTensor([5000-neg/10.0e+52,
5000-pos/10.0e+52])
        result = torch.softmax(result, 0)
        if abs(result[0].item()+result[1].item()-1) >= 1e-6:
            print('detected something')
            result = torch.FloatTensor([1,0])
        return torch.softmax(result, 0)
model = ClassificationModel('gpt2', ['Positive:'], ['Negative:'])
class_model = CustomWrapper(model)
attacker = Attacker(attack, textattack.datasets.Dataset(dataset[:10]),
AttackArgs(num_examples=10))
```

```
attacker.attack_dataset()
```

This results in the following output:

```
-------------------------------+-------+
| Attack Results                |       |
+-------------------------------+-------+
| Number of successful attacks: | 0     |
| Number of failed attacks:     | 3     |
| Number of skipped attacks:    | 7     |
| Original accuracy:            | 30.0% |
| Accuracy under attack:        | 30.0% |
| Attack success rate:          | 0.0%  |
```

```
| Average perturbed word %:      | nan%  |
| Average num. words per input:  | 17.3  |
| Avg num queries:               | 250.0 |
+-------------------------------+-------+
```

Private transformers – training LLMs using differential privacy

The complete source code for this section can be found at `https://github.com/lxuechen/private-transformers`.

Xuechen Li, Florian Tramer, Percy Liang, Tatsunori Hashimoto et al. provided private transformers to train LLMs using differential privacy.

They modified the Opacus framework, integrated it with Hugging Face's `transformers` library, and provided a **privacy engine** to train the LLMs in a privacy-preserving manner. Using this code base, they successfully fine-tuned exceptionally large pre-trained models, achieving some of the most impressive differentially private NLP results to date. In fact, certain models have exhibited performance comparable to robust non-private baseline approaches. This provides compelling empirical support for the notion that highly effective differentially private NLP models can be constructed even with relatively modest datasets. Furthermore, support for the ghost-clipping technique enables the private training of large transformers with significantly reduced memory requirements. In many instances, the memory footprint is nearly as lightweight as non-private training, with only a modest increase in runtime overhead. Private transformers currently support the following LLMs only:

- `OpenAIGPTLMHeadModel`
- `OpenAIGPTDoubleHeadsModel`
- `GPT2LMHeadModel`
- `GPT2DoubleHeadsModel`
- `BertForSequenceClassification`
- `RobertaForSequenceClassification`
- `AlbertForSequenceClassification`
- `BartForConditionalGeneration`
- `T5ForConditionalGeneration`
- `OPTForCausalLM`
- `ViTForImageClassification`
- `DeiTForImageClassification`
- `BeitForImageClassification`

Privately training Hugging Face transformers simply consists of four steps:

1. Create your favorite transformer model and optimizer; attach this optimizer to a `PrivacyEngine` instance.

2. Compute a per-example loss (1-D tensor) for a mini-batch of data.

3. Pass the loss to `optimizer.step` or `optimizer.virtual_step` as a keyword argument.

4. Repeat from *step 2*.

Example

The code shown next is designed for training a language model with privacy-preserving features. It utilizes the Hugging Face Transformers library and PyTorch. Next are the detailed steps to implement.

The following things are covered in the steps outlined next:

-   Libraries and imports

-   Dataset class

-   Loading data from a text file

-   Forward step

-   Training function

-   Running the training

1. Importing the necessary libraries and modules. These include the following:

    -   `tqdm`: A library for displaying progress bars during training.

    -   `transformers`: A library for working with transformer-based models.

    -   `torch`: The PyTorch library for **deep learning** (**DL**).

    -   `GPT2Tokenizer` and `GPT2LMHeadModel` from `transformers`: These classes provide access to the GPT-2 model and tokenizer.

    -   `PrivacyEngine` from `private_transformers`: A custom privacy engine for training the model with privacy constraints.

2. **Dataset class**: A custom `Dataset` class is defined to handle the training data. This class has the following methods:

    -   `__init__(self, texts, labels, eos_token)`: Initializes the dataset with texts, labels, and an **end-of-sequence** (**EOS**) token (`eos_token`).

    -   `__len__(self)`: Returns the length of the dataset.

    -   `__getitem__(self, index)`: Retrieves a specific text and its corresponding label at the given index.

3. **Loading data from a text file**: The `get_data_from_txt(path)` function is used to load text data and labels from a text file. Each line in the file contains a label followed by a text. This function reads the file, extracts the labels and texts, and returns them as lists.

4. **Forward step**: The `forward_step(correct_texts, wrong_texts, tokenizer, model, mismatch_loss, mismatch_weight)` function performs a forward step during training. It takes a list of correct and incorrect texts, a tokenizer, the model, and parameters for mismatch loss and mismatch weight. It tokenizes the texts, calculates the language modeling loss, and applies mismatch loss if specified. The result is a loss tensor.

5. **Training function**: The `train_llm(args_model_out, return_results, train_data, train_loader)` function trains the language model. It initializes the GPT-2 model, tokenizer, optimizer, and privacy engine. A privacy budget (epsilon) value of `0.5` is used in this example, but it can be changed to the desired privacy budget. It then iterates over training epochs, processing data in batches and calculating losses. The model is saved at the end of each epoch.

6. **Running the training**: At the end of the code, a sample dataset is loaded from a text file, and the training process is initiated using the `train_llm()` function. The function takes parameters such as the output path for saving the model, whether to return results, the training data, and the data loader.

All the preceding six steps are implemented in the following code snippet:

Source code: Privacy_Transformer.ipynb

```
!pip install transformers


!pip install git+https://github.com/lxuechen/private-
transformers.git


!pip install tqdm
from tqdm import tqdm
import transformers
import torch
from transformers import GPT2Tokenizer, GPT2LMHeadModel
from private_transformers import PrivacyEngine


class Dataset(torch.utils.data.Dataset):
    def __init__(self, texts, labels, eos_token):
        self.texts = texts
        self.y = labels
        self.eos_token = eos_token
    def __len__(self):
        return len(self.texts)
    def __getitem__(self, index):
        text = self.texts[index] + ' ' + self.eos_token
```

```
            label = self.y[index]
            return text, label


def get_data_from_txt(path: str):
    texts = []
    labels = []
    with open(path, 'r') as f:
        for line in f:
            texts.append(' '.join(line.split(' ')[1:]).replace('\n',
''))
            labels.append(int(line.split(' ')[0]))
    return texts, labels


def forward_step(texts,tokenizer, model):
    tokenized_texts = tokenizer(texts, truncation=True, max_
length=500, return_tensors='pt', padding=True).input_ids.to('cpu')

    lm_loss = model(tokenized_texts, labels=tokenized_texts).loss.
unsqueeze(dim=0)

    return lm_loss


def train_llm(train_data, train_loader, ):
    model = GPT2LMHeadModel.from_pretrained("gpt2")
    #model.parallelize()
    model.train()
    tokenizer = GPT2Tokenizer.from_pretrained("gpt2")
    tokenizer.pad_token = tokenizer.eos_token
    optimizer = torch.optim.Adam(model.parameters(),lr = 8e-6)
    args_epochs=2
    print(args_epochs)
    epsilon=0.5
    privacy_engine = PrivacyEngine(
            model,
            batch_size=1,
            sample_size=10,
            epochs=args_epochs,
            max_grad_norm=0.1,
            target_epsilon=epsilon,
        )
    privacy_engine.attach(optimizer)

    for epoch in range(args_epochs):
        total_loss = 0
```

```
            for texts, labels in tqdm(train_loader):
                lm_loss = forward_step(texts,tokenizer, model)
                optimizer.step(loss=lm_loss)
                total_loss += lm_loss.item()
        return model
train_texts, train_labels = get_data_from_txt('imdb_train.txt')
train_texts = train_texts[0:100]
train_labels =train_labels[0:100]
train_data = Dataset(train_texts, train_labels, '<|endoftext|>')
train_loader = torch.utils.data.DataLoader(train_data, shuffle=False,
batch_size=1)
pmodel = train_llm(train_data,train_loader)
print(pmodel)
Output of this program as follows:


2
0.5
training epoch 0
```

```
0%|          | 0/100 [00:00<?, ?it/s]/opt/conda/lib/python3.10/site-packages/torch/nn/modules/module.py:1344: Us
erWarning: Using a non-full backward hook when the forward contains multiple autograd Nodes is deprecated and will
be removed in future versions. This hook will be missing some grad_input. Please use register_full_backward_hook t
o get the documented behavior.
  warnings.warn("Using a non-full backward hook when the forward contains multiple autograd Nodes "
100%|██████████| 100/100 [03:27<00:00,  2.07s/it]
total language modeling loss 4.433933084011078
training epoch 1
100%|██████████| 100/100 [03:25<00:00,  2.06s/it]
total language modeling loss 3.9711666679382325

model training done!

GPT2LMHeadModel(
  (transformer): GPT2Model(
    (wte): Embedding(50257, 768)
    (wpe): Embedding(1024, 768)
    (drop): Dropout(p=0.1, inplace=False)
    (h): ModuleList(
      (0-11): 12 x GPT2Block(
        (ln_1): LayerNorm((768,), eps=1e-05, elementwise_affine=True)
        (attn): GPT2Attention(
          (c_attn): Conv1D()
          (c_proj): Conv1D()
          (attn_dropout): Dropout(p=0.1, inplace=False)
          (resid_dropout): Dropout(p=0.1, inplace=False)
        )
        (ln_2): LayerNorm((768,), eps=1e-05, elementwise_affine=True)
        (mlp): GPT2MLP(
          (c_fc): Conv1D()
          (c_proj): Conv1D()
          (act): NewGELUActivation()
          (dropout): Dropout(p=0.1, inplace=False)
        )
      )
    )
    (ln_f): LayerNorm((768,), eps=1e-05, elementwise_affine=True)
  )
  (lm_head): Linear(in_features=768, out_features=50257, bias=False)
)
```

Figure 10.6 Training Loss and Model Parameters

STOA – Privacy-preserving technologies for LLMs

The following section provides high-level SOTA research work on privacy-preserving technologies for LLMs. This is not an exhaustive list but details current trends in the research.

Prompts – Privacy: "Flocks of Stochastic Parrots: Differentially Private Prompt Learning for Large Language Models" *Research article:* `https://arxiv.org/pdf/2305.15594.pdf`

Large Language Models (LLMs) are adept at understanding contextual information; however, concerns arise regarding the privacy implications associated with the data contained within prompts. This study validates these concerns by demonstrating a simple yet highly effective membership inference attack on the data used for LLM prompts. To address this vulnerability, one option is to move away from prompting and instead focus on fine-tuning LLMs using established algorithms for private gradient descent. However, this approach sacrifices the practicality and efficiency provided by the prompting method. Therefore, the authors propose a new solution: private prompt learning. They first show the feasibility of obtaining soft prompts privately through gradient descent on downstream data. However, the challenge lies in handling discrete prompts. To overcome this, a process is devised where an ensemble of LLMs is engaged with various prompts, similar to a group of diverse parrots. A noisy vote among these LLMs privately transfers the collective knowledge of the ensemble into a single public prompt. Their results demonstrate that LLMs prompted using their private algorithms closely approach the performance of their non-private counterparts. For example, when using GPT-3 as the base model, they achieve a downstream accuracy of 92.7% on the sst2 dataset with ($\varepsilon = 0.147$, $\delta = 10\text{-}6$)-differential privacy, compared to 95.2% for the non-private baseline.

Prompts – Privacy: LLMs Can Understand Encrypted Prompt: Towards Privacy-Computing Friendly Transformers

Research article: `https://arxiv.org/abs/2305.18396`

In this study, scholars illustrated that replacing computationally and communication-intensive functions within the transformer framework with privacy-computing-compatible approximations markedly reduced the expenses linked to private inference, with only slight impacts on model effectiveness. Contrasting with the state-of-the-art Iron framework (*NeurIPS 2022*), their model inference process tailored for privacy-computing demonstrated a fivefold increase in computational speed and an 80% decrease in communication overhead, while preserving almost identical accuracy levels.

Differentially private attention computation

Research article: `https://arxiv.org/abs/2305.04701`

The attention mechanism plays a crucial role in LLMs, enabling them to selectively focus on various segments of input text. Computing the attention matrix is a well-recognized and substantial task in the LLM computation process. Consequently, determining how to offer verifiable privacy guarantees for the computation of the attention matrix is a significant research avenue. One natural mathematical concept for quantifying privacy, as found in theoretical computer science graduate textbooks, is differential privacy.

In this study, inspired by the work of Vyas, Kakade, and Barak (2023), researchers present a provable outcome that demonstrates how to differentially privately approximate the attention matrix. From a technical perspective, the results draw upon pioneering research in the realm of differential privacy as established by Alabi, Kothari, Tankala, Venkat, and Zhang (2022).

Differentially private decoding in LLMs

Research article: `https://arxiv.org/abs/2205.13621`

Researchers presented a straightforward, easily interpretable, and computationally efficient perturbation technique designed for implementation during the decoding phase of a pre-trained model. This perturbation mechanism is model-agnostic and compatible with any LLM. Their work includes a theoretical analysis demonstrating the differential privacy properties of the proposed mechanism, along with experimental results illustrating the trade-off between privacy and utility.

Differentially private model compression

Research article: `https://arxiv.org/abs/2206.01838`

Large pre-trained LLMs have demonstrated the ability to undergo fine-tuning on private data, achieving performance levels comparable to non-private models across numerous downstream NLP) tasks while ensuring differential privacy. However, these models, comprising hundreds of millions of parameters, often incur prohibitively high inference costs. Therefore, in practical applications, LLMs are frequently subjected to compression before deployment. Researchers embark on the exploration of differentially private model compression and propose frameworks capable of achieving 50% sparsity levels while retaining nearly full performance. Their study includes practical demonstrations of standard **General Language Understanding Evaluation** (**GLUE**) benchmarks using BERT models, thus establishing benchmarks for future research in this field.

Summary

In conclusion, this chapter has provided an in-depth exploration into the world of Language Models (LLMs) and the critical considerations surrounding their use, particularly focusing on privacy and security aspects. We have covered key concepts such as prompt engineering and compared open-source versus closed-source LLMs. Additionally, we delved into AI standards and terminology of attacks, highlighting NIST's guidelines and the OWASP Top 10 LLMs vulnerabilities.

Furthermore, we discussed various privacy attacks on LLMs, including real-world incidents of privacy leaks, membership inference attacks, and prompt injection attacks. These examples underscore the importance of robust privacy-preserving technologies in LLMs. We examined techniques like training LLMs using Differential Privacy with Private Transformer to mitigate privacy risks while maintaining model performance.

Overall, this chapter aims to empower readers with the knowledge and tools necessary to navigate the complexities of LLMs while safeguarding user privacy and ensuring responsible AI deployment. As the field continues to evolve, it becomes increasingly crucial to stay informed and proactive in addressing privacy concerns in LLMs. By understanding the nuances of prompt engineering, AI standards, privacy attacks, and privacy-preserving technologies, stakeholders can make informed decisions to promote trustworthy and responsible use of LLMs in various applications.

Index

`packtpub.com`

Subscribe to our online digital library for full access to over 7,000 books and videos, as well as industry leading tools to help you plan your personal development and advance your career. For more information, please visit our website.

Why subscribe?

- Spend less time learning and more time coding with practical eBooks and Videos from over 4,000 industry professionals

- Improve your learning with Skill Plans built especially for you

- Get a free eBook or video every month

- Fully searchable for easy access to vital information

- Copy and paste, print, and bookmark content

Did you know that Packt offers eBook versions of every book published, with PDF and ePub files available? You can upgrade to the eBook version at `packtpub.com` and as a print book customer, you are entitled to a discount on the eBook copy. Get in touch with us at `customercare@packtpub.com` for more details.

At `www.packtpub.com`, you can also read a collection of free technical articles, sign up for a range of free newsletters, and receive exclusive discounts and offers on Packt books and eBooks.

Other Books You May Enjoy

If you enjoyed this book, you may be interested in these other books by Packt:

Federated Learning with Python

Kiyoshi Nakayama PhD, George Jeno

ISBN: 978-1-80324-710-6

- Discover the challenges related to centralized big data ML that we currently face along with their solutions
- Understand the theoretical and conceptual basics of FL
- Acquire design and architecting skills to build an FL system
- Explore the actual implementation of FL servers and clients
- Find out how to integrate FL into your own ML application
- Understand various aggregation mechanisms for diverse ML scenarios
- Discover popular use cases and future trends in FL

Machine Learning with PyTorch and Scikit-Learn

Sebastian Raschka, Yuxi (Hayden) Liu, Vahid Mirjalili

ISBN: 978-1-80181-931-2

- Explore frameworks, models, and techniques for machines to learn from data
- Use scikit-learn for machine learning and PyTorch for deep learning
- Train machine learning classifiers on images, text, and more
- Build and train neural networks, transformers, and boosting algorithms
- Discover best practices for evaluating and tuning models
- Predict continuous target outcomes using regression analysis
- Dig deeper into textual and social media data using sentiment analysis

Packt is searching for authors like you

If you're interested in becoming an author for Packt, please visit `authors.packtpub.com` and apply today. We have worked with thousands of developers and tech professionals, just like you, to help them share their insight with the global tech community. You can make a general application, apply for a specific hot topic that we are recruiting an author for, or submit your own idea.

Share Your Thoughts

Now you've finished *Privacy-Preserving Machine Learning*, we'd love to hear your thoughts! Scan the QR code below to go straight to the Amazon review page for this book and share your feedback or leave a review on the site that you purchased it from.

https://packt.link/r/1-800-56467-8

Your review is important to us and the tech community and will help us make sure we're delivering excellent quality content.

Download a free PDF copy of this book

Thanks for purchasing this book!

Do you like to read on the go but are unable to carry your print books everywhere?

Is your eBook purchase not compatible with the device of your choice?

Don't worry, now with every Packt book you get a DRM-free PDF version of that book at no cost.

Read anywhere, any place, on any device. Search, copy, and paste code from your favorite technical books directly into your application.

The perks don't stop there, you can get exclusive access to discounts, newsletters, and great free content in your inbox daily

Follow these simple steps to get the benefits:

1. Scan the QR code or visit the link below

https://packt.link/free-ebook/9781800564671

2. Submit your proof of purchase

3. That's it! We'll send your free PDF and other benefits to your email directly

www.ingramcontent.com/pod-product-compliance
Lightning Source LLC
LaVergne TN
LVHW081512050326
832903LV00025B/1461